NOBEL SYMPOSIUM 24

Medicine and Natural Sciences

Collective properties of physical systems

Proceedings of the Twenty-Fourth Nobel Symposium
held June 12–16, 1973
at Aspenäsgården, Lerum, Sweden

Editors
BENGT LUNDQVIST and STIG LUNDQVIST

Administrative editor
VERA RUNNSTRÖM-REIO

NOBEL FOUNDATION · STOCKHOLM

ACADEMIC PRESS · NEW YORK and LONDON

A Subsidiary of Harcourt Brace Jovanovich, Publishers

Academic Press ISBN 0-12-460350-5
Library of Congress Catalog Card Number 73 – 19099

Graphic Design Dick Hallström
Printed in Sweden by
Almqvist & Wiksell, Uppsala 1974

Foreword

The idea of arranging a Nobel Symposium on the collective properties of matter formed quite some time ago. According to the first preliminary plans the symposium was to deal with collective phenomena in a very broad sense and it was even suggested that collective behaviour in social and economic systems should be included. When we were approached in the preliminary stage of discussions, such a wide scope for such a small meeting seemed a little frightening, to say the least. The Nobel Symposia are very small meetings by invitation only and the number of invited key participants is usually in the range 20–40. Nevertheless, we found it an attractive idea to arrange a symposium on collective phenomena—provided that we could restrict the programme to a few selected specialized topics of current interest in order to generate a strong interaction between the participants and produce an intensive discussion. A rather lightly sketched proposal along such lines was accepted by the Nobel Symposium Committee, and the Committee was kind enough not to ask detailed questions at the time as to how we were going to arrange things.

We feel that we need to express our apologies to all prominent scientists who could not be invited as a result of our actions. In the first place we excluded all areas which seemed to be well established and have reached a certain level of maturity or saturation irrespective of how great the individual contributions might have been. We decided firmly to concentrate on a few very recent developments which were in the focus of interest among leading scientists. Furthermore, we found it necessary to leave out entirely such large fields of physics as nuclear physics and plasma physics and concentrate mainly on collective phenomena in condensed systems. We felt that at the present time more exciting work on collective phenomena is being done in condensed matter physics than in the other main fields. Finally, even within condensed matter physics we had to select just a few topics of current interest.

The symposium was organized by a committee consisting of Lars Hedin, Lamek Hulthén, who represented the Nobel Symposium Committee, and the two of us. We wish to thank our colleagues for their valuable advice and assistance in all phases of the planning. In the detailed planning we were most fortunate to have J. W. Wilkins with us as a visiting professor during the academic year 72/73. His first suggestion was that we should just forget about the whole thing and send the money back to the sponsors, but he gradually gave up and joined us in the planning. He was the driving force in all our long discussions about topics and participants. His wide knowledge concerning almost everything going on almost everywhere, his constructive ideas and friendly criticism was a tremendous asset for us and he deserves much of the credit for the success of the symposium.

We are also grateful for some most valuable discussions during the planning period with N. H. March and P. Nozières. The choice of topics was the crucial question and J. R. Schrieffer, when visiting us after receiving his Nobel award, helped us in making the final selection of topics and gave us the courage to send out the invitations.

We would like to place on record our thanks to all the members of our solid state theory group for carrying out all the different service functions during the symposium and doing it so well. Our secretary, Mrs Agneta Connant, deserves our very special thanks for her excellent work for the symposium on top of all her regular duties.

The symposium was made possible through a major grant from the Tri-Centennial Fund of the Bank of Sweden. Additional support was provided by the Royal Academy of Sciences and the Swedish Natural Science Research Council. We gratefully acknowledge the generous support of these sponsors.

We would finally like to mention something about the Proceedings. It would of course have been ideal if all papers read at the symposium could have been published in full. However, we were aware that some participants would have found it difficult or inconvenient to prepare a full-length paper, which in some cases would have been merely modified versions of material due to appear in international journals. We therefore took a liberal attitude and accepted everything from a brief abstract with some key references, up to a full-length paper.

The discussions at the symposium were most interesting and exciting and the comments made have been edited and included in the Proceedings.

We ourselves experienced this symposium as an unforgettable scientific event, thanks to the outstanding contributions of our invited participants. We hope that this volume can convey to the reader something of the excitement felt by all the participants during the symposium week.

Göteborg, June 1973

Bengt Lundqvist — Stig Lundqvist

Contents

Effects of limited dimensionality in solids

Surface physics

High-energy excitations in solids

Liquid crystals and liquids

Biology

Sponsors

The Nobel Foundation
The Tri-Centennial Fund of the Bank of Sweden
The Royal Academy of Sciences
The Swedish Natural Science Research Council

Nobel Symposium Committee

Ramel, Stig, Chairman, Executive Director of the Nobel Foundation
Hulthén, Lamek, Professor, Member of the Nobel Committee for Physics
Fredga, Arne, Professor, Chairman of the Nobel Committee for Chemistry
Gustafsson, Bengt, Professor, Secretary of the Nobel Committee for Medicine
Gyllensten, Lars, Professor, Member of the Swedish Academy
Schou, August, Director of the Norwegian Nobel Institute (Peace)

Organizing Committee

Professor Lars Hedin
Professor Lamek Hulthén
Docent Bengt Lundqvist
Professor Stig Lundqvist

List of Participants

Ambegaokar, V, Cornell University, Laboratory of Atomic and Solid State Physics, Ithaca, N.Y. 14850, USA

Anderson, P W, Cavendish Laboratory, Department of Physics, Cambridge CB2 3RQ, UK and Bell Laboratories, Murray Hill, N.J. 07974, USA

Andersson, S, Chalmers University of Technology, Institute of Physics, S 402 20 Göteborg, Sweden

Berggren, K F, University of Linköping, Institute for Physics and Measurement Technique, S 581 83 Linköping, Sweden

Blandin, A, Faculté des Sciences, Service de Physique des Solides, 91405 Orsay, France

Blomberg, C, Royal Institute of Technology (KTH), Institute for Theoretical Physics, S 100 44 Stockholm, Sweden

Brinkman, W F, Bell Laboratories, Holmdel, N.J. 07733, USA

Cooper, L N, Brown University, Physics Department, Providence, R.I. 02912, USA

Domb, C, University of London, Kings College, Wheatstone Physics Laboratory, London WC2R 2LS, UK

Doniach, S, Stanford University, W. W. Hansen Laboratories of Physics, Stanford, Calif. 94305, USA

Dzyaloshinsky, I, Landau Institute of Theoretical Physics, USSR Academy of Science, Moscow B-334, USSR

Fisher, M E, Cornell University, Baker Laboratory, Ithaca, N.Y. 14850, USA

Fröman, A, Institute of Defence (FOA), S 104 50 Stockholm, Sweden

de Gennes, P G, Collège de France, Physique de la Matière Condensée, 11, Place Marcellin-Berthelot, 75 Paris 5e, France

Gorkov, L P, Landau Institute of Theoretical Physics, USSR Academy of Science, Moscow B-334, USSR

Grimvall, G, Chalmers University of Technology, Institute of Theoretical Physics, S 402 20 Göteborg 5, Sweden

Halperin, B I, Bell Laboratories, Murray Hill, N.J. 07974, USA

Hedin, L, University of Lund, Institute for Theoretical Physics, S 223 62 Lund, Sweden

Heeger, A J, University of Pennsylvania, Department of Physics, Philadelphia, Pa 19104, USA

Hopfield, J J, Princeton University, Department of Physics, Princeton, N.J. 08540, and Bell Laboratories, Murray Hill, N.J. 07974, USA

Hubbard, J, Atomic Energy Research Establishment, T. P. Division, Harwell, Didcot, Berks., UK

Hulthén, L, Royal Insitute of Technology (KTH), Institute for Mathematical Physics, S 100 44 Stockholm, Sweden

Josephson, B D, University of Cambridge, Cavendish Laboratory, Department of Physics, Cambridge CB2 3RQ, UK

Johansson, B, Institute of Defence (FOA), S 104 50 Stockholm 80, Sweden

Jona-Lasinio, G, University of Padova, Institute of Physics, 35100 Padova, Italy

Kohn, W, University of California at San Diego, Department of Physics, La Jolla, Calif. 92037, USA

Langreth, D C, Rutgers University, Department of Physics, New Brunswick, N.J. 08903, USA

Leggett, A J, University of Sussex, School of Mathematical and Physical Sciences, Brighton BN1 9QH, UK

Lounasmaa, O V, Helsinki University of Technology, Department of Technical Physics, SF 02150 Otaniemi, Finland

Lucas, A A, University of Liège, Institute of Physics, B-4000 Liège, Belgium

Lundqvist, B, Chalmers University of Technology, Institute of Theoretical Physics, S 402 20 Göteborg 5, Sweden

Lundqvist, S, Chalmers University of Technology, Institute of Theoretical Physics, S 402 20 Göteborg 5, Sweden

Luther, A, Harvard University, Department of Physics, Cambridge, Mass. 02138, USA

Mahan, G D, University of Oregon, Institute of Theoretical Science, Eugene, Ore. 97403, USA

March, N H, Imperial College, Department of Physics, London SW 7, UK

Martin, P C, Harvard University, Department of Physics, Cambridge, Mass. 02138, USA

Mermin, N D, Cornell University, Laboratory of Atomic and Solid State Physics, Ithaca, N.Y. 14850, USA

Niklasson, G, Chalmers University of Technology, Institute of Theoretical Physics, S 402 20 Göteborg 5, Sweden

Nozières, P, Institut Max von Laue–Paul Langevin, 38042 Grenoble, Cedex, France

Richardson, R C, Cornell University, Laboratory of Atomic and Solid State Physics, Ithaca, N.Y. 14850, USA

Schrieffer, J R, University of Pennsylvania, Department of Physics, Philadelphia, Pa 19104, USA

Sjölander, A, Chalmers University of Technology, Institute of Theoretical Physics, S 402 20 Göteborg 5, Sweden

Srinivasan, G, Chalmers University of Technology, Institute of Theoretical Physics, Fack, S 402 20 Göteborg 5, Sweden

Suhl, H, University of California, Department of Physics, La Jolla, Calif. 92037, USA

Toulouse, G, Université de Paris, Laboratoire de Physique des Solides, 91405 Orsay, France

Waller, I, University of Uppsala, Institute for Theoretical Physics, S 752 20 Uppsala, Sweden

Wendin, G, Chalmers University of Technology, Institute of Theoretical Physics, S 402 20 Göteborg 5, Sweden

Wilkins, J W, Cornell University, Laboratory of Atomic and Solid State Physics, Ithaca, N.Y. 14850, USA

Wilson, K G, Cornell University, Laboratory of Nuclear Studies, Ithaca, N.Y. 14850, USA

Zawadowski, A, Hungarian Academy of Sciences, Central Research Institute for Physics, Budapest 114, Hungary

Runnström-Reio, Vera, Administrative editor, The Nobel Foundation, c/o Karolinska Institutet, S 104 01 Stockholm 60, Sweden

Author Index

The Renormalization Group Approach, Phase Transitions and Critical Phenomena

Breakdown of Landau Theory, Fluctuations and Renormalization

K. G. Wilson

Laboratory of Nuclear Studies, Cornell University, Ithaca, N.Y. 14850, USA

This paper was based in part on a talk published previously [1] and in part on lecture IV, in [2]. The Landau theory of a critical point was reviewed briefly. It was emphasized that the magnetization density $M(x)$ involves an averaging size L; $M(x)$ is the average magnetization in a region of size L. The breakdown of Landau theory for space dimension less than 4 was attributed to L dependence for large L at the coefficients R and V in the Landau-Ginzburg interaction.

$$H = \int d^d x \{\nabla M^2(x) + RM^2(x) + VM^4(x)\}$$

L dependence is caused by statistical fluctuations of wavelength L. The graphical methods of ref. [2] were used to determine the effect of fluctuations leading to L dependence of R and V. This led to two recursion formulae determining R_{2L} and V_{2L} in terms of R_L and V_L. Solution of these recursion formulae determines the correlation length ν to be $1/(2-\varepsilon/3)$ where $\varepsilon = 4-d$. This result is valid to order ε.

References

1. Wilson, K, in AIP conf. proc no. 10, Magnetism and magnetic materials, p. 843, 1972.
2. Wilson, K G & Kogut, J, Institute for Advanced Study lecture notes (Physics repts. To be published).

Discussion

Schrieffer: (1) You mentioned that the renormalization group procedure currently being used leads to results for the critical indices which are accurate to within 5–10%. In the integrating down scheme, one neglects effective interactions between distant neighboring spins and interactions involving more than two spins at a time. While the incremental effect of these interactions may appear to be small in each stage of renormalization, could they not add up to give significant effects near the fixed point? The cumulative effect of small perturbations in each order is well known, e.g. in superconductivity theory where the pairing energy is formally of zero measure in each order, yet sums to a large value.

(2) Has the scheme been applied to the two-dimensional Ising model?

Wilson: (1) The formal idea of relevant and irrelevant variables allows one to determine in what way errors can accumulate over many iterations. One can see that these errors can be compensated by changing the values of the coefficients of relevant or marginally relevant operators in the initial interaction. Approximations of course change the location of the fixed point and the eigenvalues of the linearized equations about the fixed point; but barring the complications discussed in answer to Jona-Lasinio's question, one expects from formal topological considerations that small changes in the renormalization group transformation lead to small changes in the fixed point and associated eigenvalues. This statement is subject to one proviso: the presence of marginal or near marginal operators at a fixed point leads to instability under small changes of renormalization group transformation.

(2) Calculations are under way for the two dimensional Ising model. See my answer to P. W. Anderson's question after Halperin's talk (p. 65).

Anderson: Are there such things as droplet singularities, and how do they fit in?

Martin: I think it is very likely that the droplets are not contained in, or easily accessible to the fixed point idea. The fixed point is physically connected to what happens at arbitrarily large distances; the droplets tend to have a different origin, characteristic of a first order transition. I doubt that the asymptotic equation of state near the fixed point, the critical point, contains them. Together with the ε-expansion and the $1/n$ expansion it almost certainly doesn't. This doesn't mean that global ideas won't be useful for this problem also; just that they haven't been found yet, as far as I know.

Wilson: The droplets have a size of order ξ, the correlation length. Since the scaling equation of state on the coexistence curve is associated with a finite correlation length ξ (ξ is large compared to the lattice spacing, but still finite) there is no reason to think droplets are excluded from the renormalization group framework. To obtain the scaling equation of state from the renormalization group theory one must solve for more than the fixed point; one must follow trajectories near but not at T_c past the fixed point and on to some trivial final limit. Singularities on the coexistence curve due to droplet formation have not been found as yet but I expect the renormalization group framework with its emphasis on all possible wavelength fluctuations will be a very suitable framework for identifying cluster formation and associated singularities (see a preprint of Langer & Barron (Carnegie-Mellon University), for example).

Jona-Lasinio: When you mentioned a possible breakdown of the fixed point mechanism, what did you have specifically in mind?

Wilson: When one iterates the renormalization group equations in practice, for example the differential equations of chap. XI of the Wilson–Kogut lecture notes, one finds long range forces develop unless the spins are rescaled properly. In non-perturbative calculations one may find long range forces developing even after proper spin rescalings which never settle into a fixed point. Other breakdowns are possible also; in fact it is remarkable that all the examples studied so far have given either a fixed point or a result implying the existence of a first order transition instead of a second order transition. Alternatives not yet seen include limit cycles (see p. 68) and turbulent asymptotic behavior.

Domb: Could you supply more details of the two-dimensional calculations? You indicated that the approximations you made become worse as the dimension d moves away from 4.

Wilson: The renormalization group transformation I described here (see the Wilson & Kogut lecture notes) is useful only if one can expand about the Gaussian model. There are alternative formulations of the renormalization group which may be practical without this expansion. A specific example of such a transformation has been proposed by Niemeyer & Van Leeuwen in a Delft preprint to study the case of 2 dimensions. The general idea is that instead of integrating out a specific momentum range, one holds fixed block spins in the sense of Kadanoff, a block consisting of three or four spins of the original lattice. The hope is that in calculating the block spin interaction one can sum over only the spins of a finite size sublattice instead of an infinite lattice, with the size L of the sublattice being, say, 10 lattice sites, no matter low large the correlation length is. Niemeyer & Van Leeuwen have used an L of about 4 or 5 and their latest results (summing 21 spins) are quite good. Better calculations are possible and will be done in the next year or two; then one will know whether two-dimensional problems can be solved this way. Assuming this method works at all, the practical limitations came from limitations on the size L, for which calculations can be done. This in turn restricts the degree of nonlocality which can be tolerated in the fixed point interaction. There is no restriction on the size of non-Gaussian terms provided such terms are sufficiently local. In fact these methods work best for discrete spins, which are used for example in the Niemeyer-Van Leeuwen technique; in this case one cannot even separate Gaussian terms from non-Gaussian terms. See my Cargese lecture notes (1973, to be published) for further discussion.

General Scaling Theory for Critical Points

M. E. Fisher

Baker Laboratory, Cornell University, Ithaca, N.Y. 14850, USA

Summary

Scaling theory of the thermodynamic functions, correlations, and cumulants is developed from a general phenomenological standpoint with allowance for an exponent ω^*, the anomalous dimension of the vacuum; in dimensionality $d>4$, $\omega^*=d-4$; for the $d=3$ Ising model $\omega^*\simeq\frac{1}{12}$. The vanishing of ω^* implies "hyperscaling" relations such as $d\nu=2-\alpha$. The spectrum of critical operators and their exponents α_j, ν_j, and ω_j, are introduced by a process of operator refinement. Scaling covariance and conformal covariance are discussed in the context of non-zero ω^*. Operator product expansions are considered similarly. A general strong scaling hypothesis is formulated and shown to imply hyperscaling, i.e. $\omega^*=0$, unless the product expansion is restricted to cumulants.

The scaling theory of behavior in the vicinity of a critical point developed originally in a phenomenological manner; first for the two-point order–order correlation functions [1] by the introduction of the correlation length $\xi(T)$; then for the thermodynamic functions of temperature and field [2, 3]. Unified pictures of the scaling of both the thermodynamic and correlation functions for the order were proposed next [4, 5]. The generality of the scaling viewpoint was enlarged by the introduction of scaling for crossover phenomena of various sorts [6, 7]. More recently the concepts of an algebra of critical operators and corresponding operator expansions, has significantly enlarged our perspective [8, 10]. Finally the invention by Wilson [11] of the renormalization group and recursion relation approach to the theory of critical phenomena, and the subsequent development [12–16], have brought the whole theory into sharp focus from an essentially microscopic viewpoint. Despite the power to be gained by the renormalization group standpoint, based on the Hamiltonian of a system, it seems worthwhile to reformulate the phenomenological approach in a coherent and general way. This serves to distinguish the various hypotheses—or separate 'consequences' of the deeper approach—and to highlight the questions associated with what we term the "hyperscaling" exponent relations such as

$$d\nu = 2-\alpha, \quad 2-\eta = d(\delta+1)/(\delta-1),$$

which directly involve the spatial dimensionality d. These relations, although valid for the two-dimensional Ising model and for the spherical model when $2<d\leq 4$, seem in doubt for the three-dimensional Ising model (see e.g. review [17]) and appear to fail in all models for $d>4$.

In the following we develop the general phenomenological theory of scaling near a critical point. It is indicated how a spectrum of critical operators and corresponding critical exponents may be generated by a "refinement process". It is also shown how a consistent theory can be developed for the thermodynamic functions and cumulants which does not entail the hyperscaling relations. Rather an exponent ω^*, the "anomalous dimension of the vacuum", is introduced [18] (see also [19]). Since, for example, one finds

$$\omega^* = d-(2-\alpha)/\nu,$$

the hyperscaling relations follow only for $\omega^*=0$. We show that the principles of conformal invariance [20] and operator product expansions [8–10] can be modified to accommodate general (non-zero) ω^*. However, combination with the hypothesis of strong scaling [21, 22] (which proposes that the correlation scaling functions, which are valid for large spatial separation, can also be extrapolated back to short distances) forces the conclusion $\omega^* = 0$. Apart from special mechanisms which may effectively allow its violation [23], the formal renormalization group approach [10, 16] and calculations so far made with it [11–15] also imply hyperscaling.

In outline the subsequent development runs as follows: exponents for cumulants are defined in section I (pp 17–19); critical operators and their exponents are introduced by refinement in section II (pp 19–22); section III (pp 22–26); contains the corresponding scaling theory and includes a table of general exponent relations; scaling covariance is formulated in section IV (pp 26–28) as a preliminary to the discussion, in section V (pp 28–31), of the conformal covariance; section VI (pp 31–33) introduces the operator expansions and product reducibility; the strong scaling hypothesis is applied in section VII (pp 33–34); finally, in section VIII (pp 34–35), hyperscaling relations are shown to follow from strong scaling.

I. Cumulants

We have mentioned the scaling of the basic order–order (or two-point) correlation function, $G(\boldsymbol{r}) = \langle \Psi(\boldsymbol{0}) \Psi(\boldsymbol{r}) \rangle$, [1, 17]. More generally, the appropriate combinations of correlation functions to consider are the so-called cumulants or "connected correlations". The n-point cumulant $K_{AB\ldots N}(\boldsymbol{r}_1, \ldots \boldsymbol{r}_n)$ for a set of n local operators (or variables) $A(\boldsymbol{r})$, $B(\boldsymbol{r})$, ... $N(\boldsymbol{r})$ (which need not be distinct[1] can be generated conveniently by adding a term

$$\delta\mathcal{H}(h_A, \ldots, h_N) = -\int_\Omega h_A(\boldsymbol{r})\, A(\boldsymbol{r})\, \mathrm{d}\boldsymbol{r} - \int_\Omega h_B(\boldsymbol{r})\, B(\boldsymbol{r})\, \mathrm{d}\boldsymbol{r} - \ldots - \int_\Omega h_N(\boldsymbol{r})\, N(\boldsymbol{r})\, \mathrm{d}\boldsymbol{r} \qquad (1)$$

to the total Hamiltonian $\mathcal{H} = \mathcal{H}\{\psi(\boldsymbol{r})\}$; then we have the definition (with $\beta = 1/kT$)

$$K_{AB\ldots N}(\boldsymbol{r}_1, \ldots, \boldsymbol{r}_n) = \frac{\delta^n}{\delta\beta h_A(\boldsymbol{r}_1) \ldots \delta\beta h_N(\boldsymbol{r}_n)} \ln \mathrm{Tr}_\psi\{\exp[-\beta\mathcal{H}(h_A, \ldots, h_N)]\}, \qquad (2)$$

where it is understood that the auxiliary fields $h_A(\boldsymbol{r})$, ... are equated to zero after the final functional differentiations. When the operators $A(\boldsymbol{r})$, $B(\boldsymbol{r})$, ... commute with the Hamiltonian (or in the classical case) the first few cumulants are easily seen to be[2]

$$K_A(\boldsymbol{r}_1) = \langle A(\boldsymbol{r}_1) \rangle = \Gamma_A(\boldsymbol{r}_1),$$

$$K_{AB}(\boldsymbol{r}_1, \boldsymbol{r}_2) = \Gamma_{AB}(\boldsymbol{r}_1, \boldsymbol{r}_2) - K_A(\boldsymbol{r}_1)\, K_B(\boldsymbol{r}_2) = G_{AB}(\boldsymbol{r}_1, \boldsymbol{r}_2),$$

$$\begin{aligned} K_{ABC}(\boldsymbol{r}_1, \boldsymbol{r}_2, \boldsymbol{r}_3) = \Gamma_{ABC}(\boldsymbol{r}_1, \boldsymbol{r}_2, \boldsymbol{r}_3) &- K_{AB}(\boldsymbol{r}_1, \boldsymbol{r}_2)\, K_C(\boldsymbol{r}_3) \\ &- K_{AC}(\boldsymbol{r}_1, \boldsymbol{r}_3)\, K_B(\boldsymbol{r}_2) \\ &- K_{BC}(\boldsymbol{r}_2, \boldsymbol{r}_3)\, K_A(\boldsymbol{r}_1) \\ &- K_A(\boldsymbol{r}_1)\, K_B(\boldsymbol{r}_2)\, K_C(\boldsymbol{r}_3), \end{aligned} \qquad (3)$$

where the correlation functions Γ_A, Γ_{AB}, etc. are defined in a straightforward way as thermal expectations of $A(\boldsymbol{r}_1)$, of $A(\boldsymbol{r}_1) B(\boldsymbol{r}_2)$, etc. The general rule in forming $K_{AB\ldots N}$ is to subtract from $\Gamma_{AB\ldots N}$ the sum of products of lower order cumulants $K_{DE\ldots J}$ corresponding to all possible partitions of the points $\boldsymbol{r}_1$ to $\boldsymbol{r}_n$ and their corresponding operators. Each partition can be thought of as a "disconnected" diagram (or linear graph) of two or more "connected" components. The nth order cumulant itself then corresponds only to "completely connected" diagrams of n points. This interpretation may often be made concrete in a diagrammatic perturbation expansion; a frequently used notation, which stresses this feature, is

$$K_{AB\ldots N}(\boldsymbol{r}_1 \ldots \boldsymbol{r}_n) = \langle A(\boldsymbol{r}_1) B(\boldsymbol{r}_2) \ldots N(\boldsymbol{r}_n) \rangle_{\mathrm{C}} \qquad (4)$$

where the subscript C denotes "the connected part of". However, the concept of a cumulant function is clearly independent of perturbation theory.

At the critical point of a system we can expect the cumulants to decay slowly with separation i.e., as inverse power laws, as does the basic order–order correlation function $G(\boldsymbol{r}) = K_{\Psi\Psi}(0, \boldsymbol{r})$. In characterizing this decay it is

[1] A cumulant, such as $K_{\ldots CC\ldots C\ldots}$, with a repeated operator of multiplicity m can be alternatively defined by taking the mth order functional derivative with respect to $\beta h_C(\boldsymbol{r})$ in (2).

[2] In the general, non-commutative case the simple statistical averages like $\Gamma_{AB} = \langle A(\boldsymbol{r}_1) B(\boldsymbol{r}_2) \rangle$ in (3) must be replaced by more elaborate expressions.

natural to let all the relative distances $|\boldsymbol{r}_{ij}| = |\boldsymbol{r}_j - \boldsymbol{r}_i|$, increase in constant proportion. If, alternatively, the n points of $K_{AB\dots N}$ were to be separated into two groups, say $(\boldsymbol{r}_1 \dots \boldsymbol{r}_{j-1})$ and $(\boldsymbol{r}_j \dots \boldsymbol{r}_n)$, within each of which the relative positions remain fixed, one would, when $|\boldsymbol{r}_j - \boldsymbol{r}_i| \to \infty$, really be studying the two-point cumulant $K_{XY}(\boldsymbol{r}_1, \boldsymbol{r}_j)$ in which $X(\boldsymbol{r}_1) = A(\boldsymbol{r}_1)\dots I(\boldsymbol{r}_{j-1})$ and $Y(\boldsymbol{r}_j) = J(\boldsymbol{r}_j) \dots N(\boldsymbol{r}_n)$. Accordingly, the appropriate scaling hypothesis to make at the critical point, is $K^{c}_{AB\dots N} \approx \mathcal{K}_{AB\dots N}$, where the asymptotic scaling function satisfies the homogeneity relation

$$\mathcal{K}^{c}_{AB\dots N}(l\boldsymbol{r}_1, l\boldsymbol{r}_2, \dots, l\boldsymbol{r}_n) = l^{-\omega\{A,B,\dots,N\}}\mathcal{K}^{c}_{AB\dots N}(\boldsymbol{r}_1, \boldsymbol{r}_2, \dots, \boldsymbol{r}_n), \tag{5}$$

which introduces the decay exponent $\omega\{AB\dots N\}$.

In order to re-express this homogeneity relation in the standard scaling form it is useful (although by no means essential) to introduce a single length $R_n = R_n(\boldsymbol{r}_1, \dots, \boldsymbol{r}_n)$ which describes the scale of the configuration $(\boldsymbol{r}_1, \dots, \boldsymbol{r}_n)$. The definition of this length is fairly arbitrary although it is useful to keep it symmetric in all the $\boldsymbol{r}_i$. Two simple possibilities are

$$R_n = \left[\sum_{1\leqslant i,j\leqslant n} (\boldsymbol{r}_i - \boldsymbol{r}_j)^2\right]^{\frac{1}{2}}, \tag{6}$$

which varies analytically with the $\boldsymbol{r}_i$, and

$$R_n = \max_{1\leqslant i,j\leqslant n} |\boldsymbol{r}_i - \boldsymbol{r}_j|, \tag{7}$$

which does not. A more elaborate choice for symmetric systems is guided by consideration of the expansion of the cumulants in terms of connected diagrams of pairwise interaction "bonds", (see e.g. [24]) namely,

$$R_n = \left\{\min_{C(m)} \sum_{(i,j)\subset C(m)} |\boldsymbol{r}_i - \boldsymbol{r}_j|\right\}, \tag{8}$$

where (i, j) denotes an edge (or line, or bond) of a connected linear graph $C(m)$ $(m \geqslant n)$ formed on the n points $\boldsymbol{r}_1, \dots \boldsymbol{r}_n$ at which an *odd* number of bonds must meet, and on $(m-n)$ additional, auxiliary points at which an even number of bonds must meet. The minimization runs over all possible connected graphs (but, clearly, only "trees" of $(m-1)$ edges need be considered). (See e.g., Essam & Fisher [25] for graph theoretical definitions.) Adopting any one of the definitions we may rewrite the critical point scaling hypothesis as

$$K^{c}_{AB\dots N}(\boldsymbol{r}_1, \dots, \boldsymbol{r}_n) \approx R_n^{-\omega\{AB\dots N\}} D^{c}_{AB\dots N}(\boldsymbol{r}_1/R_n, \dots, \boldsymbol{r}_n/R_n). \tag{9}$$

For $A = B = \Psi$ (where, as before, $\Psi(\boldsymbol{r})$ denotes the order parameter) this must of course reduce to previous definitions [1, 17]; thus we have

$$\omega\{\Psi\Psi\} = d - 2 + \eta, \tag{10}$$

while $D^{c}_{\Psi\Psi}(\boldsymbol{x}_1, \boldsymbol{x}_2)$ reduces to the standard, constant critical decay amplitude, D_0. More generally, it is clear that in the thermodynamic limit the cumulant $K_{AB\dots}(\boldsymbol{r}_1, \boldsymbol{r}_2, \dots)$ should, in the bulk, depend only on the relative separations $\boldsymbol{r}_{ij}$, so that the scaling function $D^{c}_{AB\dots}(\boldsymbol{x}_1, \boldsymbol{x}_2, \dots)$, likewise depends only on the relative variables $\boldsymbol{x}_{ij} = \boldsymbol{x}_j - \boldsymbol{x}_i$ and, in particular, is simply a constant for any two-point cumulant.

By the type of arguments that lead to the basic fluctuation-susceptibility relation between $G(\boldsymbol{r})$ and the susceptibility $\chi = \chi_{\Psi\Psi}$, it is easy to see that integrals over the cumulants should yield derivatives of the free energy $F = F(h_A, h_B, \dots)$, with respect to the conjugate fields $h_A, h_B, \dots$, now taken as uniform (i.e., independent of $\boldsymbol{r}$). The first two relations are simply

$$-\frac{\partial(\beta F)}{\partial \beta h_A} = \langle A \rangle = K_A(\boldsymbol{r}) = K_A(\boldsymbol{0}),$$

$$-\frac{\partial^2(\beta F)}{\partial \beta h_A\, \partial \beta h_B} = \chi_{AB} = \int \mathrm{d}\boldsymbol{r}\, K_{AB}(\boldsymbol{0}, \boldsymbol{r}), \tag{11}$$

while in general one finds an $(n-1)$-fold integral relation,

$$-\left(\frac{\partial}{\partial \beta h_A}\right)\left(\frac{\partial}{\partial \beta h_B}\right)\cdots\left(\frac{\partial}{\partial \beta h_N}\right)(\beta F) = \chi_{AB\dots N} = \int \mathrm{d}\boldsymbol{r}_2 \int \mathrm{d}\boldsymbol{r}_3 \cdots \int \mathrm{d}\boldsymbol{r}_n\, K_{AB\dots N}(\boldsymbol{0}, \boldsymbol{r}_2, \dots, \boldsymbol{r}_n), \tag{12}$$

where $\chi_{AB\dots N}$ can be regarded as a higher order, or multiple susceptibility. Note that if an operator A is repeated with multiplicity m_A one just has m_A repeated derivatives with respect to βh_A. For the special choices $A(\boldsymbol{r}) = \Psi(\boldsymbol{r})$ the order parameter, and $B(\boldsymbol{r}) = \mathcal{E}(\boldsymbol{r})$ the energy density, we have, of course,

$$\chi_{\Psi\Psi} = \chi \text{ and } \chi_{\mathcal{E}\mathcal{E}} = C \tag{13}$$

that is, the basic susceptibility and specific heat, respectively. Naturally, we expect these generalized susceptibilities to exhibit characteristic critical behavior with suitably defined exponents, $\gamma_{AB} = \gamma\{AB\}$, $\gamma\{ABC\}$, ... (see below). Furthermore, these exponents should then be related through the generalized fluctuation-susceptibility formula (12) to the exponents $\omega\{AB\ldots\}$ describing the critical decay of the cumulants.

II. Critical Operators and their Exponents

In order to develop a theory to encompass these exponent relations (and those previously mentioned) we will formulate a more abstract and deeper approach to scaling than adopted before. As discussed in the Introduction this viewpoint derives from the idea of various authors but, in particular, the work of Patashinskii & Pokrovskii [5], Kadanoff [8], and Wilson [10], should be cited. To start we note that one may, from the basic local variables (or "fields") of operators $\Psi_\alpha(\boldsymbol{r})$ entering the Hamiltonian, form products and sums (including integrals over *short* range kernels[1], hence building up an algebra of local operators $A(\boldsymbol{r})$, $B(\boldsymbol{r})$, ... The local energy density, $\varepsilon(\boldsymbol{r})$, whose integral defines the total Hamiltonian $\mathcal{H}$ is, of course, just one such operator; however, it is the dimensionless *reduced Hamiltonian,*

$$\bar{\mathcal{H}}\{\psi\} = -\mathcal{H}\{\psi\}/k_B T = \int_\Omega \bar{\varepsilon}(\boldsymbol{r})\, d\boldsymbol{r}, \tag{14}$$

and, corresponding *reduced energy density* $\bar{\varepsilon}(\boldsymbol{r})$, which determines the basic Boltzmann factor entering the computation of all expectation values $\langle\cdot\rangle = \langle\cdot\rangle_{\bar{\mathcal{H}}}$. The specification of a reduced Hamiltonian amounts, in physical terms to the specification of all the "interactions", the precise values of all "external fields", *and* the value of the temperature.

Now 'most' reduced Hamiltonians, $\bar{\mathcal{H}}$, will be *noncritical* in the sense that the associated cumulants, in particular $K_{AB}(\boldsymbol{r})$, for all local operators will decay exponentially fast with distance.[2] Conversely, for a *critical* (reduced) *Hamiltonian,* $\mathcal{H}_c$, at least some of the cumulants will exhibit slow, power law decays.[3] To study the thermodynamics and fluctuations in the 'vicinity' of such a critical Hamiltonian we examine the response to the incremental (reduced) Hamiltonian

$$\delta\bar{\mathcal{H}} = \bar{h}_A \int_\Omega A(\boldsymbol{r})\, d\boldsymbol{r}, \tag{15}$$

where $A(\boldsymbol{r})$ is an element of the algebra of local operators, while $\bar{h}_A$ is the corresponding reduced field, which, physically, can be regarded as equal to $\beta_c h_A = h_A(kT_c)^{-1}$. The algebra contains the unit operator $I(\boldsymbol{r}) \equiv 1$, with average unity. Thus we can always subtract from $A(\boldsymbol{r})$ its critical average $\langle A\rangle_c = \langle A\rangle_{\bar{\mathcal{H}}_c}$; it is usually convenient to do this and hence assume, without further mention, that $\langle A\rangle_c \equiv 0$.

By calculating the incremental (reduced) free energy density $\bar{F}$

$$\bar{F}(\bar{h}_A) = \lim_{V(\Omega)\to\infty} \frac{1}{V(\Omega)} \ln\left[\mathrm{Tr}_\psi\{e^{\bar{\mathcal{H}}_c + \delta\bar{\mathcal{H}}}\}/\mathrm{Tr}_\psi\{e^{\bar{\mathcal{H}}_c}\}\right] \tag{16}$$

we can now define a fluctuation exponent α_A by

$$\chi_{AA} = \frac{\partial^2 \bar{F}}{\partial \bar{h}_A^2} \sim 1/\bar{h}_A^{\alpha_A}, \quad (\bar{h}_A \to 0), \tag{17}$$

or, on integrating, by

$$\frac{\partial \bar{F}}{\partial \bar{h}_A} = \langle A\rangle \sim \bar{h}_A^{1-\alpha_A}, \quad \bar{F}(h_A) \sim \bar{h}_A^{2-\alpha_A}, \tag{18}$$

from which one must normally have $\alpha_A < 1$ since $\langle A\rangle \to \langle A\rangle_c = 0$ as $\bar{h}_A \to 0$. More generally, if χ_{AA} does not diverge as $\bar{h}_A \to 0$ we may define $\alpha_A(<0)$ from the divergence of the higher order

---

[1] We must also allow coefficients, which depend on the point of locality $\boldsymbol{r}$, provided there are sufficient elements of translational invariance. Thus coefficients $\cos(\boldsymbol{q}_0\cdot\boldsymbol{r})$ or $\exp(i\boldsymbol{q}_0\cdot\boldsymbol{r})$ are needed to define the order parameter of an antiferromagnet.

[2] On a phase boundary below a critical point the cumulants are to be evaluated as limits $\zeta \to \zeta_\sigma(T)\pm$, as in the definition of spontaneous order, where ζ is the ordering field conjugate to the order $\Psi(x)$.

[3] It will, of course, be clear to the reader that we are not here giving a rigorous account of the general situation. (One might well ask here, as indeed in general, "Why not logarithmic or other nonpower-law decays?") Rather we are sketching a picture of our expectations in the same spirit as the original introduction of critical exponents and scaling behavior [1–3, 17].

susceptibilities $\chi_{AAA}, \ldots$ (as in the standard definition of singular part exponents; see [17]).

If we replace $A(\boldsymbol{r})$ by another operator $B(\boldsymbol{r})$ we can define a further exponent α_B, and so on. If $B(\boldsymbol{r})$ is a multiple of $A(\boldsymbol{r})$, one has, trivially, $\alpha_B = \alpha_A$. However, by studying examples we can see that in many other cases the exponent α_B will again be identical to α_A. Thus consider, for concreteness, a simple Ising ferromagnet (see e.g. [9, 17]) with local spin variables $\psi(\boldsymbol{r}) = s(\boldsymbol{r})$. If $A(\boldsymbol{r}) = s(\boldsymbol{r})$, the normal definition [17] of the exponent δ evidently yields $\alpha_A = \alpha_s = (\delta - 1)/\delta$. But it is clear that any "smearing" of $s(\boldsymbol{r})$ with a positive kernel $v(\boldsymbol{r} - \boldsymbol{r}')$, for example, $B(\boldsymbol{r}) = \sum_{\boldsymbol{\delta}} v(\boldsymbol{\delta}) s(\boldsymbol{r} + \boldsymbol{\delta})$ (where the $\boldsymbol{\delta}$ are the nearest neighbor lattice vectors), has an identical exponent $\alpha_B = (\delta - 1)/\delta$. Less trivially we might form $B(\boldsymbol{r})$ by adding to $s(\boldsymbol{r})$ some multiple of the energy density, $\varepsilon(\boldsymbol{r}) = \frac{1}{2} s(\boldsymbol{r}) \sum_{r'} J(\boldsymbol{r}') s(\boldsymbol{r} + \boldsymbol{r}')$; but we will again find $\alpha_B = \alpha_A$, since we know that the fluctuations of the energy are always much weaker than those of the spin, which will hence dominate on approach to the critical point. On the other hand, with $B(\boldsymbol{r}) = \varepsilon(\boldsymbol{r})$ we obtain $\alpha_B = \alpha \neq \alpha_A$, where α is the standard specific heat exponent. (This special case justifies the extension of normal exponent notation.) These examples also indicate that when the critical Hamiltonian, $\mathcal{H}_c$, has some special symmetry (in this case, invariance under $s(\boldsymbol{r}) \Rightarrow -s(\boldsymbol{r})$) one should expect operators transforming differently under the symmetry to have different exponents.

A process of *refinement* of operators or variables, leading to a *spectrum* of *critical operators* or *critical densities* Θ_j $(j = 1, 2, \ldots)$, with exponents α_j, is suggested by the above illustrations. Thus suppose two distinct operators A and B have the same exponent $\alpha_A = \alpha_B$: we may look for a linear combination $C = c_A A + c_B B$ with (positive or negative) coefficients c_A and c_B, in which the leading singularity of $\langle A \rangle \sim \langle B \rangle$, or of $\chi_{AA} \sim \chi_{BB}$, etc., is *cancelled*, revealing a weaker singularity and a new exponent $\alpha_C < \alpha_A$. Then a suitable multiple of C added to A (or to B) should lead to a refined operator, $A'(\propto B')$, whose expectation $\langle A' \rangle$, susceptibility $\chi_{A'A'}$, etc. vary more accurately as pure powers of the conjugate field than do the corresponding derivatives formed with A or B.

To see more explicitly how this might go, suppose that $g_1 = \bar{h}_{\Theta_1}$ and $g_2 = \bar{h}_{\Theta_2}$ are the fields conjugate to two critical operators Θ_1 and Θ_2 which are *well-refined* in the sense that

$$\bar{F}(g_1, 0) = f_1 g_1^{2-\alpha_1}[1 + O(g_1^{\pi_1})], \quad (g_2 \equiv 0, g_1 \to 0),$$

$$\bar{F}(0, g_2) = f_2 g_2^{2-\alpha_2}[1 + O(g_2^{\pi_2})], \quad (g_1 \equiv 0, g_2 \to 0), \tag{19}$$

where π_1 and π_2 are 'large' (at very least, not less than unity).

We will suppose that $\alpha_1 > \alpha_2$. Now the behavior of the free energy and its derivatives as a function of the single field $\bar{h}_A$ conjugate to the linear combination

$$A(\boldsymbol{r}) = a_1 \Theta_1(\boldsymbol{r}) + a_2 \Theta_2(\boldsymbol{r}), \tag{20}$$

cannot be found without further knowledge. Accordingly, let us postulate a scaling expression[1]

$$\bar{F}(g_1, g_2) \approx g_1^{2-\alpha_1} Y(g_2/g_1^{\phi_{12}}) \tag{21}$$

where, in order to reproduce (19), the "crossover exponent", ϕ_{12}, must have the value

$$\phi_{12} = (2 - \alpha_1)/(2 - \alpha_2) \tag{22}$$

while, in the usual way, the scaling function obeys

$$Y(x) \approx f_2 x^{2-\alpha_2} \quad \text{as } x \to \infty, \tag{23}$$

and, we may suppose,

$$Y(x) - f_1[1 + e_1 x + \ldots] \quad \text{as } x \to 0. \tag{24}$$

Now the effect of the field $\bar{h}_A$ (acting alone) is precisely equivalent to setting $g_1 = \bar{h}_A a_1$, and $g_2 = \bar{h}_A a_2$. On using $\alpha_2 < \alpha_1$, which implies $\phi_{12} < 1$, we thus find from (21) and (24),

$$\bar{F}(\bar{h}_A) = f_A \bar{h}_A^{2-\alpha_1}[1 + e_A \bar{h}_A^{1-\phi_{12}} + \ldots], \tag{25}$$

where $f_A = f_1 a_1^{2-\alpha_1}$ and $e_A = a_2 e_1 / a_1^{\phi_{12}}$. Now since $1 - \phi_{12} < 1 \leq \pi_1$ the operator A is clearly less refined than Θ_1 (unless $a_2 \equiv 0$) although it has the same exponent $\alpha_A = \alpha_1$. This conclusion applies equally to some other combination B, with coefficients b_1 and b_2 $(\neq 0)$. From two distinct such combi-

[1] To avoid complicating the notation by the introduction of matching scaling functions, $X_{\gtrless}(x)$, and the insertion of moduli bars, as in general needed for $x \gtrless 0$, etc. we will suppose all the relevant field variables are positive.

nations, however, we can evidently form $C = c_A A + c_B B$ such that the amplitude of $\bar{h}_C^{2-\alpha_1}$ vanishes identically; evidently, we must then have $C(\boldsymbol{r}) \propto \Theta_2(\boldsymbol{r})$ so that $\alpha_C = \alpha_2$. Then using C, we can refine A or B as before and will find $A'(\boldsymbol{r}) = A(\boldsymbol{r}) + a_C C(\boldsymbol{r}) \propto B'(\boldsymbol{r}) \propto \Theta_1(\boldsymbol{r})$. Alternatively, we could look for another combination $D = d_A A + d_B B$ in which the singular correction term in (25) vanishes ($e_D = 0$); then again we will have $D(\boldsymbol{r}) \propto \Theta_1(\boldsymbol{r})$ and $\alpha_D = \alpha_1$.

These ideas can be extended to accommodate more operators and successive further refinements, thereby building up a critical spectrum of operators (or densities) Θ_1, Θ_2, ... to any desired level. In terms of these critical operators any other operator will have a critical expansion generalizing (20). One anticipates that the Θ_j will have definite transformation properties with respect to any symmetries of the critical Hamiltonian. It is also natural to expect that operators of higher algebraic order in the basic variables $\psi_\alpha(\boldsymbol{r})$, will normally be less singular, i.e., have smaller exponents α_j. This is also true for operators formed with kernels whose spatial average (and possibly higher moments) vanish; for example, in the plane square Ising model,

$$B(\boldsymbol{r}) = s(\boldsymbol{r})[s(\boldsymbol{r}+a\boldsymbol{e}_x) - s(\boldsymbol{r}+a\boldsymbol{e}_y) + s(\boldsymbol{r}-a\boldsymbol{e}_x) - s(\boldsymbol{r}-a\boldsymbol{e}_y)],$$

where $\boldsymbol{e}_x$ and $\boldsymbol{e}_y$ are unit vectors parallel to the lattice axes.[1]

It is then reasonable to identify the most singular operator $\Theta_1(\boldsymbol{r})$, (or set of operators) with $\alpha_1 > \alpha_j$ ($j \neq 1$) as the order parameter $\Psi(\boldsymbol{r})$ [or $\underset{\rightarrow}{\Psi}(\boldsymbol{r})$]. The scheme of refinement thus provides a method of defining the order parameter which removes most of the ambiguities and arbitrariness that are normally entailed. The identifications with the ordering field, $\zeta = \zeta_\sigma(T) + \Delta\zeta$,

$$g_1 = g_\Psi = \Delta\zeta/kT \approx \Delta\zeta, \quad \alpha_1 = \alpha_\Psi = 1-(1/\delta), \tag{26}$$

follow immediately. It should be noted, however, that this definition does not necessarily ensure the presence of all the properties usually associated with the order parameter.[2] Thus, as the fluctuations in Ψ are stronger than in any other density, we may expect that 'below' or, more generally, in the vicinity of the critical point, long-range order in Ψ will set in so that there will be a phase-boundary or, more generally, surface $g_1 = g_\sigma(g_2, g_3, ...)$ on which $G_{\Psi\Psi}(\boldsymbol{r})$ does not decay to zero. On varying g_1 (or ζ) through this surface, a discontinuity in the thermodynamic average $\langle\Theta_1\rangle = \langle\Psi\rangle$ is to be expected. It is conceivable, nonetheless, that no such first order phase boundary exists, or on the other hand, that one does exist but contains, asymptotically, the g_1 axis so that $\langle\Theta_1\rangle$ is continuous through the surface while some less singular "anomalous order parameter" displays a discontinuity and long range order. At present, however, no examples of such behavior are known.

To define the critical energy density operator $\Theta_0 = \mathcal{E}(\boldsymbol{r})$ and its exponent

$$\alpha_0 = \alpha_{\mathcal{E}} = \alpha, \tag{27}$$

we may start with the energy density $\bar{\varepsilon}_c(\boldsymbol{r})$ defining the critical Hamiltonian $\mathcal{H}_c$. (We break our labelling convention $j = 1, 2, 3, ...$ here for reasons that will be apparent below.) On projecting out any contribution from the order parameter $\Psi(\boldsymbol{r})$, we find the exponent α_0 and on further refinement we obtain $\mathcal{E}(\boldsymbol{r})$. This prescription again removes much of the ambiguity in our introduction of the local energy density $\varepsilon(\boldsymbol{r})$, which in the scaling analysis should now be replaced by $\mathcal{E}(\boldsymbol{r})$. The conjugate field is, of course, almost proportional to the reduced temperature deviation since

$$g_0 = (kT_c)^{-1} - (kT)^{-1} = t/kT = \tilde{t}, \qquad t = (T-T_c)/T_c. \tag{28}$$

One might expect that there would be no other critical operators with exponents α_j lying between $\alpha_\Psi = 1-(1/\delta)$ and $\alpha_{\mathcal{E}} = \alpha$. While this seems to be the case for the standard Ising models, there is no general reason why it should be so. Indeed, analytic and numerical calculations (see [14, 15, 26]) show that in the case of isotropic systems with a vector order parameter $\underset{\rightarrow}{\Psi} = (\Psi_1, ... \Psi_n)$, critical operators like

[1] Similarly, operators formed with spatially varying coefficients with zero mean, such as $\cos(\boldsymbol{q}_0 \cdot \boldsymbol{r})$, will usually be less singular unless, as in the example of an antiferromagnet, they happen to "resonate" with some attractive term in the critical Hamiltonian.

[2] I am indebted to Professor K. G. Wilson for observations on this point.

$$\Xi_\beta(\boldsymbol{r}) = \varepsilon_\beta(\boldsymbol{r}) - n^{-1} \sum_{\gamma=1}^{n} \varepsilon_\gamma(\boldsymbol{r}),$$

$$\varepsilon_\gamma(\boldsymbol{r}) = \frac{1}{2} \int d\boldsymbol{r}'\, v(\boldsymbol{r}-\boldsymbol{r}')\, \Psi_\gamma(\boldsymbol{r})\, \Psi_\gamma(\boldsymbol{r}'), \qquad (29)$$

which break the continuous symmetry, have an exponent α_Ξ exceeding α (but less than α_Ψ).[1] For this reason, although it is natural to label the critical spectrum so that $\alpha_1 \geqslant \alpha_2 \geqslant \alpha_3 \geqslant \ldots$, we will continue to reserve the label 0 for the critical energy density.

III. General Formulation of Scaling

In terms of the critical operators Θ_j and their reduced fields g_j we may advance the general homogeneity hypothesis

$$\bar{F}(g_1, \ldots, g_j, \ldots) \approx l\bar{F}(g_1/l^{1/(2-\alpha_1)}, \ldots g_j/l^{1/(2-\alpha_j)}, \ldots), \qquad (30)$$

where, as previously, l is arbitrary. By choosing $l = g_i^{2-\alpha_i}$ we confirm the definitions of the α_i and obtain the scaling hypothesis with respect to the field g_i, as

$$\bar{F}(g_1, \ldots, g_j, \ldots) \approx g_i^{2-\alpha_i}\, Y_i(g_1/g_i^{\phi_{i1}}, \ldots, g_j/g_i^{\phi_{ij}}, \ldots), \qquad (31)$$

(As before we ignore the complications needed to accommodate all combinations of the signs of the fields.) The scaling function $Y_i(x_1 \ldots)$ is related in the standard way to $\bar{F}(x_1 \ldots)$. As in (22) the crossover exponents ϕ_{ij} are given by

$$\phi_{ij} = (2-\alpha_i)/(2-\alpha_j). \qquad (32)$$

From (32) we see that the critical field g_j "scales like" $g_i^{\phi_{ij}}$. Note that $\alpha_j > \alpha_k$ implies $\phi_{ij} > \phi_{ik}$ so that with respect to any scaling field g_i the order parameter Ψ has the largest crossover exponent. Of course, one may choose any convenient critical field for g_i but, as we know, the reduced temperature is the most natural choice (fundamentally because the inverse temperature is *always* an experimentally accessible field!). With this in mind we may drop the first subscript of the crossover exponents when $i=0$ and simply write

$$\phi_{0j} \equiv \phi_j = (2-\alpha)/(2-\alpha_j) \quad (g_0 = \bar{t},\ \Theta_0 = \mathcal{E}). \qquad (33)$$

Then for the order parameter we see that

$$\phi_1 = \phi_\Psi = \Delta = \beta + \gamma, \qquad (34)$$

where Δ is the thermodynamic gap exponent (defined e.g. in [17]). Again, we may say that "g_j scales like t^{ϕ_j}".

With the convention (33) we can define, and from the scaling hypothesis (31), evaluate, generalizations of the other thermodynamic exponents. Thus we define β_Θ by

$$\mathcal{K}_{\Theta_j} = \langle \Theta_j \rangle \sim \bar{t}^{\beta_\Theta} \sim t^{\beta_j} \quad (g_i \equiv 0, i \neq 0) \qquad (35)$$

and find

$$\beta_1 = \beta_\Psi = \beta \quad \text{and} \quad \beta_j = 2 - \alpha - \phi_j. \qquad (36)$$

Similarly, we can define the cross-susceptibility exponents via

$$\chi_{jk} = \int d\boldsymbol{r}\, K_{\Theta_j \Theta_k}(\boldsymbol{0}, \boldsymbol{r}) \sim 1/t^{\gamma_{jk}} \quad (g_i = 0, i \neq 0) \qquad (37)$$

and find

$$\gamma_{11} = \gamma_{\Psi\Psi} = \gamma \quad \text{and} \quad \gamma_{jk} = \phi_j + \phi_k - (2-\alpha), \qquad (38)$$

and the generalized Essam–Fisher relation

$$\alpha + \beta_j + \beta_k + \gamma_{jk} = 2. \qquad (39)$$

One may also (although somewhat redundantly) generalize the exponent relation in (26) by defining δ_j through

$$\langle \Theta_j \rangle \sim g_j^{1/\delta_j} \quad \text{or} \quad g_j \sim \langle \Theta_j \rangle^{\delta_j}, \quad (g_i = 0, j \neq i) \qquad (40)$$

which leads to

$$\delta_1 = \delta, \quad \delta_j = 1/(1-\alpha_j), \qquad (41)$$

and to the generalized Widom relation,

$$\beta_j + \gamma_{jj} = \beta_j \delta_j = \phi_j. \qquad (42)$$

Finally, for the generalized multiple susceptibilities defined in (12) we may write (in a slightly modified notation)

[1] For $n=3$, the three-dimensional models studied yield $\alpha_1 = \alpha_\Psi \simeq 0.8$, $\alpha_2 = \alpha_\Xi \simeq 0.3$, $\alpha_3 = \alpha_\mathcal{E} \approx -0.1$ (see [14, 26]) Note that when an operator like Ξ is added to the $\mathcal{H}_c$ we expect, after variation of t, etc. to find a crossover to a different critical Hamiltonian with different exponents [12, 14, 16].

$$\chi_{\{m_1,\ldots,m_j,\ldots\}} = \prod_j \left(\frac{\partial}{\partial g_j}\right)^{m_j} \bar{F} \sim 1/t^{\gamma\{m_1, m_2,\ldots\}},$$

$$(g_i = 0, i \neq 0), \quad (43)$$

where $m_j = 0, 1, 2, \ldots$ is the multiplicity of the critical operator Θ_j in $\chi_{ABC\ldots}$. From the scaling hypothesis (31) we find immediately

$$\gamma\{m_1, m_2, \ldots\} = \sum_j m_j \phi_j - (2 - \alpha) \quad (44)$$

which generalizes (38).

Correlation Length and Hyperscaling

In order to discuss the scaling of the correlations and cumulants we may proceed, as originally, by first introducing the correlation length $\xi(g_1, g_2, \ldots)$. As discussed in [27, 28] there are many specific definitions one can adopt [e.g. in terms of various spatial moments of $G(\boldsymbol{r})$], but we expect them all to yield the same critical exponents and exhibit the same form of scaling behavior. Correlation exponents ν_j can be defined, [generalizing the standard definitions] by

$$\xi \sim g_i^{-\nu_i}, \quad (g_j = 0, j \neq i), \quad (45)$$

with, for the order parameter and energy, the identifications

$$\nu_1 = \nu^c \quad \text{and} \quad \nu_0 = \nu, \ (g_0 = \bar{t}). \quad (46)$$

The appropriate scaling hypothesis (by analogy with the form of the thermodynamic functions) is clearly[1]

$$\xi(g_1, g_2, \ldots) \approx g_i^{-\nu_i} X_i(g_1/g_i^{\phi_{i1}}, \ldots, g_j/g_i^{\phi_{ij}}, \ldots), \quad (47)$$

whence the exponents ν_j must be related to one-another by

$$\nu_i = \phi_{ij}\nu_j \quad \text{or} \quad \nu_j = \nu/\phi_j, \quad (48)$$

where we have used (46) and the convention (33). By using the expression (32) for ϕ_{ij} we can rewrite this as

$$\frac{\nu_i}{2-\alpha_i} = \frac{\nu_j}{2-\alpha_j}. \quad (49)$$

[1] The form of the scaling function X_i must, of course depend on the particular definition of ξ which is adopted.

From these and the earlier exponent relations it is clear that a complete set of exponents for both thermodynamic and correlation scaling is provided either by (*i*) an exponent α_j for each critical operator Θ_j together with *one* correlation length exponent ν_i (say, $\nu_0 = \nu$), or (*ii*) by a correlation exponent ν_j for each operator together with one fluctuation exponent α_i (say, $\alpha_0 = \alpha$). The relations for these two choices are presented in table 1. Other intermediate combinations would, of course, also do. As in previous arguments leading to "two-exponent scaling" or "hyperscaling", we may ask if this set of exponents cannot be reduced by some relation among them. The simplest approach is to assert baldly that the critical free energy density $\bar{F}(g_i)$ [which has dimensions $(\text{length})^{-d}$ in a d-dimensional system] should vary as

$$[\xi(g_i)]^{-d} \sim g_i^{d\nu_i}.$$

This leads immediately [see (18)] to

$$d\nu_i = 2 - \alpha_i, \quad (50)$$

which generalizes the well known hyperscaling relation $d\nu = 2 - \alpha$. Note that by (49) this relation holds for all i if valid for any i: thus the relation would also follow from most of the arguments normally used to demonstrate hyperscaling in the original, less general contexts. The other hyperscaling relations, involving β, γ, etc., can be generalized by using (36) to (42). Once we accept the hyperscaling argument, either the set of fluctuation exponents $\{\alpha_j\}$, *or* the set of correlation exponents $\{\nu_j\}$ is complete.

Scaling of the Cumulants and Critical Dimensions

At this stage it is fairly straightforward to extend the scaling hypothesis (5) for the cumulants at the critical point to the whole critical neighborhood. In effect, the approach is to generalize to many variables the simple treatment used for $G(\boldsymbol{r}, T)$ [1]. For the cumulant

$$K_{\{m_1, m_2,\ldots\}}(\boldsymbol{r}_1, \boldsymbol{r}_2, \ldots; g_1, g_2, \ldots) \equiv K_{\{m_j\}}(\{\boldsymbol{r}_k\}; \{g_j\})$$

of critical operators Θ_j with multiplicities m_j we make a homogeneity hypothesis (as in [2, 7]) for the asymptotic behavior, namely,

Table 1. *General exponent relations*

Exponent ($\alpha_0=\alpha, \nu_0=\nu$)	Basic exponent set (i fixed, $j=1, 2, \ldots$) $[\nu_i; \{\alpha_j\}]$	$[\alpha_i; \{\nu_j\}]^a$	$[\omega^*; \{\omega_j\}]$
$2-\alpha_j$	$2-\alpha_j$	$\frac{\nu_j}{\nu_i}(2-\alpha_i)$	$\frac{d-\omega^*}{d-\omega_j}$
ν_j	$\left(\frac{2-\alpha_j}{2-\alpha_i}\right)\nu_i$	ν_j	$\frac{1}{d-\omega_j}$
ϕ_{ij}	$\frac{2-\alpha_i}{2-\alpha_j}$	$\frac{\nu_i}{\nu_j}$	$\frac{d-\omega_j}{d-\omega_i}$
β_j	$\left(\frac{2-\alpha}{2-\alpha_j}\right)(1-\alpha_j)$	$2-\alpha-\frac{\nu}{\nu_j}$	$\frac{\omega_j-\omega^*}{d-\omega_0}$
γ_{jj}	$\left(\frac{2-\alpha}{2-\alpha_j}\right)\alpha_j$	$\frac{2\nu}{\nu_j}-2+\alpha$	$\frac{d-2\omega_j+\omega^*}{d-\omega_0}$
δ_j	$\frac{1}{1-\alpha_j}$	$\frac{\nu_i}{\nu_j(2-\alpha_i)-\nu_i}$	$\frac{d-\omega_j}{\omega_j-\omega^*}$
ω^*	$d-\frac{2-\alpha_i}{\nu_i}$	$d-\frac{2-\alpha_i}{\nu_i}$	ω^*
ω_j	$d-\left(\frac{2-\alpha_i}{2-\alpha_j}\right)\frac{1}{\nu_i}$	$d-\frac{1}{\nu_j}$	ω_j

[a] For the exponents β_j and γ_{jj} we suppose we are given $\alpha_i=\alpha_0=\alpha$.

$$K_{\{m_j\}}(\{\boldsymbol{r}_k\};\{g_j\}) \approx \mathcal{K}_{\{m_j\}}(\{\boldsymbol{r}_k\};\{g_j\}) \quad \text{for} \quad |\boldsymbol{r}_k-\boldsymbol{r}_l| \gg R_0, \quad g_j\to 0, \tag{51}$$

where R_0 is the range of the critical interactions and where

$$\mathcal{K}_{\{m_j\}}(\{l\boldsymbol{r}_k\};\{l^{-\theta_j}g_j\}) = l^{-\omega\{m_j\}}\mathcal{K}_{\{m_j\}}(\{\boldsymbol{r}_k\};\{g_j\}). \tag{52}$$

(All formulae here and below use the multiplicity convention introduced in (43).) As usual we must relate the θ_j and $\omega\{m_j\}$ to the thermodynamic exponents. First we recover the critical point behavior by setting all $g_j=0$. Choosing $l=R^{-1}(\{\boldsymbol{r}_k\})$, with $R\equiv R_n$ defined as, for example, in (6) we reproduce (9) with the critical point 'amplitude'

$$D^c_{\{m_j\}}(\{\boldsymbol{x}_k\}) = \mathcal{K}_{\{m_j\}}(\{\boldsymbol{x}_k\};\{0\}). \tag{53}$$

Then choosing $l=g_i^{1/\theta_i}$ for a particular i, the relation

$$\mathcal{K}_{\{m_j\}}(\{\boldsymbol{r}_k\};\{g_j\}) = g_i^{\omega\{m_j\}/\theta_i}\mathcal{K}_{i\{m_j\}}(\{\boldsymbol{r}_k/g_i^{-1/\theta_i}\};\{g_j/g_i^{\theta_j/\theta_i}\}) \tag{54}$$

results. (As in the thermodynamic scaling analysis we indicate by the subscript i the fact that the function $\mathcal{K}_{i\{m\}}$ has the i-field scaling form and actually has one less independent variable than $\mathcal{K}_{\{m_j\}}$.) By comparison with the scaled fields entering the scaling forms for the free energy in (31), we must make the identification

$$\theta_j/\theta_i = \phi_{ij}. \tag{55}$$

Furthermore, if we set $g_j=0$ for all $j\neq i$, we expect all lengths to be scaled by $\xi(g_i)\sim g_i^{-\nu_i}$ [see (45)]; comparison with (54) immediately yields

$$\theta_i = 1/\nu_i. \tag{56}$$

This, of course, holds for all i and (55) then simply reproduces the relation (48).

Having made these identifications we return again to the homogeneity hypothesis (52) and introduce the overall configuration scale $R(\{\boldsymbol{r}_k\})$ by writing $r_k/g_i^{-\nu_i} = y_i(r_k/R)$ and $g_i^{\omega\nu_i} = y_i^{\omega} R^{-\omega}$ with $y_i = R/g_i^{-\nu_i}$. This yields the required generalization of the critical point postulate (9), to the critical neighborhood, namely,

$$K_{\{m_j\}}(\{\boldsymbol{r}_k\}; \{g_i\}) \approx R^{-\omega\{m_j\}} D_{i\{m_j\}}(\{\boldsymbol{r}_{kl}/R\}; R/g_i^{-\nu_i}; \{g_j/g_i^{\phi_{ij}}\}), \tag{57}$$

where we have explicitly recognized the translational invariance of the cumulants by introducing the differences $\boldsymbol{r}_{kl} = \boldsymbol{r}_l - \boldsymbol{r}_k$. The scaling function $D_{i\{m_j\}}(\{x_{kl}\}; y; \{z_j\})$ is seen to be equal to $y^{\omega}\,\mathcal{K}_{i\{m_j\}}(\{yx_{kl}\}; \{z_j\})$.

Our last task is to demonstrate that the decay exponents $\omega\{m_1, m_2, \dots\}$ are determined in terms of the fluctuation and correlation exponents α_j and ν_j. To do this we use the general fluctuation–susceptibility relation (12) and the expression (44) for the susceptibility exponents. Provided the corresponding susceptibility $\chi_{\{m_j\}}$ diverges, we may perform the required $(\Sigma_j m_j - 1)$-fold spatial integral (each being a d-fold scalar integral) by using the scaling forms (54) or (57) for $K_{\{m_j\}}$ with $g_i = g_0$, $g_j = 0$ $(j \neq i)$, and making the change of variables $\boldsymbol{x}_k = \boldsymbol{r}_k g_0^{\nu}$ and noting that R is a homogeneous function of degree unity in the variables $\boldsymbol{r}_k$. In the standard way (see e.g. [1, 17]) this leads directly to

$$\gamma\{m_1, m_2, \dots\} = \nu d(\sum_j m_j - 1) - \nu\omega\{m_1, m_2, \dots\}. \tag{58}$$

This relation is, of course, simply the analog of the standard relation $\gamma = (2-\eta)\nu$. On combining this with (44) and solving for $\omega\{m_1, \dots\}$ we may write the answer as

$$\omega\{m_1, m_2, \dots\} = \sum_j m_j\omega_j - \omega^*, \tag{59}$$

where, using earlier relations for ϕ_j,

$$\omega_j = d - (\phi_j/\nu) = d - (\phi_{ij}/\nu_i) = d - (1/\nu_j), \quad \text{(all } i\text{)}, \tag{60}$$

while

$$\omega^* = d - (2-\alpha)/\nu = d - (2-\alpha_i)/\nu_i, \quad \text{(all } i\text{)}. \tag{61}$$

The exponents $\{\omega_j\}$ together with ω^* form a complete set since on inverting these last relations, we find

$$\nu_j = \frac{1}{d-\omega_j} \quad \text{and} \quad 2-\alpha_j = \frac{d-\omega^*}{d-\omega_j}. \tag{62}$$

Other relations are listed in table 1. The exponent ω_j may be termed the critical dimension[1] of the operator Θ_j since each operatoren tering a cumulant contributes a factor $1/R^{\omega_j}$ to the critical point behavior. We may thus rewrite the full scaling hypothesis for the cumulants [by choosing $l = R^{-1}$ in (54)] as

$$K_{\dots\Theta_i\dots\Theta_j\dots} = \langle \dots \Theta_i \dots \Theta_j \dots \rangle_c \approx Q/R^{\dots+\omega_i\dots+\omega_j\dots-\omega^*}, \tag{63}$$

as $R \to \infty$, where the scaling amplitude is

$$Q = \mathcal{K}\left(\dots\frac{r_{kl}}{R}, \dots; \dots \frac{g_i}{R^{\omega_i - d}}, \dots \frac{g_j}{R^{\omega_j - d}}, \dots\right), \quad r_{kl} \to \infty, g_j \to 0. \tag{64}$$

The simplicity of this formulation is most appealing since it displays directly the transformation properties of all the critical operators and their conjugate fields with respect to a change of spatial scale. This viewpoint reflects closely that taken originally by Kadanoff [4, 30]. However, Kadanoff did not allow for the presence of the exponent ω^* since his approach led naturally to a hyperscaling theory; see also Patashinskii [31] who uses a somewhat different approach, and below. This point was noted by Domb [24] who discussed the cumulants for Ψ on the basis of a random walk analysis of the diagrammatics, which for $K_{\Psi\Psi}(\mathrm{r}) = G(\mathrm{r})$, corresponds closely to the Ornstein-Zernike approximation. Domb argued that for large d one has $\alpha, \eta \to 0$ and $\nu \to \frac{1}{2}$, which then requires the presence of an exponent equivalent to ω^* as mentioned in the Introduction.

To make contact with more familiar exponents we note the critical dimensions of the energy, namely,

---

[1] The term "anomalous dimension" has been used in a field-theoretic context Wilson [11, 29] for the analagous exponent of the vacuum expectation value of an operator product.

$$\omega_0 = \omega_{\mathcal{E}} = d - (1/\nu) \tag{65}$$

which then, typically ranges from $d-1$ to $d-2$, and of the order parameter,

$$\omega_1 = \omega_{\Psi} = d - (\beta\delta/\nu) = d - (1 - \tfrac{1}{2}\eta) - (1 - \tfrac{1}{2}\alpha)/\nu. \tag{66}$$

which usually varies from $d-2$ to $d-3$. Clearly the smaller the critical dimension, the stronger the fluctuations of the associated operator. For the plane Ising models the appropriate values are $\omega_{\mathcal{E}} = 1$, $\omega_{\Psi} = \frac{1}{8}$, and $\omega^* = 0$ (see [17]).

It is clear from the boundedness of the cumulants that the critical dimension ω_j cannot be negative. On the other hand, suppose in (63) that an operator Θ_k has a dimension ω_k *exceeding* the dimensionality d. Then in the calculation of a susceptibility $\chi_{\ldots m_k \ldots}$, the integrations $\int d^d r / r^{\omega_k}$ will converge at infinity and hence cannot contribute to the divergence of the susceptibility however large m_k. In addition, by (62) and (48), $\omega_k > d$ implies $\nu_k < 0$ and thence $\phi_k = -|\phi_k| < 0$ (assuming $\nu_0 = \nu > 0$). In this case, the scaling relation (31) for the free energy leads to

$$\bar{F}(g_1, \ldots) \approx g_0^{2-\alpha}\, Y_0(g_1/g_0^{\phi_1}, \ldots g_k g_0^{|\phi_k|}, \ldots), \tag{67}$$

which is to hold as $g_0 \sim t$, $g_1, \ldots g_k, \ldots$ approach zero; but in this limit $y_k = g_k t^{|\phi_k|}$ *always* approaches zero (in contrast to the usual scaling situation where $y_j = g_j/t^{\phi_j}$ can approach any limit or diverge to infinity). It follows that we may, asymptotically, replace y_k by zero, so that the field g_k is irrelevant to the critical thermodynamics. Thus we may term operators Θ_k for which $\omega_k > d$, *irrelevant operators* to distinguish them from the normally considered *relevant operators* for which $\omega_k < d$. In fact, the irrelevant operators may show up in the *corrections* to the asymptotic scaling laws since we may expect the scaling function $Y_0(\ldots y_k, \ldots)$ to have an expansion

$$Y_0(y_1, \ldots, y_k, \ldots) = Y_0(y_1, \ldots, 0, \ldots) + y_k Y_{0;k}(y_1, \ldots, 0, \ldots) + \ldots \tag{68}$$

so that exponents related to ω_k will appear in the correction factors for divergent derivatives of the free energy as observed by Wegner [16] within the renormalization group context.

On the borderline $\omega_k = d$ we have *marginal operators* for which the integral $\int dr/r^{\omega_k}$ is only logarithmically divergent. We may thus expect logarithmic dependencies on the critical fields to appear when such operators are present. Furthermore, as argued by Kadanoff & Wegner [32], such operators may lead to continuously variable exponents as encountered in the exact solution of the eight-vertex model by Baxter [33].

The exponent ω^* in the basic scaling hypothesis (63) can be regarded as entering the formulation naturally because in taking the thermodynamic limit in the fluctuation–susceptibility relation, *one* spatial integration is lost. (Recall that a single-variable cumulant K_A is just equal to $\langle A \rangle$ and has no spatial dependence in the thermodynamic limit; all higher order cumulants depend only on the spatial differences.) Somewhat picturesquely we may thus refer to ω^* as the *critical dimension of the vacuum.* Now from the relation (61) and the hyperscaling (or "two-exponent" scaling) formula (50) we see that the hyperscaling hypothesis may be expressed succinctly as

$$\omega^* = 0. \tag{69}$$

We mentioned the evidence on the validity of hyperscaling briefly in the Introduction. To expand in summary, one does find $\omega^* = 0$ for the two-dimensional Ising model and for the spherical model for $2 < d \leqslant 4$, although few higher order cumulants have actually been closely studied. However, the numerical evidence for the three-dimensional Ising models (see [1, 22, 28]) indicates $\omega^* \simeq 1/12$, and this seems to agree well with the best experiments determining α, β, γ, ν, and η for binary alloys [34–39]. Furthermore, the normal interpretations of the exponents $\alpha, \beta, \gamma, \delta, \eta$ and ν for the spherical model in more than four dimensions (which are consistent with standard scaling), lead unequivocally to $\omega^* = d - 4$; on the face of it, this is a clear breakdown of hyperscaling.

IV. Scaling Covariance

The basic result (63) for the scaling of the general m-point cumulant

$$K_{\{m_i\}}(\boldsymbol{r}_1, \ldots \boldsymbol{r}_n; g_1, g_2, \ldots) = \langle \Theta(\boldsymbol{r}_1) \ldots \Theta_j(\boldsymbol{r}_k) \ldots \rangle_{c\{g_i\}},$$
$$m = \sum_i m_i, \tag{70}$$

may be reformulated[1] as a covariance under a transformation which, on the one hand, changes the linear scale according to

$$\text{(a)}\quad \boldsymbol{r} \Rightarrow \boldsymbol{r}' = l\boldsymbol{r} \tag{71}$$

and, on the other hand, transforms the critical operators or densities according to

$$\text{(b)}\quad \Theta_j(\boldsymbol{r}) \Rightarrow \Theta_j'(\boldsymbol{r}') = l^{-\omega_j}\,\Theta_j(\boldsymbol{r}). \tag{72}$$

The transformation rule thus introduces the critical dimension ω_j of the operator $\Theta_j(\boldsymbol{r})$.

To see how the argument goes, we confine attention first to the critical point $g_i \equiv 0$ (all i). Then we postulate that the cumulant function transforms as

$$\text{(a)}\quad K_{\{m_j\}}(\{\boldsymbol{r}_k\};\{0\}) \Rightarrow K'_{\{m_j\}} = K_{\{m_j\}}(\{\boldsymbol{r}_k'\};\{0\}). \tag{73}$$

On the other hand, the cumulant expectation transforms according to

$$\text{(b)}\quad \langle\prod_j \Theta_j(\boldsymbol{r}_k)\rangle_C \Rightarrow \langle\prod_j \Theta_j(\boldsymbol{r}_k)\rangle_C' = \langle\prod_j \Theta_j'(\boldsymbol{r}_k')\rangle_C. \tag{74}$$

If we use (71) and (72) and on the basis of the covariance, equate $K'_{\{m_j\}}$ to $\langle\prod_j \Theta_j\rangle_C'$ we obtain

$$K_{\{m_j\}}(\{l\boldsymbol{r}_k\};\{0\}) = l^{-\Sigma_j \omega_j}\langle\prod_j \Theta_j(\boldsymbol{r}_k)\rangle_{C\{0\}}$$
$$= l^{-\Sigma_j \omega_j} K_{\{m_j\}}(\{\boldsymbol{r}_k\};\{0\}). \tag{75}$$

But on choosing $l = 1/R$, we recognize that this relation is equivalent to the original scaling formula (63) in the hyperscaling form $\omega^* = 0$ (and at the critical point $g_i = 0$). Thus the homogeneity relations are implied by covariance under the transformation of each operator as defined in (72). It should be noticed, however, that we wrote a precise transformation law in (73) for the exact cumulant $K_{\{m_j\}}$; accordingly the scaling relation (75) also appears as a precise statement; in fact, homogeneity is expected to be valid only asymptotically for $r_{kl} \gg a$; strictly, therefore, the transformation law (73), and its generalizations to be introduced below, should be understood as valid only asymptotically, or else should be expressed in terms of the asymptotic cumulants, $\mathcal{K}_{\{m_j\}}$. For notational simplicity, however, we will use the formally exact expressions in what follows, merely recalling the true asymptotic character at appropriate points.

[1] This formulation was first emphasized by Wilson [29] in a field-theoretic context; see also Polyakov [40] and Kadanoff & Ceva [9]. These authors confined themselves to the hyperscaling situation where $\omega^* = 0$.

In order to extend the formulation to accommodate the possibility of a nonzero value of ω^*, we may replace the cumulant transformation rule (73) by

$$\text{(a)}\quad K_{\{m_j\}}(\{\boldsymbol{r}_k\};\{g_i\}) \Rightarrow K'_{\{m_j\}}$$
$$= l^{-\omega^*} K_{\{m_j\}}(\{\boldsymbol{r}_k'\};\{g_i\}). \tag{76}$$

where we have also allowed for nonzero fields g_i, which will be considered below. On repeating the arguments one sees that the exponent of l in the covariance relation (75) will now be replaced by $(\omega^* - \sum_j \omega_j)$; this yields correctly the more general scaling relation.

It is instructive, with later applications in mind, to illustrate how the infinitesimal forms of the transformations can be utilized to derive equivalent results. Thus suppose

$$l = 1 + \delta l \quad (|\delta l| \ll 1) \tag{77}$$

so that the transformation of space can be interpreted as a uniform, isotropic infinitesimal stretching of the distance between any two points, namely,

$$\text{(a)}\quad \boldsymbol{r}_{ik} \Rightarrow \boldsymbol{r}_{ik}' \quad \text{with} \quad r_{ik}'/r_{ik} = 1 + \delta l,$$
$$\text{or} \quad \delta \ln|\boldsymbol{r}_{ik}| = \delta l. \tag{78}$$

The corresponding infinitesimal transformation of the operators is clearly given by

$$\text{(b)}\quad \delta\Theta_j(\boldsymbol{r}) = -\omega_j\,\Theta_j(\boldsymbol{r})\,\delta l. \tag{79}$$

For simplicity let us now consider a two-point cumulant, K_{ab}, at the critical point, that is,

$$K_{ab}(r_{12};\{0\}) = \langle\Theta_a(\boldsymbol{r}_1)\,\Theta_b(\boldsymbol{r}_2)\rangle, \tag{80}$$

where we have extended the original cumulant notation slightly in an obvious way.

We have here explicitly invoked translational and rotational invariance by displaying only the argument $r_{12} = |\boldsymbol{r}_2 - \boldsymbol{r}_1|$ (and used $\langle\Theta_j\rangle = 0$). Then from (76) we have

$$\text{(a)}\quad \delta K_{ab} = -\omega^* K_{ab}(r_{12})\,\delta l + \frac{\partial K_{ab}}{\partial \ln r_{12}}\,\delta l, \tag{81}$$

while, on the other hand using (79) yields

$$\text{(b)}\quad \delta\langle\Theta_a\Theta_b\rangle = -\omega_a\langle\Theta_a\Theta_b\rangle\delta l - \omega_b\langle\Theta_a\Theta_b\rangle\delta l$$
$$= -(\omega_a+\omega_b)K_{ab}(r_{12})\delta l. \qquad (82)$$

By the covariance hypothesis we may equate these two expressions: this produces a differential equation which is immediately integrated to give

$$K_{ab}(r_{12};\{0\}) = D^c_{ab}/r_{12}^{\omega_a+\omega_b-\omega^*}, \qquad (83)$$

which is just the expected scaling result. The critical amplitude D^c_{ab} appears as the constant of integration.

In principle, D^c_{ab} could depend explicitly on the direction of the vector $\boldsymbol{r}_{12}$, i.e., on the unit vector $\boldsymbol{e}_{12} = \boldsymbol{r}_{12}/r_{12}$. In fact, however, in the same way that we have taken for granted translational invariance in the thermodynamic limit,[1] so we expect all the (limiting) cumulants to be *rotationally invariant*. For continuum systems this can be concluded from the rotational invariance of the Hamiltonian. On the other hand, for a lattice system (which generally has only particular reflection and rotation symmetries) this argument fails. In that case we expect the cumulants to display full rotational symmetry only near the critical point and for large r_{jk}/a. This is not, of course, obvious but it is rather well borne out by experiment and by detailed model calculations and numerical studies (see e.g. [17, 28]). We will assume such asymptotic rotational invariance where needed below without further comment.

Finally, in order to rederive the homogeneity expressions away from the critical point we examine the transformations induced on the fields g_i by the scale and operator transformations. The field g_i enters the reduced Hamiltonian $\bar{\mathcal{H}}$ defining the expectation $\langle\cdot\rangle$, via

$$\delta\bar{\mathcal{H}} = \sum_i g_i \int \Theta_i(\boldsymbol{r})\,\mathrm{d}\boldsymbol{r},$$
$$\Rightarrow \sum_i g_i \int \Theta_i'(\boldsymbol{r}')\,\mathrm{d}\boldsymbol{r}' = \sum_i g_i l^{d-\omega_i} \int \Theta_i(\boldsymbol{r})\,\mathrm{d}\boldsymbol{r}. \qquad (84)$$

The effect of the transformations is thus a rescaling of the fields according to

$$\text{(b)}\quad g_i \Rightarrow g_i' = l^{d-\omega_i} g_i \qquad (85)$$

which now supplements (74). The transformation rule (76) is retained. For general $\{g_i\}$ (and ω^*) the covariance relation (75) thus becomes

$$K_{\{m_j\}}(\{l\boldsymbol{r}_k\};\{g_i\}) = l^{\omega^*-\Sigma_j\omega_j} K_{\{m_j\}}(\{\boldsymbol{r}_k\};\{l^{d-\omega_i}g_i\}). \qquad (86)$$

This is quite equivalent to the original hypothesis (63) as can be seen by putting $l = 1/R$ (as before) and also setting $g_i = g_i''/l^{d-\omega_i} = g_i''/R^{\omega_i-d}$.

V. Conformal Covariance

The scaling or homogeneity hypotheses predict and inter-relate the overall power law behavior of the cumulants and, via spatial integrations, of the free energy. However, the forms of the scaling functions or amplitudes—specifically the functions Q in (64)—are left essentially unrestricted. (But note that the finiteness of various limiting values is normally assumed implicitly.) There are two principles which yield more detailed information on the cumulants: the first, which we take up now, is the hypothesis of conformal covariance; the second, to be discussed below, is the reducibility of operator products.

Ordinary Euclidean geometry rests on the so-called principal group of transformations which is generated by translations, rotations, reflections, and dilations (or scale transformations). As we have seen, the critical point cumulants appear to be covariant under all elements of this group. Covariance with respect to dilations is, of course, just scaling covariance. Covariance with respect to rotations and translations has been discussed above. Reflection covariance follows formally from the reflection symmetries of the Hamiltonian. All the transformations of the principal group are *conformal* in that they preserve the angles between intersecting lines, or, equivalently carry any triangle into a similar triangle. (The principal group, of course, transforms lines into lines.) A general conformal transformation is only required to carry infinitesimal triangles into similar infinitesimal triangles. As is well known, the full group of conformal transformation of the plane ($d = 2$) has an intimate connection with the theory of

[1] Translational invariance of the cumulants follows formally from the translational invariance of the Hamiltonian of an infinite system.

analytic functions: indeed, every analytic function defines a conformal mapping of the complex plane (except at any singular points of the function). On the other hand, this richness is absent for $d \geqslant 3$. In that case the full conformal group can be obtained simply by adjoining to translations, rotations, reflections, and dilations, the operation of inversion with respect to the unit sphere, namely,

$$\boldsymbol{r} \Rightarrow \boldsymbol{r}' = \boldsymbol{r}/|\boldsymbol{r}|^2. \tag{87}$$

By combining this with translations and dilations we see that inversion with respect to any sphere is included in the conformal group. Such inversions carry spheres into spheres (where, as usual, planes are considered to be degenerate spheres).

A group of special conformal transformations, parametrized by a vector or, more precisely, a reciprocal vector $\boldsymbol{q}$, can be formed by taking the product of an inversion and a translation, followed by the reciprocal inversion. This yields the transformation

$$\text{(a)}\quad \boldsymbol{r}' \Rightarrow \boldsymbol{r}' = \frac{\boldsymbol{r} + |\boldsymbol{r}|^2 \boldsymbol{q}}{1 + 2\boldsymbol{r}\cdot\boldsymbol{q} + |\boldsymbol{r}|^2|\boldsymbol{q}|^2}, \tag{88}$$

which may be written less explicitly but more transparently as

$$\boldsymbol{r}/|\boldsymbol{r}'|^2 = \boldsymbol{r}/|\boldsymbol{r}|^2 + \boldsymbol{q}. \tag{89}$$

It is worth noting that in the complex plane ($z = x + iy$) this transformation becomes simply

$$z \Rightarrow z' = z/(1 + q^* z). \tag{90}$$

where $q^* = q_x - iq_y$. For the deeper study of this special group, an important role is played by an equivalence with orthogonal transformations in a space of $d+2$ dimensions with an indefinite metric [of signature $(+ \ldots +;\ + \ -)$] (see [41–43]). We will not, however, need this fact.

Following Polyakov [20] it is natural to ask whether the critical point cumulants might also be covariant under the special conformal group. Polyakov was motivated by the conformal invariance of the field-theoretic "bootstrap" equations derived previously by himself [44] and Migdal [45]. To investigate the consequences of such covariance and, indeed, to see how it may be implemented, it is appropriate to study the infinitesimal transformation with parameter $\delta\boldsymbol{q}$. [In fact there are difficulties in requiring global conformal invariance, associated with the singularities of (88) for finite $\boldsymbol{q}$.] We find

$$\delta\boldsymbol{r} = \boldsymbol{r}' - \boldsymbol{r} = |\boldsymbol{r}|^2 \delta\boldsymbol{q} - 2\boldsymbol{r}(\boldsymbol{r}\cdot\delta\boldsymbol{q}), \tag{91}$$

to first order in $\delta\boldsymbol{q}$. From this we can calculate the change in the distance between two points $\boldsymbol{r}_i$ and $\boldsymbol{r}_k$ with the result

$$\text{(a)}\quad r_{ik} \Rightarrow r'_{ik} \quad \text{with} \quad r'_{ik}/r_{ik} = 1 - (\boldsymbol{r}_i + \boldsymbol{r}_k)\cdot\delta\boldsymbol{q},$$

$$\text{so that}\quad \delta \ln|r_{ik}| = -(\boldsymbol{r}_i + \boldsymbol{r}_k)\cdot\delta\boldsymbol{q}. \tag{92}$$

On letting $\boldsymbol{r}_i \to \boldsymbol{r}_k$ we discover that the infinitesimal conformal transformation may be interpreted as an isotropic but spatially nonuniform stretching by a factor

$$l(\boldsymbol{r}) = 1 + \delta l(\boldsymbol{r}) \quad \text{with} \quad \delta l(\boldsymbol{r}) = -2\boldsymbol{r}\cdot\delta\boldsymbol{q}. \tag{93}$$

(As a matter of fact, for finite $\boldsymbol{q}$, the local stretching factor is given by the same scale factor entering (88), that is $l(\boldsymbol{r}) = [1 + 2\boldsymbol{r}\cdot\boldsymbol{q} + r^2q^2]^{-1}$.) Now by generalizing the previous expression (79) for the case of scale covariance, we see that the corresponding infinitesimal transformation of the critical operators should be

$$\text{(b)}\quad \delta\Theta_j(\boldsymbol{r}) = +2\omega_j\,\Theta_j(\boldsymbol{r})\,\boldsymbol{r}\cdot\delta\boldsymbol{q}. \tag{94}$$

To impose covariance under these transformations we will restrict ourselves to the critical point ($g_i \equiv 0$) and compute $\delta K = K' - K$ from the simple expression (73). (To avoid overburdening the notation we will hence drop the superscript c in most places.) Consider first a one-point cumulant

$$K_a^c(\boldsymbol{r}) = K_a^c = \langle\Theta_a(\boldsymbol{r})\rangle. \tag{95}$$

By translational invariance K_a is independent of $\boldsymbol{r}$. Hence on equating the increments we have

$$\delta K_a = 0 = \delta\langle\Theta_a(\boldsymbol{r})\rangle = 2\omega_a\langle\Theta_a\rangle\,\boldsymbol{r}\cdot\delta\boldsymbol{q}. \tag{96}$$

Since this must hold for arbitrary $\delta\boldsymbol{q}$ it follows that

$$K_a^c = \langle\Theta_a(\boldsymbol{r})\rangle \equiv 0 \quad \text{if} \quad \omega_a \neq 0. \tag{97}$$

This is quite consistent with our original specification that the critical operators should have vanishing expectations at the critical point, but it emphasizes that such a property is essential if conformal covariance is to hold. (One may, at this point, consistently assign the exponent $\omega^* = \omega_I = 0$ to the identity operator $I(\boldsymbol{r})$; but again, see below.)

More significant results are obtained for the two-point cumulant K^c_{ab} which, as before, must be a function only of r_{12}. Equating the increments yields [compare with (82)]

$$\partial K_{ab} = -\frac{\partial K_{ab}}{\partial \ln r_{12}} (\boldsymbol{r}_1 + \boldsymbol{r}_2) \cdot \delta\boldsymbol{q} = \delta\langle \Theta_a(\boldsymbol{r}_1)\, \Theta_b(\boldsymbol{r}_2)\rangle,$$

$$= 2\omega_a K_{ab}(r_{12})\, \boldsymbol{r}_1 \cdot \delta\boldsymbol{q} + 2\omega_b K_{ab}(r_{12})\, \boldsymbol{r}_2 \cdot \delta\boldsymbol{q}. \tag{98}$$

This relation must again hold as an identity in $\delta\boldsymbol{q}$: comparing the coefficients of $\boldsymbol{r}_1 \cdot \delta\boldsymbol{q}$ and $\boldsymbol{r}_2 \cdot \delta\boldsymbol{q}$ hence yields the pair of equations

$$\frac{\partial K_{ab}}{\partial \ln r_{12}} = -2\omega_a K_{ab}(r_{12}),$$

$$\frac{\partial K_{ab}}{\partial \ln r_{12}} = -2\omega_b K_{ab}(r_{12}). \tag{99}$$

with common solutions

$$K^c_{ab}(r_{12}) = \langle \Theta_a\, \Theta_b\rangle = D^c_{ab}/r_{12}^{2\omega_a} \quad \text{for} \quad \omega_a = \omega_b,$$
$$= 0 \quad \text{for} \quad \omega_a \neq \omega_b. \tag{100}$$

The first part of this result merely reproduces the expected power law. (The power law appears in its hyperscaling ($\omega^* = 0$) form, but at this stage the exponent ω^* could have been absorbed into the definition of ω_a.) On the other hand, the second part asserts the orthogonality of the critical fluctuations (for operators of differing critical dimensions). This consequence of conformal covariance was first observed by Migdal [46] for the field-theoretic situation. In certain cases (for example, an Ising model ferromagnet) some orthogonality relations such as this are certainly to be expected on the grounds of known symmetries. Thus the spin $s(\boldsymbol{r})$ and energy density $\varepsilon(\boldsymbol{r}) = \frac{1}{2}s(\boldsymbol{r}) \sum_{\boldsymbol{\delta}} s(\boldsymbol{r} + \boldsymbol{\delta})$ must be orthogonal. In general, however, the result is new and surprising. It should be recalled, however, that in the foregoing discussion of scaling and conformal covariance we have, for convenience, used the "equals" symbol in the homogeneity relations rather than the symbol for "asymptotically equals". Thus one must be prepared for corrections to (100), at least for small r (of order a) but also, possibly, for large r in the form of higher order inverse powers of r. (Such powers, however, might be susceptible to removal by the refinement process.)

The full strength of the conformal covariance hypothesis becomes apparent in the case of a three-point cumulant as observed by Polyakov, [20]. By translational and rotational invariance K^c_{abc} must be a function only of r_{12}, r_{23}, and r_{31}. On the one hand, we then have

$$\text{(a)} \quad \delta K_{abc} = -\frac{\partial K_{abc}}{\partial \ln r_{12}} (\boldsymbol{r}_1 + \boldsymbol{r}_2) \cdot \delta\boldsymbol{q} - \frac{\partial K_{abc}}{\partial \ln r_{23}} (\boldsymbol{r}_2 + \boldsymbol{r}_3) \cdot \delta\boldsymbol{q} - \frac{\partial K_{abc}}{\partial \ln r_{31}} (\boldsymbol{r}_3 + \boldsymbol{r}_1) \cdot \delta\boldsymbol{q} \tag{101}$$

while, on the other hand,

$$\text{(b)} \quad \delta\langle \Theta_a\, \Theta_b\, \Theta_c\rangle = 2\omega_a K_{abc} \boldsymbol{r}_1 \cdot \delta\boldsymbol{q} + 2\omega_b K_{abc} \boldsymbol{r}_2 \cdot \delta\boldsymbol{q} + 2\omega_c K_{abc} \boldsymbol{r}_3 \cdot \delta\boldsymbol{q}. \tag{102}$$

As before we may equate the coefficients of $\boldsymbol{r}_1 \cdot \delta\boldsymbol{q}$ in these expressions thereby obtaining a set of three simultaneous partial differential equations. By combining equations, the variables are easily separated and one obtains the solution

$$K^c_{abc}(r_{12}, r_{23}, r_{31}) \approx \frac{D^c_{abc}}{r_{12}^{\omega_a+\omega_b-\omega_c}\, r_{23}^{\omega_b+\omega_c-\omega_a}\, r_{13}^{\omega_c+\omega_a-\omega_b}}, \tag{103}$$

where D^c_{abc} is a constant amplitude. We have replaced the previous equals sign by $\approx$ for the reasons mentioned.

This is a remarkable result since it gives the complete (asymptotic) form of the three-point cumulant of any three critical operators. It is, as expected, quite consistent with the scaling form (63) when hyperscaling is valid (so that $\omega^* = 0$). This is because the argument was based on the rule (73); (see further below). More strikingly, the predicted form has been verified by explicit calculations for the plane square Ising model with the various operators lying on a common line [47–49]. In particular, one can check the behavior of the order–order–energy correlation function—this entails the identification, at least to a sufficient degree of refinement, of the order parameter and the energy with their usual forms in terms of the spin operators. This success compels one to consider conformal covariance seriously in general systems.

Conformal covariance also yields restrictions on the form of the higher order cumulants, although the results are not as definitive as for the three-point cumulants. For the four-point cumulant analogous arguments lead to

$$K^{c}_{abcd}(r_{12}, r_{13}, r_{14}, r_{23}, r_{24}, r_{34}) \approx \frac{r_{12}^{\omega_c+\omega_d} r_{34}^{\omega_a+\omega_b}}{r_{13}^{\omega_a+\omega_c} r_{24}^{\omega_b+\omega_d} r_{14}^{\omega_a+\omega_d} r_{23}^{\omega_b+\omega_c}} \times Q_{abcd}\left(\frac{r_{12}r_{34}}{r_{13}r_{24}}, \frac{r_{12}r_{34}}{r_{14}r_{23}}\right). \quad (104)$$

As it should be, this expression is invariant, up to a redefinition of the residual amplitude function $Q_{abcd}(X, Y)$, under permutation of the points $\boldsymbol{r}_k$. Again the result is fully consistent with the previous hyperscaling relations. Note that the variables X and Y entering the amplitude function (whose form is otherwise undetermined) are the cross-ratios formed from the four points $\boldsymbol{r}_1$, $\boldsymbol{r}_2$, $\boldsymbol{r}_3$, and $\boldsymbol{r}_4$. It is easy to check, using the infinitesimal transformation (92), that the cross-ratios are conformal invariants. For m points there are $\frac{1}{2}m(m-3)$ independent cross-ratios which will appear similarly in conformally covariant expressions for the m-point cumulants.

Finally, it remains to show that conformal covariance can be formulated in a way that allows for $\omega^* \neq 0$. This possibility was not considered by Polyakov or Migdal. To do so, we must extend the general transformation rule (76) to the case of a nonuniform stretching $l(\boldsymbol{r})$ [as specified in (93)]. A natural approach is to replace l in the original formulation by the geometric mean,

$$\tilde{l}_m = \tilde{l}_m(\boldsymbol{r}_1, \boldsymbol{r}_2, \ldots, \boldsymbol{r}_m) = \left[\prod_{k=1}^{m} l(\boldsymbol{r}_k)\right]^{1/m}, \quad (105)$$

of the values of l evaluated at the m points $\boldsymbol{r}_k$ entering the cumulant $K_{\{m_j\}}$. The general transformation rule for the cumulants then becomes

$$\text{(a)} \quad K_{\{m_j\}} \Rightarrow K'_{\{m_j\}} = \tilde{l}_m^{-\omega^*} K_{\{m_j\}}(\{\boldsymbol{r}'_k\}; \{g_i\}), \quad (106)$$

where $m = \sum_i m_i$, as before, and $\tilde{l}_m(\{\boldsymbol{r}'_k\})$ is defined as above.

When this definition is used to compute the increment in $K_{\{m_j\}}$, an additional term $-(\omega^*/m)(-2\boldsymbol{r}_k \cdot \delta\boldsymbol{q})K_{\{m_j\}}$ is obtained for each argument $\boldsymbol{r}_k$. Each such term combines directly with the corresponding increment obtained by transforming the operator $\Theta_j(\boldsymbol{r}_k)$. The net effect is hence the replacement of ω_a, ω_b, etc. in the previous results by $\omega_a - (\omega^*/m)$, $\omega_b - (\omega^*/m)$, etc. For the one-point cumulant we thus obtain (97) again but with the condition $\omega_a \neq \omega^*$. (In view of this it is no longer consistent to regard the unit operator $I(\boldsymbol{r})$ as a member of the set of critical operators.) The orthogonality (100) of the two-point cumulants still applies; but when $\omega_a = \omega_b$, the cumulant decays with the standard scaling exponent

$$\omega\{\Theta_a\, \Theta_b\} = 2(\omega_a - \tfrac{1}{2}\omega^*) = 2\omega_a - \omega^* = \omega_a + \omega_b - \omega^*. \quad (107)$$

In the same way, the conformally covariant expression (103) for the three-point cumulant now becomes

$$K^{c}_{abc}(r_{12}, r_{23}, r_{31}) \approx \frac{D^{c}_{abc}}{r_{12}^{\omega_a+\omega_b-\omega_c-\frac{1}{3}\omega^*} r_{23}^{\omega_b+\omega_c-\omega_a-\frac{1}{3}\omega^*} r_{13}^{\omega_c+\omega_a-\omega_b-\frac{1}{3}\omega^*}} \quad (108)$$

This is equivalent to the original hyperscaling result multiplied by $(r_{12}r_{23}r_{31})^{\frac{1}{3}\omega^*}$. Similarly for general ω^* the four-point cumulant hyperscaling expression (104) should be multiplied by $(r_{12}r_{34}r_{13}r_{24}r_{14}r_{23})^{\omega^*/6}$.

We conclude that conformal invariance of the cumulants is consistent with standard homogeneity ($\omega^* \neq 0$) although the formulation is, perhaps, not very elegant.

VI. Product Expansions and Reducibility

The process of refinement of operators expounded earlier, leads to the expansion of an arbitrary local operator $A(\boldsymbol{r})$ in terms of critical operators, namely,

$$A(\boldsymbol{r}) \approx a_I I(\boldsymbol{r}) + \sum_j a_j\, \Theta_j(\boldsymbol{r}), \quad (109)$$

where $I(\boldsymbol{r})$ is the unit operator. Note that in general the coefficients a_j may have tensor character. Taking expectation values at the critical point ($g_i \equiv 0$) leads to $a_I = \langle A \rangle_c$ and demonstrates the necessity for including $I(\mathbf{r})$. The degree of approximation of the expansion depends on the set of critical operators retained; this set should include at least all the relevant operators. The expansion is expected to reproduce the dominant singularities of the average $\langle A(\mathbf{r})\rangle$ as a function of a field g_i, and hence is valid with fixed coefficients a_j in the immediate critical vicinity. In order to reproduce higher order singularities and any analytic background or analytic factors, it is necessary to allow the coefficients a_I and a_j themselves, to develop a dependence on the field g_i. One might hope that this dependence will in general be analytic, or at most weakly singular; however, we will avoid further consideration of this question[1] by restricting attention to the most dominant singularities and ignoring analytic contributions.

Now one way of producing new local operators is to take the product of two nearby critical operators, say, $\Theta_a(\mathbf{r}_1)$ and $\Theta_b(\mathbf{r}_2)$. We may associate the resulting product P_{ab} with the midpoint

$$\mathbf{r}_0 = \tfrac{1}{2}(\mathbf{r}_1 + \mathbf{r}_2). \tag{110}$$

Then by implementing a refinement process we can 'reduce' the product and write an operator product expansion [8, 9, 29]

$$\begin{aligned} P_{ab}(\mathbf{r}_0) &= \Theta_a(\mathbf{r}_0 + \tfrac{1}{2}\mathbf{r})\,\Theta_b(\mathbf{r}_0 - \tfrac{1}{2}\mathbf{r}), \\ &\approx p_I^{ab}(\mathbf{r})\, I(\mathbf{r}_0) + \sum_j p_j^{ab}(\mathbf{r})\,\Theta_j(\mathbf{r}_0), \end{aligned} \tag{111}$$

where $\mathbf{r} = \mathbf{r}_2 - \mathbf{r}_1$, and, as mentioned before, the p_j^{ab} will usually be tensors. When the operators Θ_a and Θ_b are spatial scalars (which is the only case we have so far considered explicitly) the expansion coefficients p_j^{ab} will depend only on the separation $r = |\mathbf{r}|$. Taking critical point expectations and recalling that $\langle \Theta_j \rangle = 0$, yields

$$\begin{aligned} p_I^{ab}(r) &= \langle \Theta_a(0)\,\Theta_b(\mathbf{r})\rangle_c = K_{ab}^c(r) \\ &\approx \frac{D_{ab}^c}{r^{\omega_a + \omega_b - \omega^*}}. \end{aligned} \tag{112}$$

Although this relation holds formally for all r it is reasonable to expect that the expansion (111) is actually valid only for r small, that is of order a microscopic length a; in other words the expansion may well fail for $r/a \gg 1$.

Let us now take the expectation value of the expansion (111) in the presence of a field g_i (with $g_i = 0$ for all $j \neq i$). This yields

$$\begin{aligned} \langle \Theta_a\,\Theta_b \rangle &= K_{ab}(r; g_i) + \langle \Theta_a\rangle \langle \Theta_b \rangle, \\ &\approx p_I^{ab}(r) + \sum_j p_j^{ab}(r)\,\langle \Theta_j \rangle. \end{aligned} \tag{113}$$

However from the thermodynamic scaling hypothesis (31) we have

$$\langle \Theta_j \rangle \approx C_{ij}\, g_i^{\phi_{ij}(1-\alpha_j)} = C_{ij}\, g_i^{\nu_i(\omega_j - \omega^*)}, \tag{114}$$

where C_{ij} is a fixed amplitude and the second equality follows from the scaling relations summarized in table 1. The expansion thus predicts the nature of the thermodynamic singularities in the two-point cumulant at short range [$r = O(a)$]: specifically, as $g_i \to 0$, we have

$$\begin{aligned} K_{ab}(r, g_i) \approx K_{ab}^c(r) &+ \sum_j E_{ij}^{ab}(r)\, g_i^{\nu_i(\omega_j - \omega^*)} \\ &- C_{ia}\, C_{ib}\, g_i^{\nu_i(\omega_a + \omega_b - 2\omega^*)}, \end{aligned} \tag{115}$$

with the amplitudes

$$E_{ij}^{ab}(r) = C_{ij}\, p_j^{ab}(r). \tag{116}$$

This result should be compared with the conclusion [21, 22 28] that the order–order correlation function $G(r; t) = K_{\Psi\Psi}(r; t)$ should, for fixed small r, contain the energy singularity $t^{1-\alpha}$. In the present formulation that prediction follows by taking $g_i = g_0 \sim t$ and examining the term $j = 0$ for which $\Theta_c = \mathcal{E}$ and $\nu_i(\omega_j - \omega^*) = \phi_{00}(1 - \alpha_0) = 1 - \alpha$. In this general context, however, we see that a $t^{1-\alpha}$ term is to be expected in any two-point cumulant.

We may, similarly, obtain the behavior of a three-point cumulant K_{abc} as its argument r_{12} becomes short range, i.e., as two points come close together. To this end multiply the product expansion (111) by $\Theta_c(\mathbf{r}_3)$ and take the critical point expectation value. With a notation consistent with (110) we find

$$K_{abc}^c(r_{12}, r_{23}, r_{31}) \approx \sum_j p_j^{ab}(r_{12})\, K_{jc}^c(r_{03}). \tag{117}$$

for $r_{12} = O(a)$ and $r_{23} \simeq r_{31} \simeq r_{03} \gg a$. The condition $r_{03} \gg a$ is needed since otherwise three operators

---

[1] The problem has been studied by K. G. Wilson in the context of field theory (see Wilson [29] section IV) who concludes that the $a_j(g_i)$ should be analytic except for the possible presence of logarithmic factors. The question can also be discussed on the basis of the renormalization group approach.

would be close together and we would have to make a fresh, triple product expansion with coefficients $p_j^{abc}(r_{12}, r_{23}, r_{31})$. This relation thus indicates the decay of the cumulant as $1/r_{03}^{\omega_c+\omega_f-\omega^*}$ as $r_{03}\to\infty$, where ω_f is the smallest exponent for which p_f^{ab} does not vanish.

VII. Strong scaling

At the cost of what is really a new hypothesis, we may also use the operator product expansion to obtain information on the cumulant scaling functions. Specifically we will demonstrate how the relations (115) and (117) yield results for the scaling functions $D_{ab}(w)$ and $D^c_{abc}(x, y, z)$ which enter the homogeneity relations

$$K_{ab}(r;g_i)\approx D_{ab}(rg_i^{\nu i})/r^{\omega_a+\omega_b-\omega^*}, \qquad (g_j=0, j\neq i), \quad (118)$$

and

$$K^c_{abc}(r_{12}, r_{23}, r_{31}) \approx D^c_{abc}\left(\frac{r_{12}}{R}, \frac{r_{23}}{R}, \frac{r_{31}}{R}\right)\Big/ R^{\omega_a+\omega_b+\omega_c-\omega^*}. \quad (119)$$

Note that, since R is at our disposal, $D^c_{abc}(x, y, z)$ is a homogeneous function of its arguments which, in turn, must satify the triangle inequalities $x+y\geqslant z$ etc. Now these homogeneity relations are expected to be valid for $r, r_{12}, r_{23}, r_{31}\gg a$; conversely the product expansions are to be valid for r and r_{12} of order a. We will nevertheless postulate that in the limits $rg_i^{\nu i}\to 0$ and r_{12}/r_{23}, $r_{12}/r_{31}\to 0$, the scaling expressions should asymptotically reproduce the product expansions; in other words there should be a region of overlapping validity. Such a postulate we term a *strong scaling hypothesis*. (This terminology was first introduced in a somewhat less general context by Ferer, Moore & Wortis [22].)

On this basis let us compare (117) and (119) for large r_{12} *and* large r_{03}/r_{12}: we find

$$\frac{D^c_{abc}(r_{12}/r_{03}, 1, 1)}{r_{03}^{\omega_a+\omega_b+\omega_c-\omega^*}}\approx p_f^{ab}(r_{12})\frac{D^c_{fc}}{r_{03}^{\omega_f+\omega_c-\omega^*}}, \quad (120)$$

where f is the label of most singular operator, Θ_f, appearing in the expansion of $\Theta_a\,\Theta_b$, subject to $D^c_{fc}=D_{fc}(0)\neq 0$. (In case distinct operators Θ_{f_1}, Θ_{f_2}, ... with equal exponents ω_f appear, one must obviously take a sum over f_i here and in the following equations.) The two sides of this relation can be consistent only if

$$p_f^{ab}(r)\approx P_f^{ab}/r^{\omega_a+\omega_b-\omega_f} \quad \text{as} \quad r\to\infty, \quad (121)$$

where P_f^{ab} is a constant amplitude, and if at the same time,

$$D^c_{abc}(x, 1, 1)\approx P_f^{ab}\,D^c_{fc}\,x^{\omega_f-\omega_a-\omega_b} \quad \text{as} \quad x\to 0. \quad (122)$$

Evidently the strong scaling postulate determines both the asymptotic forms of the operator product expansion coefficient and the behavior of the three-point scaling function when an argument is small. (Note that owing to the homogeneity and the triangle inequalities only one argument of $D^c_{abc}(x, y, x)$, can be small at a time.)

By letting Θ_c run through all the critical operators we conclude, in general, that each expansion coefficient $p_j^{ab}(r)$ should have the form (121). Notice, in addition, that these results are independent of the value of ω^* since this cancels out in (120). The same conclusion regarding the expansion coefficients is reached by comparing the scaling form (118) with the field-dependent expansion (115). Consider first the case where $\langle\Theta_a\rangle$ or $\langle\Theta_b\rangle$ vanishes even when g_i varies (so that $C_{ia}C_{ib}=0$); this will typically follow from some symmetry of the system: consider, for example, the order–order correlations in an Ising model in zero field above T_c. On cross multiplying by $r^{\omega_a+\omega_b-\omega^*}$ and putting $w=rg_i^{\nu i}$ we obtain

$$D_{ab}(w)\approx D^c_{ab}+\sum_j E_{ij}^{ab}(r)\,r^{\omega_a+\omega_b-\omega_j}\,w^{\omega_j-\omega^*}, \quad (123)$$

where, if strong scaling is correct, the right hand side must be independent of r. This means that

$$E_{ij}^{ab}(r)\sim 1/r^{\omega_a+\omega_b-\omega_j}, \quad (r\to\infty), \quad (124)$$

which is precisely equivalent to the previous results (116) and (121). The singular variation of the scaling function as $w\to 0$ (or $g_i\to 0$ at fixed r) is thus predicted to be

$$D_{ab}(w)\approx D^c_{ab}+\sum_j C_{ij}\,P_j^{ab}\,w^{\omega_j-\omega^*}. \quad (125)$$

Such a prediction was discussed originally [21] for the order–order correlation function which is expected to exhibit an energy-derived singularity as discussed above. As mentioned, one may also expect nonsingular background terms varying as w, w^2, ...

This, and similar predictions have been tested successfully for the square lattice Ising model, particularly by Kadanoff & Ceva [9].[1]

VIII. Strong scaling, hyperscaling and conformal invariance

We noted that the derivation of the formula (121) for the expansion coefficients was independent of the value of ω^*. The same is true for the scaling function formulae (122) and (125). In the latter case, however, this fact was subject to the proviso $C_{ia} C_{ib} = 0$, which enabled us to drop the last term in (115), namely

$$-C_{ia} C_{ib} g_i^{\nu_i(\omega_a+\omega_b-2\omega^*)}.$$

If the argument is repeated with this term present the matching relation (125) becomes

$$D_{ab}(w) \approx D^c_{ab} + \sum_j C_{ij} P_j^{ab} w^{\omega_j-\omega^*} - r^{\omega^*} C_{ia} C_{ib} w^{\omega_a+\omega_b-2\omega^*}. \qquad (126)$$

But the two sides of this formula can be compatible only if $\omega^* = 0$, so that the residual dependence on r drops out. Thus in the absence of some very special cancellations (arising, perhaps, from higher order corrections to the expressions for $p_j^{ab}(r)$ and $\langle\Theta_j\rangle$) *the strong scaling postulates together with the product expansion hypothesis, implies the hyperscaling relation* $\omega^* = 0$. When hyperscaling is valid (126) merely provides an extension of the previous result (125).

This conclusion is not just a consequence of having introduced a non-vanishing field g_i. Thus we might, alternatively, have examined the four-point cumulant

$$K^c_{abcd}(r_{12}, r_{13}, r_{14}, r_{23}, r_{24}, r_{34}) = \langle\Theta_a(r_1)\,\Theta_b(r_2)\,\Theta_c(r_3)\,\Theta_d(r_4)\rangle_c - K^c_{ab}(r_{12})\,K^c_{cd}(r_{34}) - K^c_{ac}(r_{13})\,K^c_{bd}(r_{24}) - K^c_{ad}(r_{14})\,K^c_{bc}(r_{23}), \qquad (127)$$

at the critical point. Then arguments, paralleling those used to derive (122) lead to

$$D^c_{abcd}(x, 1, y, 1, y, z) \approx \sum_j P_j^{ab}\, D_{jcd}(1, z, y)\, x^{\omega_j-\omega_a-\omega_b} - r_{03}^{\omega^*}\,\mathcal{D}_{abcd}(y), \qquad (128)$$

as $x \to 0$ where

$$\mathcal{D}_{abcd}(y) = \frac{D_{ac}\,D_{bd}}{y^{\omega_b+\omega_d-\omega^*}} + \frac{D_{ad}\,D_{bc}}{y^{\omega_a+\omega_d-\omega^*}}. \qquad (129)$$

Once again, consistency is ensured only if $\omega^* = 0$. (To avoid the possibility that the amplitudes D_{ab}, D_{ac}, ... vanish leading to $\mathcal{D}_{abcd} \equiv 0$ we may simply take $a = b = c = d$.)

This intimate connection between hyperscaling and the strong scaling postulate could be broken if one insisted that the product expansion was not to be taken literally (in the way that we have done by equating direct expectation values of both sides) but rather is to be interpreted as a relation between cumulants [so replacing $\langle\cdot\rangle$ by $\langle\cdot\rangle_c$]. With this convention the offending last terms in (126) and (128) could not arise and the value of ω^* would remain free. While such a formulation would retain the predictive power of the product expansion procedure, it robs the principle of its simplicity and directness. On the other hand we are, of course, at liberty to retain a general value for ω^* by discarding the strong scaling hypothesis.

Finally, we may bring the hypothesis of conformal invariance into play. In the first place it implies the orthogonality of the fluctuations [see (100)] so that in the matching relation (120) for the three-point cumulant we have $D^c_{fc} = 0$ unless $\omega_f = \omega_c$. Secondly, the three-point function at the critical point has the definite hyperscaling form (103) or more general form (108). The matching relation for r_{12}/a and r_{12}/r_{03} large, thence becomes

$$\frac{D^c_{abc}\, r_{12}^{\omega_c-\omega_a-\omega_b+\frac{1}{2}\omega^*}}{r_{03}^{2\omega_c-\frac{3}{2}\omega^*}} \approx p_f^{ab}(r_{12})\,\frac{D^c_{fc}}{r_{03}^{2\omega_c-\omega^*}}, \qquad (130)$$

[1] Furthermore Kadanoff [8] also demonstrated that knowledge (i) of all the relevant critical operators and their transformation properties under rotation, and (ii) of the "selection rules" that determine which expansion coefficients vanish identically, is sufficient in the case of the plane square Ising model, to determine the critical exponents. Whether this is so more generally can only be conjectured at present; but unless independent means can be found for determining or guessing the selection rules the approach would not, in any case, be practicable.

where D^{c}_{abc} is now a fixed amplitude. By examining the exponents of r_{03} on each side we see that this relation also is inconsistent unless $\omega^* = 0$. Thus, if conformal invariance is valid, the product expansion and the strong scaling postulate imply hyperscaling even when attention is confined to the critical point and to two- and three-point cumulants only. Conversely, if hyperscaling is valid, identifying the functions of r_{12} in the matching relation leads, once again, to the expansion coefficient formula (121) [recall that $\omega_f = \omega_c$]. In addition the three-point amplitude must be given by

$$D^{c}_{abc} = P^{ab}_{f} D^{c}_{fc}, \quad (\omega_f = \omega_c), \tag{131}$$

where a sum over f may be needed in case there are a number of operators of dimensions ω_c. [If there is only one such operator we may replace f by c.] It is easy to check by writing the conformal formula for K_{abc} in scaling function form, that this relation is just a special case of the previous prediction (122). Conformal invariance is thus quite consistent with the strong scaling hypothesis.

The consistency may be further cross-checked by using the conformally invariant expression (104) for the four-point cumulant to substitute for $D^{c}_{abcd}(x, 1, y, 1, y, z)$ in the matching relation (128). After some algebra this leads to an expansion for the four-point scaling function $Q_{abcd}(X,X)$ for small, namely

$$Q_{abcd}(X,X) \approx \sum_j P^{ab}_{j} D_{jcd} X^{\omega_j - \omega_a - \omega_b - \omega_c - \omega_d} - (D_{ac} D_{bd} + D_{ad} D_{bc}) X^{-\omega_a - \omega_b}, \tag{132}$$

where it must be remembered that D_{kl} vanishes identically unless $\omega_k = \omega_l$. As in the previous applications, then, strong scaling yields information on the limiting behavior of the scaling function.

I am indebted to Professor Kenneth G. Wilson and Professor David Jasnow for comments and discussion. The support of the National Science Foundation, in part through the Materials Science Center at Cornell University is gratefully acknowledged.

References

1. Fisher, M E, J math phys 1964, 5, 944.
2. Widom, B, J chem phys 1965, 43, 3392, 3398.
3. Domb, C & Hunter, D L, Proc phys soc 1965, 86, 1147.
4. Kadanoff, L P, Physics 1966, 2, 263.
5. Patashinskii, A Z & Pokrovskii, V L, Zh eksp teor fiz 1966, 50, 439 (Soviet Phys JETP 1966, 23, 292).
6. Fisher, M E, Phys rev lett 1966, 16, 11.
7. Riedel, E K & Wegner, F, Z phys 1969, 225, 195.
8. Kadanoff, L P, Phys rev lett 1969, 23, 1430.
9. Kadanoff, L P & Ceva, H, Phys rev B 1971, 3, 3918.
10a. Wilson, K G, Phys rev 1969, 179, 149.
10b. — Ibid D 1970, 2, 1473.
11. — Ibid B 1971, 4, 3174, 3184.
12. Wilson, K G & Fisher, M E, Phys rev lett 1972, 28, 248.
13. Wilson, K G, Phys rev lett 1972, 28, 540.
14. Fisher, M E & Pfeuty, P, Phys rev B 1972, 6, 1889.
15. Wegner, F, Phys rev B 1972, 6, 1891.
16. — Ibid 1972, 5, 4592.
17. Fisher, M E, Rept progr phys 1967, 30, 615.
18. — Notes for the 1972 MIT summer school on phase transitions and critical phenomena.
19. Stell, G, Phys rev lett 1970, 24, 1343.
— Phys lett 1968, 27 A, 550.
— Phys rev B 1972, 5, 981.
— in Critical phenomena (ed M S Green). Proc Enrico Fermi summer school course 51, p. 188. Academic Press, New York, 1971.
20. Polyakov, A M, Soviet phys-JETP lett 1970, 12, 381.
21. Fisher, M E, in Critical phenomena (ed M S Green & J V Sengers) p. 108. NBS misc publ no. 273, 1966.
22. Ferer, M, Moore, M A & Wortis, M, Phys rev lett 1969, 22, 1382.
23. Fisher, M E, in Proc conf on renormalization group, Temple University, May, 1973.
24. Domb, C, Phys rev lett 1968, 20, 1425.
25. Essam, J W & Fisher, M E, Rev mod phys 1970, 42, 271.
26. Pfeuty, P, Fisher, M E & Jasnow, D, AIP conf proc on magnetism and magnetic materials, 1972, AIP, 1973.
27. Fisher, M E, Phys rev 1969, 180, 594.
28. Fisher, M E & Burford, R J, Phys rev 1967, 156, 583.
29. Wilson, K G, Phys rev 1969, 179, 1499.
30. Kadanoff, L P, in Critical phenomena (ed M S Green). Proc Enrico Fermi summer school, course 51, p. 118. Academic Press, New York, 1971.
31. Patashinskii, A Z, Soviet phys-JETP 1968, 26, 1126.
32. Kadanoff, L P & Wegner, F J, Phys rev B 1971, 4, 3989.
33a. Baxter, R J, Phys rev lett 1971, 26, 832.
33b. — Ann phys N.Y. 1972, 70, 193.
33c. — Ibid 1972, 76, 1, 25, 48.
34. Als-Nielsen, J & Dietrich, O, Phys rev 1967, 153, 706, 711, 717.
35. Norvell, J C & Als-Nielsen, J, Phys rev B 1970, 2, 277.
36. Als-Nielsen, J, Phys rev 1969, 185, 664.
37. Guttman, L, Schnyders, H C & Arai, G J, Phys rev lett 1969, 22, 517.
38. Guttman, L & Schnyders, H C, Phys rev lett 1969 22, 520.

39. Ashman, J & Handler, P, Phys rev lett 1969, 23, 642.
40. Polyakov, A M, Soviet phys JETP 1970, 30, 151.
41. Kastrup, H A, Phys rev 1966, 150, 1189.
42. Mack, G & Salam, A, Ann phys NY 1969, 53, 174.
43. Ferrara, S, Gatto, R & Grillo, A F, Conformal algebra in space-time and operator product expansions. Springer, Berlin, 1972.
44. Polyakov, A M, Soviet phys JETP 1969, 28, 533.
45. Migdal, A A, Soviet phys JETP 1969, 28, 1036.
46. — Phys lett 1971, 37 B, 98.
47. Kadanoff, L P, Phys rev 1969, 188, 859.
48. Pink, D A, Phys rev 1969, 188, 1032.
49. Mikulinskii, M A, Soviet phys JETP 1970, 31, 991.

Discussion[1]

Mermin: When you say the continuously variable exponents of the Baxter model can be understood as arising from a line of fixed points and an operator that is marginal all along the line, do you mean only that such structure leads to the Baxter type of behavior, or, more specifically, that, perhaps in some approximation, the renormalization group of the Baxter model can explicitly be seen to have a line of fixed points.

Fisher: The answer to your question is contained in the work of Kadanoff & Wegner. They showed firstly that the Baxter model could be fruitfully expressed as a pair of independent square lattice Ising models coupled with a four-spin (energy-energy) term. Since one knows the exact properties of the Ising lattices one can use perturbation theory in the four-spin coupling. First one sees that this operator is indeed marginal; then one shows within (low order) perturbation theory that it remains marginal (as I said was required). One can also estimate the other exponents correctly. Thus the situation is fully consistent with the picture of a line of fixed points; however, an explicit calculation of the renormalization group for the Baxter model has not been made.

Doniach: In the two-dimensional Heisenberg model it appears that there is a line of singularities below the critical point. Does this correspond to two fixed points or to a line of fixed points? In the latter case what is the parameter determining where one is on the line? (the η exponent appears to be temperature dependent in this case).

[1] This Discussion followed the lecture presented orally by Professor Fisher; it does *not* refer to the foregoing printed paper.

Fisher: To my knowledge the problem of the two-dimensional Heisenberg model has not been investigated by the renormalization group approach (and perhaps one may still reasonably doubt that there is a nonzero critical temperature!). However, if we accept some of the arguments that have been given, which indicated that the exponent η varies continuously with T, then we would conclude that the region below T_c should be described by a line of fixed points rather than an isolated fixed point. (Presumably for dimensionality $d > 2$ there is an isolated fixed point.)

Domb: May I comment on the apparent failure of the scaling relation $d\nu = 2 - \alpha$ in three dimensions? I wonder if this might not be due to the washing out of the lattice structure at the beginning of the Wilson theory. This would then be associated with the failure of strong scaling near the origin. We have known for some time of such a failure for self avoiding walks where the origin is a peculiar point. Mackenzie has recently demonstrated that a second scaling length must be introduced for this problem.

Fisher: I do not believe that the lattice structure is "washed out" at the beginning of the Wilson renormalization group theory. On the contrary, it is explicitly represented by the momentum cut-off and by fourth and higher order powers of the momentum in the Hamiltonian. However, by explicit calculation, at least for small $\varepsilon = 4 - d$ and for large n, one finds that these effects are irrelevant. More generally, as I have tried to show explicitly in my talk (the contributed paper does not cover the renormalization group aspects of the lecture), if a fixed point Hamiltonian is found with a spatially rescaling renormalization group then the hyperscaling relation $d\nu = 2 - \alpha$ (or $\omega^* = 0$) is necessarily involved. Some ways in which this hyperscaling may, nevertheless, be effectively violated can be devised within the renormalization group context. In particular a "dangerous" irrelevant variable may rescale the whole singular free energy. This mechanism (described at the recent conference in Philadelphia) seems to be responsible for the violation of hyperscaling in more than four dimensions. However, it is by no means clear that the numerical evidence for the three-dimensional Ising model can be explained along such lines.

March: I wonder about the applicability of your approach to the specific case of liquid-vapor critical point. I have two worries:

(1) You seem to assume exponential decay of correlation functions, when one is away from the critical point. But in the case of a liquid like argon, with van der Waals interactions, the pair correlation function at the triple point already approaches its limiting asymptotic form as an inverse power of distance (r^{-6}).

(2) There is a formula for the critical exponent η, due to Green & Gunton, which appears to depend in an essential way on the detailed pair interaction between atoms.

Fisher: I do not believe that the points you raise are any cause for concern in the application of renormalization group ideas to the liquid-vapor critical point. In regard to (1), I only mentioned the exponential decay of correlation away from criticality to make matters concrete with the case of short range interactions in mind. As a matter of fact Ma, Nickel and myself [1] and, later, J. Sak [2] have studied the situation of short range forces plus a long range attractive potential of form const.$/r^{d+\sigma}$. For $\sigma > 2$ (as applicable to van der Waals interactions) one finds that the critical exponents do not change. When $2-\sigma > \eta_0$ (where the subscript denotes the short range value) however, the exponents do change.

As regards the Green & Gunton expression for η I think the dependence on the pair potential is a mirage. If we express the net pair correlation function $G(r)$ in terms of the Hamiltonian in the standard way and, as originally, define η by

$$\eta = 2 - d - \lim_{r\to\infty} [\ln G_c(r)/\ln r]$$

then again one might feel η depended on the pair interactions. However we know from many models that this is not, in fact, so. Fluids should be no different.

References

1. Fisher, M E, Ma, Sa, S & Nickel, B G, Phys rev lett 1972, 29, 917.
2. Sak I V, Phys rev B, 1973, 8, 281.

Generalized Renormalization Transformations

G. Jona-Lasinio

Istituto di Fisica dell'Università, Padova, Italy

In his famous essay, The structure of scientific revolutions [1] Thomas Kuhn defines a paradigm as an achievement in science which shares two characteristics: it is "sufficiently unprecedented to attract an enduring group of adherents away from competing modes of scientific acitivity". "Simultaneously" it is "sufficiently open ended to leave all sorts of problems for the redefined group of practitioners to resolve". And to clarify further this concept he adds "paradigms gain their status becuse they are more successful than their competitors in solving a few problems that the group of practitioners has come to recognize as acute". "The success of a paradigm is at the start, largely a promise of success discoverable in selected and still incomplete examples. Normal science consists in the actualization of that promise." "If the paradigm represents work that has been done once for all, what further problems does it leave to the united group to resolve?"

If one looks back at the development of renormalization group ideas in the past few years one has the impression that many of the above requirements are indeed fulfilled. No doubt a new group of initiates has been engaged in rather esoteric activities like going to a fixed point, eliminating irrelevant variables, rescaling everything at hand and last but not least, escaping momentarily (so people say or hope), into 3.99 dimensions. There is also no doubt that the renormalization group is largely a promise of success and many problems are still left for the practitioners to resolve. So in order that Kuhn's definition be fulfilled a long period of "normal science" should start and promises should be actualized.

The early developmental stages of a new approach are often characterized by a competition between a number of distinct views. What differentiates the various trends is not so much the aim and scope of the research but rather a kind of incommensurability which in theoretical physics is likely to be connected also with the linguistic frameworks used by different people.

Apparently, as far as our subject is concerned, one can recognize two major "schools": we may call them the followers of Gell-Mann & Low and the followers of Wilson. The second school may be is now the most conspicuous. I am no exception to the rule and until some time ago I liked to consider myself an active member of the Gell-Mann & Low school.

But times change and we would like to see the route to normal science open. What I would like to do in this report is to offer a contribution towards the reunification of various points of view. Possibly, what I am going to present is only a linguistic improvement which could however facilitate the setting up of certain problems. To explain the origin of my reasoning perhaps I would like to make a few remarks on personal experience in this field.

Although may be the idea was in the air, to my knowledge the first explicit connection between the renormalization group and critical phenomena, more specifically scaling, was made by Di Castro and myself early in 1969 [2]. The guiding principle of our work had been the following: by examining the renormalization group equations a la Gell-Mann & Low, as formulated for example in Bogoliubov and Shirkov book, we were struck by the fact that these equations can be read as generalized homogeneity relationships for the Green's functions, the only parameter scaling in a complicated way being the coupling constant. (In this connection I should mention the fact that only much later we read the Gell-Mann and Low paper which contains ideas much closer to the problems encountered in critical phenomena.) It was then obvious to us that if some mechanism could be found which would asymptotically (i.e. in the critical region) eliminate the explicit dependence on the coupling, the usual scaling laws would follow. This was essentially the content of our first paper on the subject. At that time we were trying to establish analogies with problems in field theory like infrared divergences and Regge pole behavior, where power laws are present—our main concern was to justify scaling. Then the first of a series of major works of Wilson [3] appeared and the concept of fixed point came into focus.

The renormalization group equations and the assumption that the properties of the critical point are connected with a fixed point provides today a very convincing, although still essentially qualitative, picture of critical phenomena. In fact there are problems of major importance which require further analysis. For example to my knowledge no direct proofs exist, even in simplified models, that the critical point corresponds to a fixed point of the renormalization group equations. Then there is another aspect, to which I will dedicate most of this paper, which in my opinion deserves our attention.

Until now I have referred essentially to the renormalization group à la Gell-Mann & Low (also the paper of Wilson cited above used the method of GM–L). However in these years Wilson has proposed a very successful and rather unconventional approach which he calls the modern version of the renormalization group [4]. As far as critical phenomena are concerned, notably the ε-expansion, the old and the new version of the renormalization group give exactly the same results [5]. The questions then spontaneously arise: are the two renormalization groups entirely equivalent? What is the connection between the two? Do they represent the only possible candidates for what we might reasonably call a renormalization transformation? Can we find other possibilities?

As I said, in this paper I will address myself mainly to these latter questions to which some preliminary partial answers can be given without too much labour. In my opinion the importance of these problems resides in the fact that if many different equivalent renormalization transformations exist, the critical point has to be associated with the properties of a class of transformations rather than with those of a special transformation. In other words, what is physically relevant is the class of renormalization transformations as a whole. But if many transformations exist, then one may be confronted with the existence of several classes of renormalization groups, two groups being "equivalent" if they belong to the same class and "non equivalent" if they belong to different classes. I am using the word equivalent in the sense of physical equivalence; the precise mathematical definition will be given later.

If this situation arises one must find physical arguments to choose one class rather than the other.

As we shall see, in principle one may be confronted with a great variety of possibilities, and a considerable amount of additional analysis is likely to be needed for a thorough clarification of the nature of the critical point. For example the role of boundary conditions is still insufficiently understood.

The ideas developed in this paper were inspired by the previous work of the author in field theory and are also based on partly unpublished work. In the next section some concepts will be summarized which were introduced in earlier papers.

Basic Ideas on Generalized Action Principles

In the series of papers [6–9] I am referring to, an effort was made to formulate quantum field theory in terms of c-number generalized action principles whose functional arguments are a finite number of the time ordered Green's functions. It would be long to describe why it is interesting to work with such formulations. Here we shall only say that the variational framework seems appropriate for the search of non-perturbative methods of solution. An example of this was given in reference [8]. Another important advantage is the fact that once a stationary functional is at hand, the most convenient set of variables can be chosen due to the invariance of the associated Euler equations under arbitrary changes of variables. This property was exploited in our previous work to obtain a new form of the renormalized equations of motion in which the renormalization constants appeared as Lagrange multipliers associated to subsidiary conditions of the variational principle.

The technique to construct generalized action

principles is the functional Legendre transformation. Suppose that the Lagrangian of our system is specified by a set of parameter functions $\mu_n(x_1, x_2, \ldots, x_n)$ describing many particle interactions in a parameter space $\{\mu\}$. Generalizing Schwinger [10] one can introduce the generating functional M

$$e^{iM(\{\mu\})} = \langle 0^+ | 0^- \rangle \tag{1}$$

where the right hand side represents the transition amplitude from the vacuum state in the far past to the vacuum state in the far future, in the presence of the sources $\{\mu\}$. We define now a set of conjugate functions to the $\{\mu\}$

$$\frac{1}{n!} w_n(x_1, \ldots, x_n; \{\mu\}) = \frac{\delta M}{\delta \mu_n(x_1, \ldots, x_n)} \tag{2}$$

If eq. (2) can be inverted, a multiple Legendre transformation allows us to switch to a new functional $W(\{w\})$

$$W(\{w\}) = M - \sum \frac{1}{n!} \int \mu_n w_n \tag{3}$$

and to the equations conjugate to (2)

$$\frac{1}{n!} \mu_n(x_1, \ldots, x_n; \{w\}) = -\frac{\delta W}{\delta w_n(x_1, \ldots, x_n)} \tag{4}$$

The system (4) can be interpreted in two ways. If the w_n are known they express the interaction parameters in terms of the Green's functions they produce. However in general the situation is exactly the opposite: the μ_n are given and one wants to calculate the corresponding w_n. Eq. (4) then represents a system of non-linear equations for the w_n and they should be written in this case.

$$\frac{1}{n!} \mu_n(x_1, \ldots, x_n) = -\frac{\delta W}{\delta w_n(x_1, \ldots, x_n)} \tag{5}$$

It is now trivial to recognize that the system (5) can be associated to a variational principle. In fact the functional

$$A(\{w\}, \{\mu\}) = W(\{w\}) + \sum \frac{1}{n!} \int \mu_n w_n \tag{6}$$

where the $\{w\}$ and the $\{\mu\}$ are treated as independent variables, is stationary with respect to variations of the $\{w\}$ and its associated Euler equations are the system (5). $A(\{w\}, \{\mu\})$ will be called a generalized action.

Some remarks are here in order. It is clear that we could have performed the Legendre transformation with respect to a subset of the $\{\mu\}$. In such a case W would be a mixed functional of some w's and some μ's. For example by performing a single Legendre transform with respect to μ_1, we would get a functional

$$W^{(1)} = \Gamma = \Gamma(\varphi, \mu_2, \mu_3, \ldots) \tag{7}$$

where φ is the vacuum expectation value of the quantum field ϕ. Γ is a very important quantity because its derivatives with respect to φ are the one particle irreducible vertex functions

$$\Gamma = \sum_{n=0}^{\infty} \frac{1}{n!} \int \Gamma_n(x_1, \ldots, x_n) \varphi(x_1) \ldots \varphi(x_n) \tag{8}$$

Γ was introduced by the author [6] for the study of spontaneous breakdown of symmetries.

If we perform a double transform with respect to μ_1 and μ_2 we obtain a functional

$$W^{(2)} = W^{(2)}(\varphi, \Delta, \mu_3, \mu_4, \ldots) \tag{9}$$

where Δ is the one particle propagator.

As a second remark we observe that with every Legendre transformation the solution of the dynamical problem is split into two steps: (*a*) the construction of the corresponding W; (*b*) the solution of the variational equations. In a previous paper we argued that the essential non-analyticities of field theory are already contained in the expression of $W^{(2)}$ and we gave an explicit example. We shall come back to this later.

For the purposes of the present paper, which are mainly illustrative, it will be enough to deal with the functional Γ. Γ depends on φ, the $\{\mu\}$ and then on a scale parameter Λ in terms of which we fix its absolute value. Since Γ is an extensive quantity there will be also a parameter L fixing the size of the system. In the end we shall not be interested in the dependence on L because the limit of infinite L has to be taken.

We are now in a position to define renormalization transformations. The usual GM–L transformations lead from φ, $\{\mu\}$, Λ to another set φ', $\{\mu'\}$, Λ' such that the value of Γ remains unchanged

$$\Gamma(\varphi, \{\mu\}, \Lambda) = \Gamma(\varphi', \{\mu'\}, \Lambda') \tag{10}$$

Clearly the GM–L renormalization group is not the only possibility. Wilson has given us another example. In the following we shall define any transformation which satisfies (10) a generalized renormalization transformation. We shall explain later the peculiarities of the GM–L and Wilson transformations among the generalized renormalization transformations.

Equivalent Transformations

In the previous section we gave the definition of generalized renormalization transformations (g.r.t.). They will now be considered in greater detail. We first observe that the equation

$$\Gamma(\varphi, \{\mu\}, \Lambda) = \text{const.} \quad (1)$$

defines a surface S_Γ in the space of the variables φ, $\{\mu\}$, Λ. A g.r.t. can then be characterized as a vector field over S_Γ. The problem of studying renormalization transformations mathematically can be formulated as the problem of studying the flows over S_Γ [11]. To each flow there corresponds a transformation T. It is useful to introduce some terminology from the theory of dynamical systems. Let T and T' be diffeomorphisms (i.e. differentiable one-to-one transformations) acting on a manifold M. T and T' are said differentiably conjugate if there exists a diffeomorphism h acting on M such that

$$hT = T'h \quad (2)$$

Transformations which are conjugate in the above sense have a very important property: if p is a fixed point of T, and therefore $h(p)$ a fixed point of T', the derivatives DT and DT' in p and $h(p)$ respectively, have the same eigenvalues. We recall that DT and DT' are linear operators in the tangent spaces at p and $h(p)$.

To prove this proposition take the derivative of (2)

$$Dh(T(q))DT(q) = DT'(h(q))Dh(q) \quad (3)$$

For $q = p$, because of $T(p) = p$, (3) shows that DT and DT' are related by a similarity transformation and have therefore the same eigenvalues.

It may happen that two differentiable transformations T and T' are related by a transformation h which is not a diffeomorphism but only a homeomorphism (i.e. one-to-one continuous transformation). In such a case T and T' are said topologically conjugate. Topological conjugacy means that T and T' have qualitatively the same orbit structure. The proposition proved above however in general does hot hold.

We now consider the GM–L and Wilson transformation. In terms of their action on φ they are defined by

$$\text{GM-L} \quad \varphi(x) \to Z\varphi(x)$$

$$\text{Wilson} \quad \varphi(x) \to \int Z(x, x')\,\varphi(x')\mathrm{d}x' \quad (4)$$

In the GM–L case Z does not depend on x, i.e. it does not change the weight of the different degrees of freedom. In the Wilson case $Z(x, x')$ is the Kernel of a non local transformation which in certain limiting cases reduces to a pure scale transformation. The non locality is what makes Wilson transformation a very efficient calculational tool. For what follows it is important to remark that both transformations are linear in φ. This means that also the form of the functional Γ is essentially invariant under these groups as far as the φ dependence is concerned.

Let us consider now the Wilson group acting on the parameters $\{\mu\}$. I will not describe the many steps involved in the construction of the transformation: this is now part of the common knowledge and in any case several good expositions are available. The reader is referred in particular to the lucid recent presentation of Ma [12]. What is of interest to us is that the critical behaviour is determined by the behaviour of this transformation near a fixed point. In formulas if μ^* is such that

$$R_W(\mu^*) = \mu^* \quad (5)$$

(R_W is the Wilson group acting on the submanifold of the μ variables), the critical indices are determined by the eigenvalues of $DR_W(\mu^*)$. But then the proposition we proved earlier in this section tells us that any other transformation R' which is differentiably conjugate to R_W will describe the same critical behaviour. In this case we shall say that R' is equivalent to R_W.

Then what about topological conjugacy? We think that this weaker equivalence property will describe in general those cases in which different transformations correspond to slightly different relevant physical conditions. However, more complicated cases may arise, as we shall explain later.

We now go back to the group action on φ. It is clear that in general, there is no reason why the action on φ induced by R' should be linear. We shall devote the remaining part of this section to an analysis and interpretation of this circumstance.

If the group action on φ is non linear, the derivatives of Γ, i.e. the one particle irreducible vertex function, are mixed by the transformation. To be more specific we consider as an illustration the simplified case in which the law according to which φ changes is of the form [13]

$$\varphi(x) \to \varphi(x)\, Z(\varphi(x)) \tag{6}$$

where $Z(\varphi)$ is a local function of φ. In this case the n-point vertex transforms according to the rule

$$\begin{aligned}\Gamma_n \to Z^n(0)\, \Gamma_n &+ \sum Z_1\, \Gamma_{n-1}\, \delta \\ &+ \sum Z_2\, \Gamma_{n-2}\, \delta\cdot\delta + \ldots \\ &+ \sum Z_{n-2}\, \Gamma_2\, \underbrace{\delta\cdot\delta\ldots\delta}_{n-2}\end{aligned} \tag{7}$$

where the constants Z_i are related to the derivatives of Z at $\varphi = 0$. Eq. (7) is best understood if interpreted diagrammatically. Take as an example the 6-point vertex and assume that originally only the Γ_n with even n were different from zero.

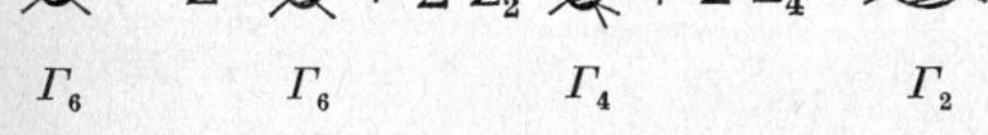

(8)

We see that the δ-functions in (7) induce new couplings by attaching additional lines to the elementary couplings of the original Lagrangian. The Σ symbol indicates summation over all the possible ways of attaching the new lines.

So far so good. But then someone might ask the very legitimate question: can you suggest any interesting application of the more general renormalization transformations?

My suggestion is that these general transformations may shed new light on the problem of irrelevant variables. One may argue as follows. Given a certain problem of critical behaviour usually one has reasons to think that some of the μ are irrelevant parameters. It is then conceivable to construct a transformation that, at each iteration, in some sense minimizes the effect of these irrelevant parameters by inducing suitable cancellations in the vertex functions. The irrelevancy of these variables should manifest itself in the fact that the non linear transformation on φ is non singular near the fixed point. In other words we expect that for $\varphi \to 0$

$$\varphi \to Z\varphi + Z_2\varphi^3 + \ldots \tag{9}$$

and in the end only the linear term is important. Said differently the new transformation is differentiably conjugate to R_W.

If the argument we have described is correct, this would give substance to a formulation of universality given by Kadanoff. Kadanoff [14] has conjectured that near the critical point the irrelevant variables can be eliminated, i.e. set equal to 0, by rescaling the relevant ones. The g.r.t. should be able to accomplish this. What we are saying is essentialy the following: the location of the fixed point is not important and in principle it can be chosen where it is more convenient. In particular one can set all the irrelevant couplings equal to zero, according to Kadanoff's picture. This choice, in general, will give rise to a non linear transformation of the order parameter φ. In this connection we observe that non linear transformations on the $\{\mu\}$ may correspond to linear transformations on φ (as in the GM–L and Wilson cases) while arbitrary linear substitutions on the $\{\mu\}$ will induce non linear transformations on φ.

We conclude this section with a remark which may be useful in certain cases. Once a given renormalization transformation has been chosen, eq. (1) can be written in differential form [15]

$$\mathrm{d}_R \Gamma = 0 \tag{10}$$

where d_R means that the variations are made along the transformation. This equation is obviously invariant under arbitrary invertible, differentiable changes of variable. By appropriate changes of variable one can introduce a whole class of formally equivalent systems described by functionals Γ'. Given a particular problem, a particular form Γ' might be especially convenient.

In a recent work in collaboration with Di Castro & Peliti, a unified derivation of the scaling laws was presented using the renormalization group equation in the form (10) for the GM–L group [16].

Outlook

We first summarize what we have achieved. Starting from the generalized action functional which generates the one particle irreducible vertex functions, we have defined g.r.t. as those transformations which leave the action invariant in value. They include as a particular case the GM–L and Wilson renormalization groups.

We have then introduced the notion of equivalent transformations. They are transformations which are connected by an arbitrary diffeomorphism and have the property that their derivatives have the same eigenvalues in corresponding fixed points. This leads to immediate generalizations of the already known renormalization groups and we have pointed out the possible role of these transformations to the effect of eliminating irrelevant variables. The generalized transformations usually will act non linearly on the vacuum expectation value of the field, φ.

All this raises several questions. For example:

(1) How large is the class of transformations equivalent to that of Wilson and which act linearly on φ?

(2) In principle transformations may exist which are not equivalent to Wilson and yet they have a similar orbit structure. What we have in mind is a situation in which different ways of constructing the transformation, for example by modifying the cut off procedure in Wilson type transformations, lead to non uniform results. These transformations would represent alternative candidates for the description of critical points. If such a situation should arise, on what basis has a choice to be made?

(3) The renormalization groups studied so far have proven very effective in studying those limiting situations in which scale invariance is expected. However there may exist g.r.t. which have a very different orbit structure and which might turn useful in different physical problems. Then the question: is it possible to classify the flows on S_Γ?

Some further comments are in order. The ideas discussed in this paper were expressed within the variational formulation of field theory. As we mentioned on p. 39, this formulation originated in the attempt of solving field theory beyond perturbation theory. In a previous paper it was shown [8] that a great deal of the expected non analyticity of field theory is already present in the structure of the generalized action principle obtained by performing a double Legendre transformation. In fact it was found in the ϕ^3 theory (but the same applies to the ϕ^4 interaction) that the action functional $W^{(2)}(\varphi, \Delta)$ is not analytic in Δ. To exhibit the non analytic character of the solutions one has to keep enough Feynman diagrams so that the divergent character of perturbation theory is maintained. The advantage of working with the generalized actions is that only skeleton diagrams have to be considered and this reduces considerably the number of graphs in the game.

The idea of g.r.t. can be easily extended to $W^{(2)}(\varphi, \Delta)$ by considering non linear transformations on φ and Δ. Then, given a particular problem, the question is to construct by appropriate transformations an effective action which includes the essential features of the physical situation. In this connection an analogy may be helpful. In a sense renormalization transformations are like canonical transformations in analytical mechanics. Finding an appropriate canonical transformation may be the essential step towards the solution of a classical problem. Similarly, using a good renormalization transformation may open the way to the understanding of an otherwise intractable problem in field theory or statistical mechanics. I do not believe that there is a recipe good for all cases.

During the preparation of this report I benefited from several conversations with C. Di Castro and L. Peliti. Of course they have no responsibility for the specific views expressed here.

References

1. Kuhn, T S, The structure of scientific revolutions, 2nd enlarged edn. University of Chicago Press, 1970.
2. Di Castro, C & Jona-Lasinio, G, Phys lett 1969, 29 A, 322.
3. Wilson, K G, Phys rev 1971, D3, 1818.
4. An extended exposition of these ideas is contained in Wilson, K J & Kogut, J, The renormalization group and the ε expansion. Preprint 1972. To appear in Phys repts. In this paper one finds also an exhaustive list of references.
5. Di Castro, C, Lett n cim 1972, 5, 69 and ref. [16].
6. Jona-Lasinio, G, N cim 1964, 34, 1790.

7. Dahmen, H D & Jona-Lasinio, G, N cim 1967, 52A, 807.
8. — Ibid 1969, 62A, 889.
9. Dahmen, H D, Jona-Lasinio, G & Tarski, J, N cim 1972, 10A, 513.
10. Schwinger, J, Proc natl acad sci US 1951, 37, 452, 455.
11. E.g. Smale, S, Bull Am math soc 1967, 73, 747.
12. Ma, S, The renormalization group and the large n limit. 1973. Preprint.
13. In a different context transformations of this kind were introduced by Zumino, B, in Lectures in elementary particles and quantum field theory. MIT Press, 1970.
14. Kadanoff, L P, in Proc int school of physics E Fermi, LI Corso. Academic Press, New York, 1970.
15. Coleman, S & Weinberg, E, Phys rev 1973, D7, 1388.
16. Di Castro, C, Jona-Lasinio, G & Peliti, L, Variational principles, renormalization group and Kadanoff's universality, 1973. Preprint.

Discussion

Kohn: I believe you said that different renormalization transformations could give rise to different critical exponents and that one would then have to use "physical considerations" to resolve the ambiguity. This seems to me rather vague and, if it happened, I would seriously worry about the use of the renormalization method.

Jona-Lasinio: I have shown that there exists a variety of renormalization transformations relevant for the description of critical points. Some of them will be strictly equivalent in the sense that they will have the same eigenvalues at corresponding fixed points. Other transformations will be equivalent in a weaker sense because they will show only the same orbit structure, for instance the same number of fixed points, the same stability properties, etc., but their derivatives will not necessarily have the same eigen values at the fixed points.

My idea is that in the latter case different transformations will correspond in general to slightly different relevant physical conditions. Changing the dimensionality near 4 dimensions is an example of the second situation. I mentioned as another possible example changes in the boundary conditions, if these turn out to be relevant. These cases are clear. However, in principle one may think of situations in which different indices are obtained by constructing the transformation differently from Wilson (by using e. g. different cut off procedures) and the difference is not immediately connected with the change of a relevant variable.

These considerations are not at all academic. When Fisher in his talk showed the connection between the renormalization group and critical phenomena, the essential argument he used was the circumstance that near a fixed point you get scaling. However, one would like to be able to prove directly that the critical point corresponds to a fixed point of the renormalization groups used today. So one has to face the problem.

Fisher: Maybe I could add a further reply to Professor Kohn's question. Certainly one would generally hope and expect that different renormalization groups would, if they described the same class of critical points, yield the same critical exponents. As a matter of fact, one can check this explicitly in the simple (but still nontrivial) examples of various one-dimensional Ising models. Here $T_c = 0$ and the exact solution is, of course, available. In work by David Nelson and myself we find, as anticipated, that different groups do yield identical critical behavior. However they describe the *approach* to the critical region differently, although always exactly. It may happen, however, that a badly chosen, "unfocussed", group exhibits *no* fixed point for some class of critical point. In such circumstances one clearly learns little. (I should mention that Leo Kadanoff has independently studied some renormalization group aspects of the linear Ising chain.)

More generally Franz Wegner has recently started to investigate the whole question more systematically (as reported at the recent conference in Philadelphia). In brief, he shows that under continuous and, I believe, differentiable transformations of a differential renormalization group the physically significant exponents associated with a fixed point do not change. However certain non-physical exponents, which essentially describe only changes of field variables, may alter.

Lastly, as I interpret Professor Jona-Lasinio's remarks, he is saying that if different exponents were found, one would have to reexamine the physics/mathematics more closely to find the origin of the differences and hence chose the formulation most accurately reflecting the physical situation of interest.

Developments in the Theory of Critical Phenomena

G. Toulouse

Laboratoire de Physique des Solides, Université Paris-Sud, Centre d'Orsay, 91405 Orsay, France

Summary

An exploration of the (n,d) plane is described (n = order parameter dimensionality, d = space dimensionality). Exact results for ($n = -2$, d arbitrary) and for ($d = 1$ and 0, n arbitrary) are presented. An internal symmetry property is discussed.

According to our present knowledge of critical phenomena in second order phase transitions, the critical exponents appear to be, quite generally, universal functions of two parameters, the dimensionality d of the system and the number n of components of the order parameter. For long-range forces, an additional parameter sets in; for simplicity, unless specifically stated, only short-range forces will be discussed here.

Let us consider then an (n, d) table. The presently physical region lies somewhere inside the domain $0 \leqslant n \leqslant 3$, $0 \leqslant d \leqslant 3$, although critical properties of systems with higher n values will undoubtedly be soon investigated. Two frontiers have recently received considerable attention, namely the line $d = 4$ and its vicinity $(4-d)$ small [1], the limit $n \to \infty$ and its vicinity $1/n$ small [2]. In an attempt to achieve the surrounding of the physical region, exact solutions for $n = -2$ (d arbitrary) [3] and for $d = 1$, $d = 0$ (n arbitrary) [4] have been obtained.

Considering the (n, d) table as a Mendeleeff table of critical phenomena, it is then possible to make precise predictions on the trends of variations of the critical temperature (for a given system) and of the critical exponents as functions of n and d; by interpolation, the critical exponents of systems not yet observed experimentally can be predicted accurately.

The quest for the boundaries of the continuity domain is also motivated by the observation that boundaries are good starting points for expansions, such as the $(4-d)$ and $1/n$ expansions. Besides, the solution for $n = -2$ has led to the discovery of internal symmetry properties in the (n, d) plane [5], introducing connections between apparently widely different systems, and paving the way for a new relation between exponents. Another motivation for this exploration of the (n, d) plane is the substantiated hope that various properties, such as disorder effects, may be understandable on a domain basis.

Some of the main results of a study undertaken in collaboration with Roger Balian are reviewed here, together with some new results.

Solution for $n = -2$ (d arbitrary) [3]

For $n = -2$, the free energy (in zero field) and the two-point correlation function are unmodified by the interaction between fluctuations. As a consequence, no renormalization of the critical temperature occurs and the exponents γ, ν, η take classical values. These results are true for arbitrary d; for $d < 4$, Josephson scaling law is satisfied and $\alpha = 2 - \frac{1}{2}d$.

There exists by now four different proofs of these results [3, 6]; the unpublished one, suggested by C. de Dominicis, makes use of auxiliary field techniques to calculate functional integrals of a ϕ^4 field. The partition function appears then as a new functional integral of a determinant elevated at power $-\frac{1}{2}n$; continuation in n is then straightforward and the results for $n = -2$ follow.

The progression toward negative values of n was initiated by de Gennes [7] who gave a physical meaning to the case $n = 0$, shown to describe the self avoiding walk or polymer excluded volume problem. The case $n = -2$ corresponds (as far as critical properties are concerned) to the random walk problem and to ideal Bose condensation at constant pressure.

The main merit, so far, of the solution for $n = -2$ has been to lead to the discovery of a symmetry which is discussed below. An expansion in terms of $(n+2)$ is a desirable development, considering the early asymptotic character of the $(4-d)$ expansion [8]. Such an expansion in $(n+2)$ requires, however, a knowledge of the anisotropic part of the four point correlation function, which does not exist so far for arbitrary d.

Symmetry of Critical Exponents

Two preliminary observations will help introducing this internal symmetry:

(i) exact solutions exist for $n = -2$ (ideal Bose condensation at constant pressure) and for $n = \infty$ (ideal Bose condensation at constant volume); the two corresponding sets of exponents are related by a transformation $\mathcal{T}$ such that

$$(1-\alpha)(1-\alpha') = \frac{\eta'}{\eta} = \frac{\delta'}{\delta} = 1 \qquad \frac{\beta'}{\beta} = \frac{\gamma'}{\gamma} = \frac{\nu'}{\nu} = \frac{1}{1-\alpha}$$

note that $\mathcal{T}^2 = 1$ and that the four scaling laws are conserved under $\mathcal{T}$; transformation $\mathcal{T}$ appears in a number of exactly soluble perturbed models, which exhibit the so-called renormalization of critical exponents by hidden variables [9]; (ii) for fixed $d\,(2<d<4)$, as functions of n, exponent α has one zero and exponent η has one maximum; in the $(4-d)$ expansion, up to second order for α (third order for η), it appears that the zero of α and the maximum of η are on the same line $\bar{n}(d)$, which passes by $(n=4,\ d=4)$ with a tangent of slope 1/4; independent information tells us that $\alpha = 0$ for $(n=1,\ d=2)$ and $\alpha \simeq 0$ for $(n=2,\ d=3)$.

Initial Conjecture [5]

These two observations led to our initial conjecture, which can be stated as follows: to one point (n, d) in the (n, d) plane, there corresponds another point $(n', d' = d)$ given by

$$(n+2)(n'+2) = (\bar{n}(d)+2)^2$$

such that the two sets of critical exponents are related by transformation $\mathcal{T}$.

Using this conjecture, the results of the preceding observations are derived as immediate consequences. The conjecture is verified by the $(4-d)$ expansion up to second order (third order for η); it holds for long range forces with an $n(d)$ varying continuously as the range varies, despite the discontinuity in the critical exponents expansion, as noticed by J. Winter.

From the point of view of renormalization by hidden variables, transformation $\mathcal{T}$ applies if the unperturbed system has $\alpha > 0$; the perturbation due to the hidden variables is then seen to be equivalent to the addition of $(n'-n)$ degrees of freedom (note that $(n'-n)$ is then a positive number, as it should). This equivalence gives a more precise meaning to the vague concept of hidden variables.

Subsequent calculation of terms of third order in $(4-d)$ (fourth order for η) has shown that the initial conjecture was violated [8]. The line of zeroes for α and the line of maxima for η have the same sign of curvature but different curvature radii at $(n=4,\ d=4)$. The initial conjecture can be thought of then as a sort of broken symmetry. A revised version, in agreement with all known results, will be given now, showing corrections to the initial conjecture of order η, which is a very small number even for $4-d=1$, i.e. in three dimensions.

Revised Version of the Symmetry

To one point (n, d) with a set of exponents characterized by the values of η and $(1-\alpha)$, there corresponds another point $(n',\ d'=d)$ with $(\eta'=\eta,\ 1-\alpha')$ such that

$$(n+2)(n'+2) = C(d, \eta)$$

$$(1-\alpha)(1-\alpha') = \prod(d, \eta)$$

Thus d and η are constant in the correspondence; C and $\prod$ are obtained by expansions in $\varepsilon = 4-d$ and η as:

$$C(d, \eta) = 36 - 48\varepsilon + 12\varepsilon^2(1+8x) + 96(7+18x)\eta$$

$$\prod(d, \eta) = 1 + 4\varepsilon\eta(\tfrac{1}{3} + 4x)$$

where

$$x = \frac{\zeta(3)}{2} \sim 0.601$$

For completeness, note that the line ($\alpha = 0$) is given by

$$n = 4 - 4\varepsilon + (\tfrac{1}{2} + 7x)\varepsilon^2,$$

whereas the line (η maximum) is given by

$$n = 4 - 4\varepsilon + (\tfrac{5}{6} + 11x)\varepsilon^2,$$

up to second order in ε.

From this revised version of the symmetry, a relation between η, α and d is easily derived; for given η and d, two values of $(1 - \alpha)$ are possible; the relation is then binomial in $(1 - \alpha)$:

$$(1 - \alpha)^2 - \sum(d, \eta)(1 - \alpha) + \prod(d, \eta) = 0$$

$\prod(d, \eta)$ has been given above; $\sum(d, \eta)$ is obtained as:

$$\sum(d, \eta) = 2 + \frac{\varepsilon^2}{4 - 2\varepsilon} - 4\eta(3 - 4\varepsilon(\tfrac{1}{3} + x)).$$

This relation between α, η and d is a sort of fifth scaling law; a sixth one would have to involve n and would give a complete determination of the set of critical exponents.

Phase Transitions for One-dimensional Systems [4]

The theorem on the absence of phase transitions in one dimension is violated, even for short-range forces, if n, the number of components of the order parameter, is made smaller than one. As a matter of fact, the variation of critical properties as functions of n is as follows. For $-2 < n < 1$, a phase transition occurs at non zero temperature, with a critical regime characterized by exponents independent of n, $\nu = \gamma = \eta = \alpha = 1$. At $n = -2$, a new phase comes into play, the point (T_c, $n = -2$) is a triple point and the critical exponents take the values $\nu = \frac{1}{2}$, $\gamma = 1$, $\eta = 0$, $\alpha = 2 - \frac{1}{2}d = \frac{3}{2}$, in agreement with the general theory for $n = -2$, discussed above. For $n < -2$, the transition is of a different character without divergence of the correlation length (first order transition).

These results have been obtained by solving the one-dimensional Ginzburg–Landau–Wilson (GLW) model. The solution is based on a known correspondence between this classical statistical model and an equivalent quantum mechanical problem. The unique space variable being considered as a time variable, the n components of the order parameter being considered as cartesian coordinates of a point in a n-dimensional space, the GLW free energy density may then be considered as a sort of Lagrangian density and the partition function as a Wiener-Feynman path integral. One is thus led to solve the Schrödinger equation for a central anharmonic potential in n dimensions.

$$(-\tfrac{1}{2}\Delta_n + V(\varrho))\,\psi_j(\boldsymbol{\rho}) = \varepsilon_j\,\psi_j(\boldsymbol{\rho})$$

where $V(\varrho)$ has the functional form of the local free energy, function of the amplitude of the order parameter. The correspondence with the one-dimensional phase transition problem implies that the free energy F is proportional to the ground state eigenvalue of this Hamiltonian and that the correlation length ξ is inversely proportional to the energy gap between the ground state eigenvalue and the eigenvalue of the first excited state appropriately coupled to the ground state by non zero matrix elements of the coordinates operators:

$$F \sim \varepsilon_0 \qquad \xi \sim \frac{1}{\varepsilon_1 - \varepsilon_0}$$

It is then manifest that a phase transition in the one-dimensional GLW model is associated with a crossing of levels in the corresponding central anharmonic quantum oscillator problem.

Before considering the particular anharmonic potential of interest here, a general theorem will be presented: for any central potential, the eigenvalues of the isotropic n-dimensional quantum oscillator depend on l, angular momentum, and n, dimensionality, only through the combination $l + n/2$. This is a universality property of the quantum problem, which shows that analytic continuation in n is closely related to the Regge program of analytic continuation in l. For an anharmonic potential, and for special values of the sum $l + n/2$, additional universality properties can be proved, such as the existence of eigenvalues independent of the anharmonicity.

The study of the various types of level crossing and of their existence domains leads to a detailed description of the corresponding (T, n) phase diagram, for one-dimensional systems. It can be summarized as follows: (i) for $n>1$, the ground state is always the nodeless s-state, no crossing occurs, and $\xi \sim 1/T$; (ii) for $n=1$, the first excited p state is only separated from the ground s state by a tunnel matrix element, $\xi \sim e^{-1/T}$; (iii) for $n<1$, s–p crossing occurs, the first excited p state taking advantage of an effectively attractive centrifugal potential; this leads to a finite T_c rising initially with infinite slope as function of $(1-n)$; (iv) for $n<0$, crossing of levels of same symmetry may occur, the eigenvalues becoming complex after getting degenerate: complex eigenvalues lead to an oscillating correlation function in the phase transition problem; this type of accident occurs at temperatures inferior to the s–p crossing T_c, for $-2<n<0$; (v) for $n=-2$, the s–p and s–s crossings occur simultaneously and a triple point is obtained.

Another one-dimensional model, the classical Heisenberg model considered by Stanley [10], exhibits also a phase transition for $n<1$ [4], with critical exponents identical to those of the GLW model. Unfortunately, the transition temperature of the classical Heisenberg model becomes infinite for $n=0$, and this divergence prevents continuation to negative values of n. Nonetheless, it is interesting to note that universality appears to hold for different one-dimensional models, as far as critical exponents are concerned. In this respect, one should note also that the self avoiding walk problem, which is exactly soluble in one dimension and corresponds to $n=0$, gives also the same critical exponents.

Zero-dimensional Systems [4]

An exact solution of the GLW model in zero dimension has also been obtained. A phase transition occurs for $n<0$, with T_c rising initially with infinite slope as function of $(-n)$. For $n<0$, the critical exponents are $\gamma=1$, $\alpha=2$.

These last results, together with the results on the absence of phase transitions for $(n=\infty, d<2)$ and for $(d=1, n \geqslant 1)$, help to fix the boundaries of the domain in the (n, d) plane inside which no phase transition is possible.

References

1. Wilson, K G, Phys rev lett 1972, 28, 240.
2. Brézin, E & Wallace, D J, Phys rev 1973, B7, 1967.
3. Balian, R & Toulouse, G, Phys rev lett 1973, 30, 544.
4. — Ann phys. Submitted for publication.
5. Toulouse, G, Compt rend acad sci 1973, 276B, 193.
6. Fisher, M E, Phys rev lett 1973, 30, 679.
7. de Gennes, P G, Phys lett 1972, 38A, 339.
8. Brézin, E, le Guillou, J C, Zinn-Justin, J & Nickel, B G, Phys lett 1973, 44A, 227.
9. Fisher, M E, Phys rev 1968, 176, 257.
10. Stanley, H E, Phys rev 1969, 179, 570.

From Self-Avoiding Walks to Real Polymer Chains

C. Domb

Wheatstone Physics Laboratory, King's College, Strand, London WC2R 2LS, UK

The topic which I am presenting seems unrelated to what has been previously presented in this session. I shall therefore briefly outline some of the problems with which polymer specialists have been concerned. When these problems are identified in the terminology of critical behaviour we shall find that the concepts of scaling, universality etc. can be of great help.

The earliest model of a polymer chain was a simple random walk, but it proved convenient to assume that the individual steps were not of fixed length but followed a Gaussian distribution of mean square length b^2. This is because each step of the walk is meant to correspond to many units of the chain. The geometrical properties of the chain can then be readily determined from the statistics of a random walk, and, for example, the mean square end-to-end length is given by

$$\langle R_n^2 \rangle = Nb^2 \tag{1}$$

However, it was soon realized that intra-molecular forces must play an important part and hence an intra-molecular potential $V(\boldsymbol{R}_i - \boldsymbol{R}_j)$ was introduced acting between the end points $\boldsymbol{R}_i$, $\boldsymbol{R}_j$ of the ith and jth steps of the walk. The potential $V(\boldsymbol{r})$ is similar to the normal inter-molecular potential between molecules of a gas, and consists of a short range "hard core" repulsion with a longer range attraction. By analogy with a condensing gas one might expect to find a phase transiton at sufficiently low temperatures corresponding to the folding up of the chain. The detailed exploration of this phase transition is very difficult since it is even more complicated than the condensing gas transition. In the same manner as for the condensing gas we can make progress by using a lattice model, but this will be considered elsewhere [1].

Because of the difficulty of dealing with the original problem the polymer workers put forward a simpler model which is more amenable to detailed treatment. The intra-molecular forces are replaced by a pseudo-potential $w\delta(|\boldsymbol{R}_i - \boldsymbol{R}_j|)$ where w represents a type of "second-virial coefficient",

$$w = -\int [1 - \exp(-\beta V(\boldsymbol{r}))]\, d\boldsymbol{r}. \tag{2}$$

It is clear that for $w > 0$ (net attraction) the chain will fold up completely into one unit and this corresponds to the phase transition. We shall be concerned with $w < 0$ (net repulsion), and a problem of central interest is to calculate the expansion factor arising from the intra-molecular forces,

$$\alpha_N^2(w) = R_N^2(w)/Nb^2, \tag{3}$$

as a function of N and w.

Following the cluster-integral procedure for a condensing gas Zimm [2] attempted to derive a series expansion for $\alpha_N^2(w)$ in powers of w,

$$\alpha^2 = 1 + A_1 z + A_2 z^2 + A_3 z^3 + \ldots, \tag{4}$$

where

$$z = -(3/2\pi b^2)^{3/2} w N^{\frac{1}{2}}. \tag{5}$$

So far three coefficients in this series have been evaluated [3]. When we examine this series in detail there are a number of unusual features.

(*a*) For a normal virial expansion we expect to obtain a thermodynamic limit as $N \to \infty$; the bulk properties are proportional to N and other terms (e.g. surface effects of order $N^{2/3}$) have a negligible effect for large N. Hence we obtain an expansion for these bulk properties as a power series in w independent of N. However, in the series (4) N enters in the expansion parameter; the larger N is the smaller the region in w for which the series can be used.

(*b*) In the series (4) A_n is not exact but represents the leading term in a power series in $N^{\frac{1}{2}}$. For example, the true coefficient of w^3 in the expansion (4) is

$$a_3 N^{3/2} + b_3 N + c_3 N^{\frac{1}{2}} + d_3 + \dots \quad (6)$$

The approximation of ignoring all but the leading term yields the "two parameter function". Although this is valid in the limit $N \to \infty$, $w \to 0$, $N^{\frac{1}{2}}w$ finite, it is not clear for what range of values of N, w this will provide a reasonable approximation.

(*c*) We may conclude from the collapse when $w > 0$ that the series (4) cannot have a finite radius of convergence and must therefore be asymptotic.

There seemed no possibility of using the virial series (4) to make direct calculations of α^2 for values N, w of practical interest, and hence (as in the case of fluids) a number of approximations of "mean field" type were put forward. The first and best known of these, due to Flory [4],

$$\alpha^5 - \alpha^3 = \frac{3\sqrt{3}}{2} z \quad (7)$$

can be regarded as the "van der Waals' equation" for a single polymer chain (the random chain with no intra-molecular forces corresponding to the ideal gas). Surprisingly, this equation gives the asymptotic behaviour

$$\alpha^2 \sim N^{1/5} \quad (8)$$

for large N, and there is good evidence from self-avoiding walks both from direct enumerations [5] and from the ε-expansion for $n = 0$ [6] that this is correct. However, an expansion of (7) for small z does not even give the first term of (4) correctly. Hence a variety of attempts have been made to find an improved equation. We shall consider the two most significant of these which give the correct exponent (8), and the correct value for the first term of the expansion (4). Flory & Fisk [7] modified the right-hand side of (7), obtaining

$$\alpha^5 - \alpha^3 = \frac{9\sqrt{3}}{14} z \quad (9)$$

We shall refer to eq. (7) as F1 and eq. (9) as F2. Recently Alexandrowicz [8] and Kurata [9] have independently put forward the formula

$$\alpha^5/5 + \alpha^3/3 - 8/15 = 4z/3 \quad (10)$$

which we shall refer to as A–K.

It is a great help in identifying the nature of this problem to take a corresponding lattice model of a random walk on a lattice with an interaction $w\delta_{ij}$. We must then average over all walks of N steps with the weight function

$$\pi(1 + w\delta_{ij}). \quad (11)$$

$w = 0$ corresponds to a random walk ($n = -2$ in the ε-expansion from the work of Balian & Toulouse [10] and Fisher [11]; when $w = -1$ all configurations with any contacts are eliminated and we obtain a self-avoiding walk ($n = 0$[6]). We see now that in critical phenomena terminology we are endeavouring to find the "crossover scaling function" referred to by Fisher previously between $n = -2$ and $n = 0$. We can understand the difference (*a*) noted above from a normal virial expansion which is an analytic expansion.

We can set up a perturbation expansion in w for any lattice [12] (and this can be applied equally to the Gaussian continuum model) for finite N in which the coefficients are exact and not asymptotic in N; thus we can test the two parameter approximation. In fact we find that if we introduce a constant f_0 for each lattice and define a new parameter

$$u = f_0 w N^{\frac{1}{2}}, \quad (12)$$

the two parameter function is the same function of u for all lattices. Although this result has been checked only for the first three terms of the perturbation expansion, it is exactly what we should expect from our ideas on the universality of scaling functions. Not surprisingly f_0 is inversely proportional to the volume of unit cell of the lattice (w has the dimensions of a volume). Hence if we estimate α^2 as a function of N, $f_0 w$ for a number of lattices and study the region of large N and small $f_0 w$ we should obtain the same

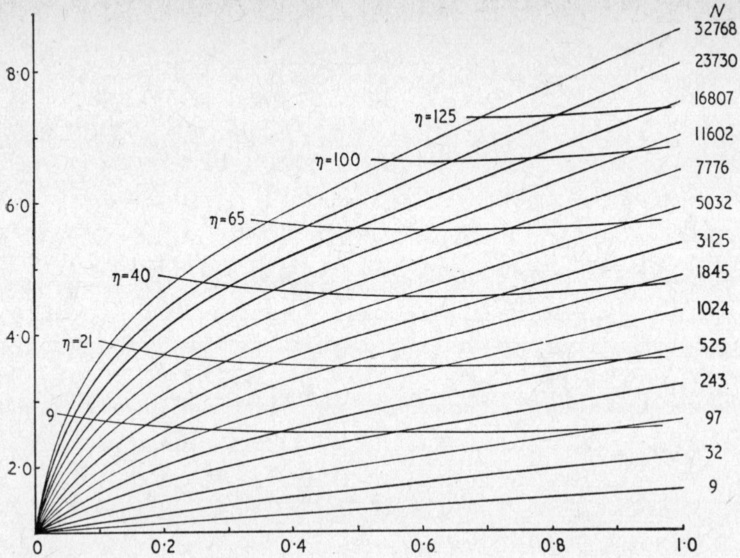

Fig. 1. *Abscissa: w; ordinate:* α^2.

Expansion factor α^2 for a chain on the SC lattice. Lines of constant $\eta(=N^{\frac{1}{2}}w)$ would be horizontal in two parameter approximation.

result for all lattices and this should provide a good estimate of the two parameter function. Even for a single lattice if we extrapolate to the region of large N and small w we should obtain a fair estimate of this function.

Detailed calculations have been undertaken in collaboration with my graduate student, A. J. Barrett of King's College, London (virial coefficients for lattice models) and Dr Melvin Lax of the Weizmann Institute, Israel (exact enumerations for random walks with contacts). To obtain asymptotic estimates near $w=-1$ from the exact enumerations the smoothness postulates [13] was invoked, and the enumerations were fitted to the formula

$$\alpha_N^2(w) \sim A(w)\,N^{1/5}[1+b(w)/N]. \qquad (13)$$

This formula was found satisfactory from $w=-1.0$ to $w=-0.5$, and the region near $w=0$ was determined from the virial series; the intermediate region could readily be covered by interpolation.

Estimates of $\alpha_N^2(w)$ obtained in this way for the simple cubic lattice are shown in fig. 1. From these curves it is possible to assess the validity for this lattice of the two parameter approximation. If the approximation were correct the lines of constant $\eta(=-wN^{\frac{1}{2}})$ would be horizontal. In fact it will be seen that the deviations from horizontal are relatively small.

An estimate of the two parameter function based on these calculations is shown in fig. 2 (labelled DBL). Fig. 2*a* covers the range of small z (0–12) and fig. 2*b* the range of large z (0–250). The resulting curve is compared with the approximations mentioned above (F1, F2, and A–K). It will be seen that the original Flory formula F1 comes closest to the DBL estimate. The error for small z is rapidly corrected and in the range $3<z<10$ the two curves are very close. However for larger z deviations become more apparent. By contrast although both F2 and A–K start correctly, they rapidly deviate in opposite directions from the DBL estimate and this deviation increases as z increases.

Having established this general procedure of passing from a self-avoiding walk to a more realistic model of a polymer chain it should be possible to undertake similar calculations for other properties of interest like the radius of gyration, the probability distribution of end-to-end length, and the pair-correlation function between two different points of the chain.

The convergence of different lattices to the two-parameter function as $N\to\infty$, $w\to 0$ for a particular value of the parameter $f_0 N^{\frac{1}{2}}w$ is illustrated in fig. 3 for the simple cubic, body-centred cubic, and face-centred cubic lattices. To

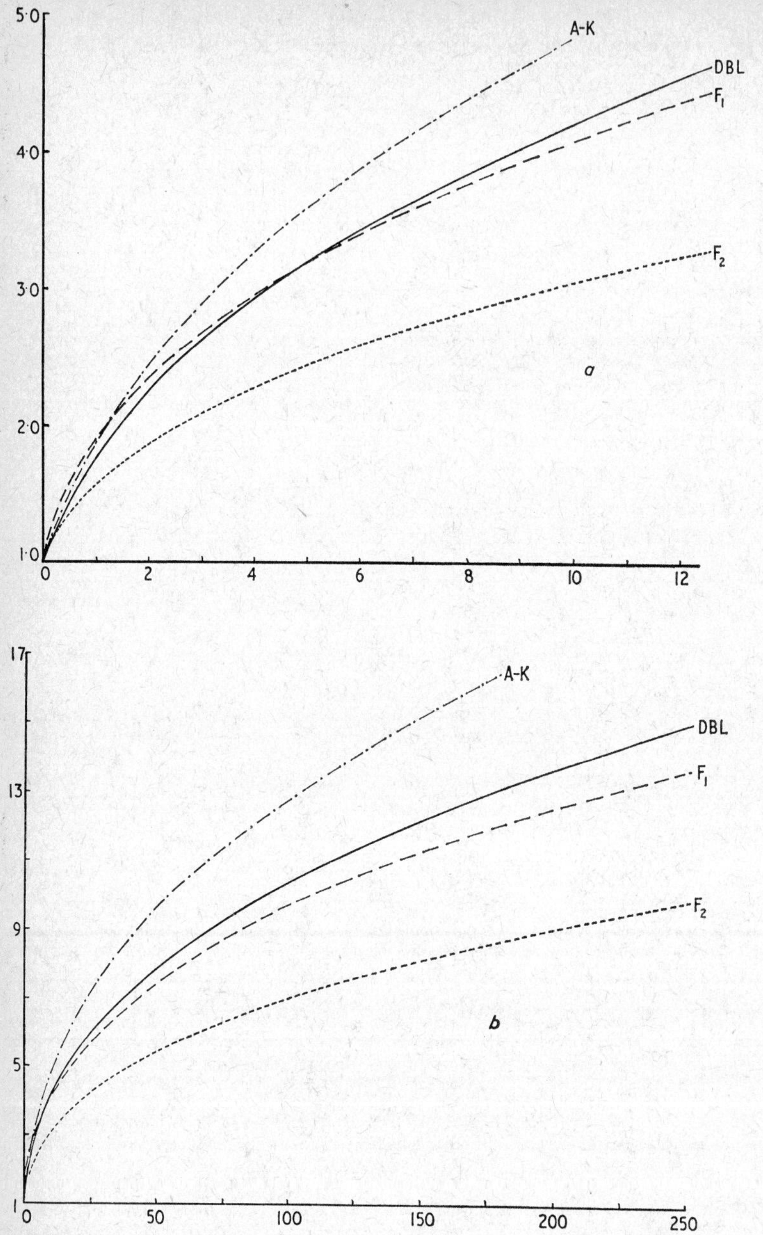

Fig. 2. *Abscissa: z*; *ordinate:* α^2.

Comparison of the present estimate of the two parameter function (DBL) with closed form approximations F_1 = Flory, 1949, F_2 = Flory & Fisk, 1966, A–K = Alexandrowicz, 1968; Kurata, 1968. (*a*) *z*, 0–12; (*b*) *z*, 0–250.

obtain an improved estimate of the two parameter function, results for all available lattices should be incorporated in this manner for different values of the parameter $f_0 N^{\frac{1}{2}} w$. Calculations by A. J. Barrett following this approach and using suitable interpolation formulae are currently in progress.

This work has been supported (in part) by the US Department of the Army through its European Office.

References

1. Domb, C, Proc van der Waals conf. 1973. To be published in Polymers, 1974.
2. Zimm, B H, J chem phys 1946, 14, 164.

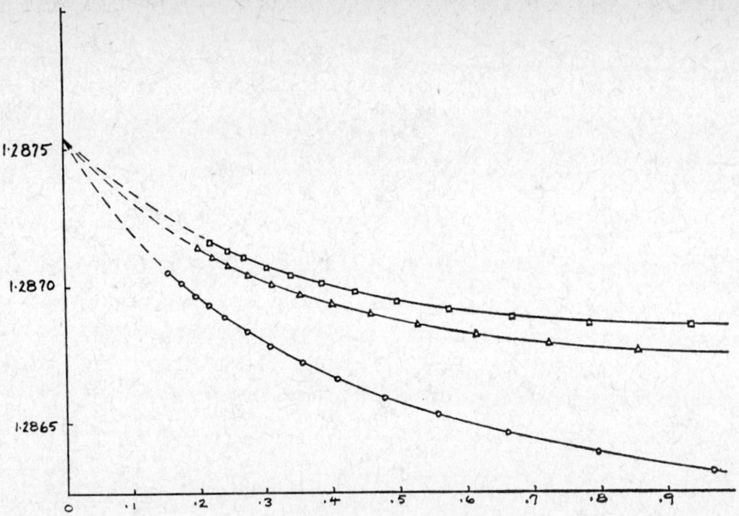

Fig. 3. *Abscissa:* $-w \times 10^2$; *ordinate:* $\alpha^2 \cdot u = 1$. ○, s.c.; △, b.c.c.; □, f.c.c.

Convergence of different lattices to the two parameter function (*a*) SC; (*b*) BCC; (*c*) FCC.

---

3. Yamakawa, H, Modern theory of polymer solutions, p. 89. Harper & Row, New York, 1971.
4. Flory, P F, Principles of polymer chemistry, p. 600. Cornell Univ Press, Ithaca, N.Y., 1953.
5. Domb, C, Adv chem phys 1969, 15, 229.
6. de Gennes, P, Phys lett 1972, A38, 339.
7. Flory, P J & Fisk, S, J chem phys 1966, 44, 2243.
8. Alexandrowicz, Z, J chem phys 1968, 49, 1599.
9. Kurata, M, J polymer sci 1968, A26, 1607.
10. Balian, R & Toulouse, G, Phys rev lett 1973, 30, 544.
11. Fisher, M E, Phys rev lett 1973, 30, 679.
12. Domb, C & Joyce, G S, J phys 1972, C5, 956.
13. Griffiths, R B, Phys rev lett 1970, 24, 1479.

Key words

Self-avoiding walk, polymer, two-parameter function, mean field

Aspects of Time-Dependent Critical Phenomena

B. I. Halperin

Bell Laboratories, Murray Hill, N. J. 07974, USA

Summary

A brief review of the current theory of dynamic critical phenomena is presented.

Along with the static phenomena discussed in some of the previous lectures at this conference, the dynamic or time-dependent properties of a system exhibit a variety of singular behaviors when the system passes through a critical point. In the present lecture, I will give a brief review of the present status of the theory of dynamic critical phenomena. Four different ideas will be introduced, and the successes and failures of these ideas will be discussed:

(1) The "Dynamic Scaling Hypothesis", first used by Ferrell et al. [1, 2], and later generalized and reformulated by P. C. Hohenberg and myself [3, 4].

(2) The "conventional" theory of dynamic critical behavior, introduced by Van Hove [5].

(3) The various "mode-mode coupling" approaches, used by Fixman [6], Kawasaki [7], Kadanoff, Swift & Laramore [8, 9], Wegner & Riedel [10–12], Résibois & De Leener [13], Ferrell & Perl [14, 15], and others.

(4) Some recent calculations using Wilson's expansion approach, which have been carried out by P. C. Hohenberg, S. Ma, and myself [16, 17] and by Suzuki & Igarashi [18].

We shall assume throughout this article that the static critical properties of the system in question can be understood in terms of the renormalization group and scaling ideas developed by Kadanoff [19], Wilson, Fisher, and others which are discussed on pp 14 and 16.

1. Definitions

Among the time-dependent properties considered in the theory of dynamic critical phenomena are time-dependent correlation functions for fluctuations in the system, the various transport coefficients, and the linear response functions for time-dependent external perturbations. These may be contrasted with "static" properties of the system such as the thermodynamic functions, equal time or instantaneous correlation functions, and linear response to time-independent (but possibly spatially inhomogeneous) perturbations. We shall here work with frequency-dependent correlation functions defined by

$$C_A(\boldsymbol{k}, \omega) = \int e^{-i\boldsymbol{k}\cdot\boldsymbol{r}+i\omega t} \langle \delta A(\boldsymbol{r}, t)\, \delta A(0,0)\rangle \, d\boldsymbol{r} dt, \quad (1)$$

where $\delta A(\boldsymbol{r}, t)$ is the fluctuation of the variable A at point $\boldsymbol{r}$ and time t about its average value, and the angular brackets denote the expectation value in thermal equilibrium.

2. Dynamic Scaling Hypothesis

The correlation function $C_A(\boldsymbol{k}, \omega)$ obeys the dynamic scaling hypothesis, if it has the following properties:

(A) Exactly at the critical point, for wave vectors $\boldsymbol{k}$ small compared to the inverse of any microscopic distances, $C_A(\boldsymbol{k}, \omega)$ has the form

$$C_A(\boldsymbol{k}, \omega) \sim 2\pi \bar{\omega}_k^{-1} C_A(\boldsymbol{k}) f_A(\omega/\bar{\omega}_k), \quad (2a)$$

$$\bar{\omega}_k \equiv \text{const} \cdot k^z \quad (2b)$$

where $C_A(\boldsymbol{k})$ is the Fourier transform of the equal time correlation function for the variable A, z is an unspecified exponent, and f_A is a function having unit area, and depending on ω only through the ratio $\omega/\bar{\omega}_k$. The dynamic scaling hypothesis supposes further that at T_c, f_A is a well behaved function of its argument (e.g., no δ-functions) with a characteristic width of order unity. Thus the spectrum of $C_A(\boldsymbol{k}, \omega)$, at fixed $\boldsymbol{k}$, has a single characteristic frequency, $\bar{\omega}_k$.

(B) For temperatures near to T_c, the dynamic scaling hypothesis requires that the function $C_A(\boldsymbol{k}, \omega)$ have a form similar to (2), but now the functions depend in an essential way on the ratio $k/\varkappa$, where $\varkappa$, the inverse coherence length, goes to zero at the critical point. (Along the "coexistence curve", or the "critical isochore" $\varkappa$ varies as $(T-T_c)^\nu$.) For example, in any given direction away from the critical point, the characteristic frequency $\bar{\omega}$ is hypothesized to have the form

$$\bar{\omega}_{k,\varkappa} = \text{const} \cdot k^z \, \Omega(k/\varkappa) \tag{3}$$

where Ω depends on k only through the ratio $k/\varkappa$. In the limit $k/\varkappa \to 0$, the characteristic frequency will have the form

$$\bar{\omega} \sim \varkappa^z (k/\varkappa)^x \tag{4}$$

where the exponent x is determined by macroscopic, or "hydrodynamic" considerations, which will be discussed further below. Typically, x has the value 0, 1 or 2, depending on the variable A.

The dynamic scaling hypotheses were developed in refs [1–4] by analogy to the more familiar static scaling laws, primarily as unproved postulates, that were to be tested by experiment. It was recognized, however, that if Kadanoff's renormalization picture of static critical phenomena [19] could be generalized to include dynamic phenomena, then the dynamic scaling laws would follow. Somewhat more precisely, in terms of Wilson's renormalization ideas (p. 14), we may say that if one defines a renormalization group for the "equation of motion" of a system, in which one successively eliminates explicit consideration of "short-wavelength" fluctuations, and if the "equation of motion" of a system at its critical point approaches a fixed point under the renormalization transformations, and if the fixed point is suitably well-behaved (e.g. has the proper stability requirements, has all parameters finite, etc.) then the dynamic scaling hypothesis must hold.

In almost all cases, as far as is known, for systems with short range forces and dimensionality $d \geq 2$ but less than 4, the dynamic scaling hypothesis works when A is the order parameter of the system at the critical point. (An exception will be given in Section 7, p. 59 however.) The dynamic scaling hypothesis frequently does not work for other choices of A, however. In particular, for complicated systems with several low-frequency dynamic modes (such as the liquid-gas system), there will often be several characteristic frequencies which go to zero as different powers of k for $T = T_c$. Whereas the order parameter normally couples only to the lowest of these modes, other variables may couple to higher frequency modes in varying proportions, and may have correlation functions with several characteristic frequencies, or with frequencies varying with different exponents z in different regions of temperature and wave vector [8].

3. Hydrodynamic Considerations

We shall consider only the simplest possible situation, where the variable A relaxes to its equilibrium value via a single, non-propagating mode. If the variable A is conserved by the equation of motion of the system (i.e., $\int A(\boldsymbol{r}, t)\,\mathrm{d}\boldsymbol{r}$ independent of time) then at any temperature other than T_c, the relaxation of A, at sufficiently long wavelengths, is described by a diffusion equation. The characteristic relaxation rate of A may then be written

$$\bar{\omega} = \frac{\lambda_A k^2}{\chi_A} \tag{5}$$

Here, λ_A is the "transport coefficient" for the variable A, and χ_A is the "susceptibility" describing the linear response of the system to an applied field conjugate to A. The most familiar example of this equation is the law of relaxation of temperature fluctuations. Here the transport coefficient is the thermal conductivity λ_E, and the susceptibility is the specific heat per unit volume at constant pressure, ϱC_p:

$$\bar{\omega}_E = \frac{\lambda_E k^2}{\varrho C_p} \tag{5a}$$

Note that (5) corresponds to the case $x = 2$, in (4).

If the variable A is not conserved, then the relaxation to equilibrium is expected to occur at a finite rate even at $\boldsymbol{k} = 0$, and we may write

$$\bar{\omega} = \frac{\Lambda_A}{\chi_A} \tag{6}$$

where Λ_A is described as the "kinetic coefficient" for the variable A. This corresponds to $x = 0$ in (4). More generally, if the system also has one or more diffusive modes, the correlation function $C_A(\boldsymbol{k}, \omega)$ for a nonconserved variable A can have one part with a finite characteristic frequency of form $\bar{\omega} \propto k^0$ and can have another part which couples to the diffusive modes, and has characteristic frequency $\bar{\omega} \propto k^2$ [20, 21]. In many cases of interest, however, the coupling of a particular variable A to the diffusive modes may be absent, because of symmetry or other considerations.

4. The Conventional Theory

The "conventional theory" of critical slowing down, first proposed by Van Hove, states that the transport coefficients λ_A, and by extension, kinetic coefficients Λ_A, remain finite and nonzero at the critical point. In support of this theory, one tries to argue that the transport and kinetic coefficients are determined by scattering events on the interatomic length scale. Since instantaneous correlation functions at short distances do not diverge at the critical point, one argues that the transport and kinetic coefficients will also not be drastically affected by passage through the critical point.

If the variable A is the order parameter ψ of the system, then χ_A diverges near T_c as $\varkappa^{-2+\eta}$, where the exponent η is positive but close to zero. If the relaxation of ψ is described by a single nonpropagating mode, the Van Hove theory therefore predicts a critical slowing down of the order parameter fluctuations at long wavelengths near T_c, with

$$\bar{\omega} \propto \varkappa^{2-\eta} k^2, \quad \text{if } \psi \text{ is conserved,} \tag{7a}$$

$$\bar{\omega} \propto \varkappa^{2-\eta}, \quad \text{if } \psi \text{ is not conserved.} \tag{7b}$$

This is consistent with the dynamic scaling hypothesis, provided that

$$z = 4 - \eta, \quad \text{for } \psi \text{ conserved,} \tag{8a}$$

$$z = 2 - \eta, \quad \text{for } \psi \text{ not conserved.} \tag{8b}$$

The notion that transport coefficients might behave like short-wavelength, instantaneous correlation functions, which was used to justify the Van Hove theory, can be carried one step further. Although short-wavelength correlation functions are expected to remain finite at T_c, the temperature derivative of such a correlation function is expected to show singularities proportional to the specific heat. Thus as an extension of the conventional theory, one would predict a singularity given by

$$\frac{d\lambda_A}{dT} \propto |T - T_c|^{-\alpha}, \tag{9}$$

where α is the specific heat exponent (cf eq. 18, p. 60). The constant of proportionality may be positive or negative, but should have the same sign above and below T_c. This type of behavior for a transport coefficient near T_c, was first predicted by Mannari [22], Fisher & Langer [23], and Nabutovskii & Patashinskii [24], for the case of the electrical conductivity of a metal near a ferromagnetic critical point. Similar behavior is expected theoretically for the electrical conductivity very near the critical point of an antiferromagnet, or of an order–disorder transition in an alloy [24–26]. The experimental situation generally tends to support these predictions, but is not conclusive.

From a purely theoretical viewpoint, I know of no arguments to make me disbelieve the detailed predictions of the conventional theory for the electrical conductivity in the above cases. There are a variety of other examples of transport coefficients, in these and other systems, which may well obey the conventional theory. Nevertheless, there are also a great number of situations in which both theoretical considerations and experimental results clearly rule out the conventional theory. We discuss some of these systems in sections 5–7, pp 57–58.

5. Systems with a Propagating Mode below T_c

The isotropic Heisenberg ferromagnet and antiferromagnet, along with superfluid helium and several other systems, have the property that in the ordered phase, below T_c, the order parameter fluctuations at long wavelengths are dominated by a propagating mode (spin waves, or second sound in He) which does not exist in the disordered phase above T_c. First consider the situation above T_c. The order parameter in the ferromagnet is conserved by the Hamiltonian, and relaxes according to a diffusion equation above T_c. Thus if the Van Hove theory were correct, and the spin-transport coefficient were nondivergent above T_c, this would imply $z = 4 - \eta$, if dynamic scaling were to hold as well. In the superfluid and in the Heisenberg antiferromagnet, the order parameter is not conserved, and the conventional theory of relaxation above T_c leads to $z = 2 - \eta$.

The situation below T_c is quite different, however. The frequencies of the spin waves or second sound below T_c are entirely determined by hydrodynamic considerations in terms of equal time correlation functions, or "static" properties of the system, and the temperature dependence of these quantities can be expressed in terms of $\varkappa$ using only the usual static scaling laws [21, 27]. Specifically one finds that at long wavelengths below T_c the order parameter oscillates with a real frequency

$\bar{\omega} \sim \varkappa^{\frac{1}{2}-\frac{1}{2}\eta} k^2$ isotropic ferromagnet,

$\bar{\omega} \sim \varkappa^{\frac{1}{2}} k$ isotropic antiferromagnet, or helium.

(In the case of helium, we are neglecting a possible weak divergence in the specific heat, which enters the hydrodynamic formula.) If the dynamic scaling theory is to describe the situation below T_c as well as at and above T_c, it then follows that we must have

$$z = \tfrac{5}{2} - \tfrac{1}{2}\eta \quad \text{(ferromagnet)}, \tag{10a}$$

$$z = \tfrac{3}{2} \quad \text{(antiferromagnet, superfluid)}, \tag{10b}$$

and that above T_c, in contradiction to the conventional theory, we must have

$$\lambda_\psi \sim \varkappa^{-\frac{3}{2}+\frac{1}{2}\eta} \quad \text{(ferromagnet)}, \tag{11a}$$

$$\Lambda_\psi \sim \varkappa^{-\frac{1}{2}+\eta} \quad \text{(antiferromagnet, helium)}, \tag{11b}$$

In a similar way, application of the dynamic scaling hypothesis to the characteristic frequency of energy fluctuations in helium, which are also dominated by second sound below T_c, leads to the prediction of a diverging thermal conductivity above T_c [1]:

$$\lambda_E \sim \varkappa^{-\frac{1}{2}} \quad \text{(helium)} \tag{12}$$

Experiments in helium and in nearly isotropic magnetic systems favor the dynamic scaling hypothesis, with its divergent transport and kinetic coefficients and not the conventional theory [28, 29].

Calculations of the "mode–mode" coupling type have also been used to obtain results (11) and (12), by considerations which involve dynamic properties only at or above T_c, making no explicit reference to the propagating modes below T_c. It is not possible to give here a detailed description of the various mode–mode coupling calculations, but perhaps a few words about one of these calculations will give the spirit of the approach [7]. Briefly, in order to calculate the divergence of the spin transport coefficient λ_ψ in the Heisenberg system, one considers the contribution to the spin-current at wavevector $\boldsymbol{k}$ arising from nonlinear coupling to two or more "intermediate" spin fluctuations, the sum of whose wavevectors is also $\boldsymbol{k}$. The magnitude of the contribution to λ_ψ is determined by the magnitude of the "vertex" coupling the spin current to the intermediate fluctuations, by the form of the appropriate instantaneous correlation functions, and by the characteristic relaxation rates of the intermediate modes. It turns out that in this case the largest contribution comes from intermediate fluctuations whose wavevectors are of the order of k or $\varkappa$, whichever is smaller, and not of the order of the reciprocal interatomic spacing, as would have been required for the validity of the conventional theory. The magnitudes of the vertices and instantaneous correlation functions are estimated by means of static scaling laws, and the relaxation rates are to be determined self-consistently. When proper care is exercised in estimating the interaction vertices, one indeed winds up with the results (10) and (11) quoted above, including the small terms $\frac{1}{2}\eta$ in the exponents [7].

6. Liquid-gas and Phase Separation Critical Points

Divergent transport coefficients for the order parameter are also found at the liquid-gas critical point and at the critical point for phase separation of a binary liquid. In both these cases the order parameter is conserved and is dominated by a nonpropagating, diffusive mode at long wavelengths both above and below the transition. Thus the conventional theory leads to no immediate inconsistency with the dynamic scaling hypothesis in these systems, if we chose $z = 4 - \eta$. However, a mode–mode analysis shows that the conventional theory is not correct [7, 8, 14, 15].

In the liquid gas transition, the order parameter may be taken to be $\varrho - \varrho_c$, where ϱ is the density of the fluid and ϱ_c is the value of ϱ at the critical point. Long wavelength density fluctuations near T_c are dominated by entropy fluctuations at constant pressure, whose relaxation is given by the thermal diffusion formula (5a). The order parameter for the binary liquid case may be taken to be $x - x_c$, where x is the relative concentration of the two components. The characteristic rate of relaxation of the concentration fluctuations is given by

$$\bar{\omega} = \frac{\lambda_x k^2}{\chi_x} \tag{13}$$

where λ_x is the transport coefficient for relative diffusion of the two components, and χ_x is the susceptibility for concentration changes, which may be defined as the derivative with respect to x of the difference in chemical potentials of the species. For our purposes we may neglect cross coupling between the species diffusion and thermal conduction modes. The susceptibilities (ϱC_p) and χ_x diverge as $\varkappa^{-2+\eta}$ near T_c in the two respective cases. In both cases the mode–mode approach predicts that the transport coefficient diverges approximately as $\varkappa^{-1}$, and the dynamic scaling hypothesis can be satisfied only if $z \approx 3$. Experiments on both cases are consistent with these results [30–32].

The physical reason for the divergent transport coefficients can be understood by means of the following argument [33], which we shall present for the case of the binary liquid. Let us divide the liquid into regions the size of the coherence length each having a volume $\varkappa^{-3}$, and let $\delta\bar{x}$ be the deviation of the concentration in a given volume $\varkappa^{-3}$ from the average value in the sample as a whole. The fluctuations in $\delta\bar{x}$ are given by the equipartition law as

$$\langle(\delta\bar{x})^2\rangle \approx T\chi_x \varkappa^3, \tag{14}$$

Suppose one imagines applying a field $\boldsymbol{F}$ which exerts a force proportional to the deviation in relative concentrations $\delta\bar{x}$. Then the liquid in any given volume $\varkappa^{-3}$ will tend to drift in the field at a velocity $\boldsymbol{v}$ such that the applied force $f_{\text{app}} = \delta\bar{x}\varkappa^{-3}\boldsymbol{F}$ is balanced by the viscous drag on the volume, which is given by Stokes' law as:

$$f_{\text{visc}} \approx -\eta^* \boldsymbol{v} \varkappa^{-3} \cdot \varkappa^2, \tag{15}$$

where η^* is an effective viscous coefficient. It is then a simple matter to solve for $\boldsymbol{v}$, and add up the contributions of all the volumes $\varkappa^{-3}$ to get the transport coefficient

$$\lambda_x \approx T\chi_x \cdot \varkappa^3/\eta^* \varkappa^2 \sim \varkappa^{-1+\eta}/\eta^*. \tag{16}$$

It follows that either λ_x diverges or η^* diverges, or both coefficients diverge. (Note that the viscosity is itself a transport coefficient.) The quantity η^* differs from the true viscosity η_0 in that the former refers to motion on a scale of relatively short lengths and times, while η_0 refers to the limit of long wavelengths and low frequencies. In fact detailed mode–mode coupling analysis of viscous transport shows that both η_0 and η^* are at most weakly divergent.

Formula (16) for the divergence of λ_x, is usually derived more formally using the various mode–mode coupling techniques [7, 8, 15]. From these points of view, the contribution to λ_x that we have considered would be described as arising from the product of a concentration fluctuation and a viscous relaxation mode.

Note the importance of the fact that the viscous force per unit volume is proportional to the inverse square of the characteristic length for gradients in the velocity, which accounts for the final factor $\varkappa^2$ in (15). From a hydrodynamic point of view, this dependence is a result of the law of conservation of momentum for the fluid, which therefore plays a vital role in the divergence of λ_x.

In closing this section, let me mention one additional area, in which the mode–mode coupling approach has been used extensively—the prediction of anomalies in the ultrasonic attenuation at the critical points of the gas-liquid system, at magnetic critical points, and at the superfluid transition in He [8, 9, 34]. We shall not discuss these calculations here.

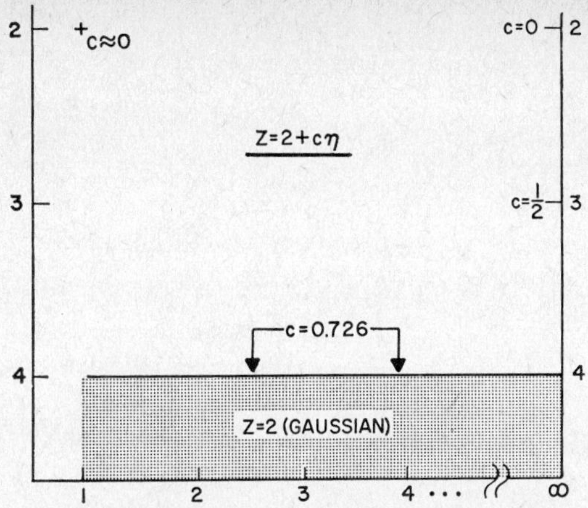

Fig. 1. *Abscissa: n; ordinate: d.* Dynamic critical exponent z for the generalized kinetic Ising model with neither energy nor order parameter conserved (case A). Ordinate and abscissa are lattice dimensionality d and order parameter dimensionality n. Values of z at boundaries come from expansions in $(4-d)$ or n, computed in [13]. Value at $d=2$, $n=1$ comes from high temperature series expansion for the two-dimensional kinetic Ising model [27].

7. The Kinetic Ising Models and Related Systems

We have now seen several examples where the conventional theory of dynamic critical phenomena breaks down because of a divergence in transport or kinetic coefficients. Recent theoretical work by P. C. Hohenberg, S. Ma and myself [16, 17] shows that there are also systems where the conventional theory breaks down because the kinetic coefficient for the order parameter goes to zero as $T \to T_c$. (We have not found any examples where a true transport coefficient, referring to the relaxation of a conserved quantity in the extreme long wavelength limit at fixed $T \neq T_c$, goes to zero as $T \to T_c$.)

The models we have considered are continuum generalizations of the kinetic Ising models, first introduced by Glauber [35]. In the kinetic Ising models, the spin system is supposed to interact with a "bath", which causes spins to flip in a stochastic manner. The spin-flip probabilities in turn are chosen consistent with the principle of detailed balance, so that the system tends to the equilibrium ensemble of the Ising model, at some specified temperature T. Consistent with this requirement, there are still many possibilities. One may choose the transition probabilities so that the order parameter (spin) of the system is conserved or in a manner that the total spin is not conserved. Similarly one may define the model so that energy is conserved, and therefore there is a slow, long-wavelength mode associated with thermal diffusion, or one may define the model so that energy is not conserved. The latter case corresponds physically to an infinite heat bath at each lattice site, or to a bath with an infinite thermal conductivity. We designate the four cases by

Case A, ψ not conserved, E not conserved;
Case B, ψ conserved, E not conserved;
Case C, ψ not conserved, E conserved;
Case D, ψ conserved, E conserved.

Note that the conventional theory for the order parameter relaxation predicts $z=2-\eta$ for cases A and C, and $z=4-\eta$ for cases B an D. Analyses based on the mode–mode coupling approach have suggested that the conventional theory should be correct in these models [9, 12, 36]. In addition, because the transition rates in these systems may be expressed in terms of a Hermitian matrix (purely "relaxational" dynamics) it is possible to prove certain rigorous inequalities for the dynamics. In particular, one can prove rigorously that the transport or kinetic coefficients do not diverge in these models [37, 38].

The critical behavior of these models has been studied using Wilson's renormalization group and expansion methods, in which the critical exponents and scaling functions are considered to be continuous functions of the lattice dimensionality d and the order parameter dimensionality n. The results are as follows:

For case B, expansions have been carried out to order $1/n$, for $2<d<4$, and to order ε^2 ($\varepsilon \equiv 4-d$), for $0<n<\infty$. Results confirm the conventional theory. Renormalization group arguments suggest that the conventional theory should hold to all orders in ε.

For case D, only preliminary analysis has been carried out, but it appears that the conventional theory should apply for the relaxation of the order parameter.

For case A, calculations have again been carried out to order ε^2 or $1/n$. The results are indicated in fig. 1. The exponent z is found to be greater than the conventional results by an amount of order ε^2 or $1/n$, which is the same order as the exponent η. It is convenient to write

$$z = 2 + c\eta \tag{17}$$

so that $c = -1$ corresponds to the conventional theory. We find in contrast that $c = 0.726$ independent of n, in the limit of $\varepsilon \to 0$. In the limit $n \to \infty$, we find that c is a function of d which has the value $c = \frac{1}{2}$ at $d = 3$, $c \to 0.726$, as $d \to 4$, and $c \to 0$ as $d \to 2$. For $d > 4$, the model behaves in a trivial Gaussian manner near T_c, and obeys the conventional theory, with $z = 2$ and $\eta = 0$.

The point $d = 2$, $n = 1$ (2-dimensional kinetic Ising model) had been studied earlier by Yahata & Suzuki [39] using extrapolations of high-temperature series expansions. Within the accuracy of their calculations, they found $z = 2$ ($c = 0$), which already implied the failure of the conventional theory. The result $z = 2$ is also consistent with recent Monte Carlo calculations [40]. Unfortunately, it is difficult to be certain about one's estimates of the accuracy of exponents derived from high temperature expansions or Monte Carlo calculations, and thereby to exclude absolutely the result $c = -1$ ($z = 1.75$) for the 2-dimensional kinetic Ising model from these results.

The scaling function f, in eq. (2) has been calculated to lowest nontrivial order in $1/n$ or ε, for $T = T_c$, and the dynamic scaling behavior at T_c has been confirmed in detail to those orders. Recently Suzuki & Igarashi [18] have explicitly calculated the dynamic critical exponent for $T > T_c$, and have checked that dynamic scaling applies here too.

The relaxation of the nonconserved energy has been explicitly calculated in case A only to first order in ε or zeroth order in $1/n$ [17, 41]. To these orders, the characteristic relaxation rate of the singular part of the energy correlation function has the same critical exponent as that for the order parameter. Renormalization group arguments indicate that this should continue to hold to all orders in ε.

Finally we turn to case C. Let us first consider the relaxation rate for the energy. In the limit of long wavelengths, at any fixed $T > T_c$, the energy relaxation is given by the thermal diffusion formula (5*a*). Furthermore, as mentioned above, the true transport coefficient λ_E remains finite and nonzero as $T \to T_c$, in the system under consideration. The specific heat entering (5*a*) may be written

$$\varrho C_p \sim A|T - T_c|^{-\alpha} + B, \tag{18}$$

where A and B are constants. The specific heat exponent α is generally close to zero, but may be positive or negative. If we consider only the larger of the two terms in (18), in the limit $T \to T_c$, we may write

$$\varrho C_p \propto \varkappa^{-\tilde{\alpha}/\nu} \tag{19}$$

where

$$\tilde{\alpha} = \alpha \quad \text{if } \alpha > 0$$
$$\tilde{\alpha} = 0 \quad \text{if } \alpha < 0. \tag{20}$$

From the form of the diffusion equation above T_c, we are led to define the characteristic exponent for energy relaxation as

$$z_E = 2 + \tilde{\alpha}/\nu. \tag{21}$$

Note that if dynamic scaling holds for the energy relaxation, then at T_c we would have

$$\bar{\omega}_E \propto k^{2+\tilde{\alpha}/\nu}. \tag{22}$$

Let us now consider the exponent z for the relaxation rate of the order parameter. The simplest possible "phase diagram", consistent with the results of the renormalization group analysis [17] and of the ε and $1/n$ expansions, is shown in fig. 2. The exponent z has different analytic behaviors in the various regions indicated, and cannot be continued from one region to another. In the unshaded region of fig. 2, the exponent z is exactly the same as for case A, which we continue to write as $z = 2 + c\eta$. This unshaded region is in fact defined by the inequality

$$2 + c\eta > 2 + \tilde{\alpha}/\nu. \tag{23}$$

Throughout this region the characteristic relaxation rate for the energy at T_c for case A is already slower than the energy relaxation rate that would be predicted by (22). It is not surprising, therefore, that imposition of energy conservation as in model C, leads to no change in the energy relaxation rate at T_c, and no change in exponent

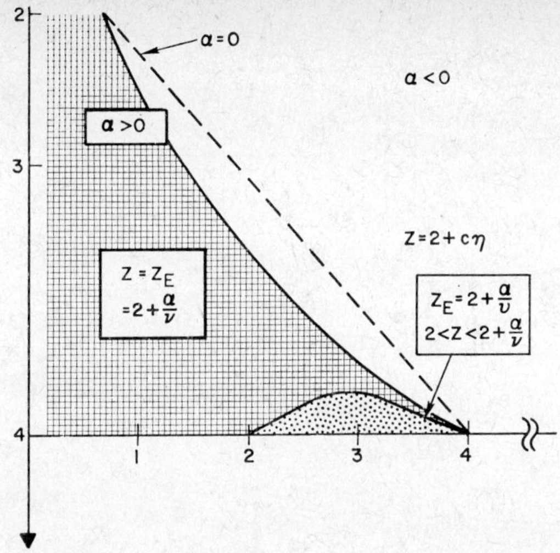

Fig. 2. *Abscissa: n; ordinate: d.* Simplest possible "phase" diagram for the dynamic critical behavior of the generalized kinetic Ising model, where energy is conserved but the order parameter is not (case C). The exponents z and z_E refer to relaxation of the order parameter and energy respectively, the function c is the same as in fig. 1, and α is the specific heat exponent. In constructing this figure it has been assumed that $c \geqslant 0$ throughout the region of interest, as is true at the boundaries $n \to \infty$ or $d \to 4$. The two shaded regions lie entirely within the region where the specific heat diverges ($\tilde{\alpha} = \alpha > 0$). The region near $d = 2$, $n = 1$ was drawn under the assumption that the result $c = 0$ obtained in [33], for the two-dimensional Ising model, holds exactly.

of the spin relaxation. On the other hand, in that portion of the $d - n$ plane where

$$2 + c\eta < 2 + \tilde{\alpha}/\nu, \tag{24}$$

the energy relaxation rate for case C, predicted by (22) is slower than the relaxation rate of the order parameter (and of the energy) in case A. We might expect, then, that imposition of energy conservation would reduce the various relaxation rates, and that the exponent z would be increased. In the cross hatched region of fig. 2, the effect of energy conservation is sufficiently strong so that z is increased all the way to $2 + \tilde{\alpha}/\nu$. In this region we find that dynamic scaling holds for both the order parameter and the energy, and that the characteristic frequencies of the two quantities are the same. In the dotted region of fig. 2, the effect of energy conservation is not sufficiently strong to increase z to the value $2 + \tilde{\alpha}/\nu$, although z is increased above the value $2 + c\eta$ appropriate to case A. In this region we find that the energy and order parameter each obey dynamic scaling, at and above T_c, but with different characteristic exponents. Below T_c, however, the energy and order parameter are linearly coupled. From hydrodynamic considerations one expects that a fraction of the weight of the spin correlation function will relax with the (slower) characteristic frequency of the energy [20, 21]. It therefore appears that in the dotted region ($2 < n < 4$ and d close to 4) there is a breakdown of dynamic scaling for the order parameter, below T_c.

The existence of the three regions described near $\varepsilon = 0$ is apparent in the series expansion results for z to first order in ε, and can be obtained by an approximate recursion relation for the equations of motion of the system similar to the calculation discussed by Wilson, for the static case, in his earlier lecture (p. 14). For $2 < n < 4$, and ε near zero, the thermal conductivity parameter is zero at the fixed point of the recursion relation. The nonexistence of a single characteristic relaxation rate for the energy and spin, in the dotted region of fig. 2, is thus signalled by the fact that the fixed point of the renormalization group is singular.

8. Anisotropic Heisenberg Systems

Consider a three-dimensional anisotropic Heisenberg ferromagnet or antiferromagnet, in which the coupling constants for the three components of the spin are all different. In this system there is no conservation law other than conservation of energy, and the only low frequency mode at long wavelengths, for $T \neq T_c$, is the thermal diffusion mode. Since the spin has a single easy direction, the order parameter is characterized by $n = 1$, and the static exponents are expected to be the same as those of the Ising model. Similarly we expect that the dynamic critical properties should be the same as those of the kinetic Ising model with energy conservation, i.e., of case C of the previous section, for $d = 3$, $n = 1$.

In uniaxial Heisenberg ferromagnet, the couplings of the x and y components of the spin are equal but smaller than the coupling of the z component. The order parameter is the z component of the spin, which is now conserved by the Hamiltonian. We expect that the dynamic critical behavior should be the same as that of case D of the previous section.

In the uniaxial antiferromagnet, the order parameter is the z-component of the staggered magnetization which is not conserved. In addition to the energy, there is now a second conserved quantity, the z-component of the total magnetization. Under repeated applications of the renormalization group, the (nonlinear) coupling of the order parameter and the energy to long wavelength fluctuations of the z-component of the magnetization becomes vanishingly small, however, and we do not expect this variable to affect the critical dynamics. Thus the uniaxial antiferromagnet should have the same critical behavior as case C above.

In real magnetic systems, the thermal conductivity due to phonons is generally much larger than that of the isolated spin system. In the experimentally obtainable range of temperatures, the extrapolated thermal relaxation rate given by $\lambda_E \varkappa^2/\varrho C_p$ may be much faster than the characteristic rate of relaxation of the order parameter. Thus one has not attained the critical region described by case C (cross hatched region of fig. 2) in which the two relaxation rates are comparable. The exchange of energy between the lattice and spins may be sufficiently slow so that one cannot describe the dynamics as in case A, in which the system freely interacts with a heat bath of infinite thermal conductivity. At the same time, energy exchange with the lattice may be too fast to consider the spin system as thermally isolated from the lattice, so that case C could be applied to the spin system alone. The observable dynamics may therefore be intermediate between case A and case C.

9. Universality

Motivating the discussion in the previous sections is the notion of "universality" for dynamic critical phenomena. As in the case of static critical phenomena, "universality" does not mean that all systems have the same critical exponents, but that there exist large classes of systems with the same critical behavior. For example all Ising models of a given lattice dimensionality d, with short range forces, are supposed to have the same static critical behavior, regardless of the details of the underlying lattice. In the language of the renormalization group, details of the lattice are irrelevant variables, which disappear from the problem near T_c. Renormalization group ideas applied to dynamic critical phenomena clearly imply that most of the details of the dynamics are also irrelevant near T_c, and that large classes of systems will show the same dynamic critical behavior. It is clear from the above examples however that a class of systems showing the same static critical behavior may be divided into several classes of dynamic behavior. For example the four cases discussed in Section 7 (pp 59–61) all correspond to systems with the same static properties, for a given n and d. Similarly, as a more familiar example, the classical isotropic Heisenberg ferromagnet and antiferromagnet have identical equilibrium properties, but have very different time-dependences, both in the hydrodynamic and critical regimes.

As in the static case, a complete classification of all possible systems has not been undertaken, but we think we know how to assign at least the most commonly occurring types of systems to their appropriate critical behavior classes. We are always assuming, of course, that the renormalization group ideas are correct in the first place. As was mentioned by the earlier speakers, there exist large enough discrepancies between the predictions of the renormalization group and scaling ideas and the results of certain experiments and high temperature series expansions, so that one should keep open the possibility that some of our ideas about static critical phenomena may be incorrect. Presumably, if this is the case, many of our ideas about dynamic critical phenomena will also have to be revised.

I have benefited over the years from conversations with many researchers on the topic of dynamic critical phenomena, among whom have been P. C. Martin, M. E. Fisher, L. P. Kadanoff, G. Ahlers, G. Thomas, and K. G. Wilson. The contributions of P. C. Hohenberg and S. Ma to my understanding of this field require special mention here. The analysis in Sections 7 and 8 of this lecture, in particular, was entirely the result of collaboration with them.

References

1. Ferrell, R A, Menyhárd, N, Schmidt, H, Schwabl, F & Szépfalusy, P, Phys rev lett 1967, 18, 891.
2. — Ann phys 1968, 47, 565.
3. Halperin, B I & Hohenberg, P C, Phys rev lett 1967, 19, 700.
4. — Phys rev 1969, 177, 952.
5. Van Hove, L, Phys rev 1954, 93, 1374.
6. Fixman, M, J chem phys 1967, 47, 2808 (and references therein).
7. Kawasaki, K, Ann phys (NY) 1970, 61, 1 (and references therein).
8. Kadanoff, L P & Swift, J, Phys rev 1968, 166, 89.
9. Laramore, G & Kadanoff, L P, Phys rev 1969, 187, 619.
10. Wegner, F, Z phys 1968, 216, 433.
11. — Z Phys 1969, 218, 260.
12. Riedel, E & Wegner, F, Phys lett 1970, 32 A, 273.
13. Résibois, P & DeLeener, M, Phys rev 1969, 178, 806 and 819.
14. Ferrell, R A, Phys rev lett 1970, 24, 1169.
15. Perl, R & Ferrell, R A, Phys rev lett 1972, 29, 51.
16. Halperin, B I, Hohenberg, P C & Ma, S, Phys rev lett 1972, 29, 1548.
17. Halperin, B I, Hohenberg, P C & Ma, S. To be published.
18. Suzuki, M and Igarashi, G, Prog theor phys 1973, 49, 1070.
19. Kadanoff, L P, Physics 1966, 2, 263. See also Wilson, K G & Kogut, J Phys repts. In press (and reference therein).
20. Heller, P, Int j magn 1970, 1, 53.
21. Halperin, B I & Hohenberg, P C, Phys rev 1969, 188, 898.
22. Mannari, I, Phys lett 1968, 26 A, 134.
23. Fisher, M E & Langer, J S, Phys rev lett 1968, 20, 665.
24. Nabutovskii, V M & Patashinskii, A Z, Fiz tverd tela 1968, 10, 3121. (English translation: Soviet phys—Solid State 1969, 10, 2462.)
25. Marčelja, S, J phys C 1971, 4, L26.
26. Halperin, B I & Thomas, G A. Unpublished.
27. Josephson, B D, Phys lett 1966, 608, 21.
28. Hohenberg, P C, in Critical phenomena (ed M S Green) Proc 51st Enrico Fermi Summer School, Varenna, Italy. Academic Press, New York, 1971.
29. Kahn, R & Parette, G, J phys (Paris) 1971, 32, suppl. C1, 523 (and references therein.)
30. Swinney, H L & Henry, D L. To be published in Phys rev A.
31. Goldburg, W I & Pusey, P N, J phys (Paris) 1972, 33, suppl. C1, 105.
32. Chu, B, Lee, S P & Tscharnuter, W, Phys rev 1973, A7, 353 (and references therein).
33. Arcovito, G, Faloci, C, Roberti, M & Mistura, L, Phys rev lett 1969, 22, 1040.
34. Luthi, B, Moran, T J & Pollina, R J, J chem phys soc 1970, 31, 1741 (and references therein).
35. Glauber, R, J math phys (NY) 1963, 4, 294.
36. Kawasaki, K, Prog theor phys 1968, 40, 706.
37. Kawasaki, K, Phys rev 1966, 148, 375.
38. Kadanoff, L P & Swift, J, Phys rev 1968, 165, 310.
39. Yahata, H & Suzuki, M, J phys soc Jap 1969, 27, 1421.
40. Stoll, E, Binder, K & Schneider, T. Phys rev B. In press.
41. Suzuki, M, Phys lett 1973, 43 A, 245.

Discussion

Martin: What can be said about situations in which spin dynamics determine the relaxation? How does one go from the kinetic Ising model to, say, the Heisenberg model in which the dynamics is governed by a Hamiltonian.

Halperin: If you are talking about the isotropic Heisenberg ferromagnet or antiferromagnet, then one cannot connect the system with a kinetic Ising model. In particular, the Heisenberg systems have propagating modes below T_c, and according to the mode–mode coupling theory or dynamic scaling theory, have divergent transport coefficients above T_c. The kinetic models, regardless of the value of n, have no propagating modes, and cannot have any divergences in transport coefficients. We have not been able to apply the ε-expansion or $1/n$-expansion to the Heisenberg models, but I believe that the mode–mode theory should be correct in that case.

If one is interested in anisotropic Heisenberg systems, then I believe we can make a connection with the kinetic models. For example, in the completely anisotropic ferromagnet or antiferromagnet, there are no propagating modes at long wavelengths above or below T_c, and the only hydrodynamic mode away from T_c is thermal diffusion. I expect this model to have the same critical properties as case C in our kinetic Ising calculation, for $n = 1$. The short wavelength spin interactions play the role of the bath in the kinetic Ising model. The details of the bath are presumably "irrelevant" to the critical behavior—only the mode structure at long wavelengths and low frequencies should be relevant.

Domb: One of the results of numerical estimates from series expansion is that δ is very close to 5 both for the self avoiding walk ($n = 0$) and Ising model ($n = 1$). Also, we know that for $n = \infty$ (spherical model) δ is exactly 5. This has led to the suggestion that δ might be exactly 5 for all short range force models and Brout has published a paper recently (Phys lett 1973, 43 A, 363) in which he puts forward a field theoretical interpretation of this result. Could anyone please comment on this paper?

Martin: I think it's impossible to accept that $\delta = 5$ for all n in these dimensions since there is no reason to doubt the $1/n$ expansions. For large n they give

$$\delta \cong 5 - \frac{16}{\pi^2 n} + O\left(\frac{1}{n^2}\right)$$

Wilkins: As we reach the end of the day some of us on the outside of the fields would appreciate a summing up of where we are. Let me start off by asking how well the ε-expansion is working. Ken Wilson referred to the theory as a 5–10% one and I have heard the ε-expansion referred to as asymptotic. What do the experts believe and what can they prove?

Wilson: How good is the ε expansion? The answer has three parts.

(1) For standard numbers, for example γ for the Ising case in 3 dimensions, one gets 1.16 to order ε, 1.24 to order ε^2 and 1.20 to order ε^3, compared to the best series value 1.25. Other exponents show similar results. In general one seems to get an accuracy to about 5% or 0.05, whichever is larger, for any standard exponent.

(2) The ε expansion is most likely an asymptotic expansion, and has already been demonstrated for the approximate recursion formula. However higher order calculations plus Padé-type abracadabra may produce convergent results. If enough different black magic techniques give the same results the convergence would even be credible.

(3) Since the ε expansion is calculated by graphical methods one can calculate many more details of critical behaviour than can easily be calculated by series, e.g. details of dipolar effects (Fisher & Aharony) and dynamics.

(4) The ε expansion gives definite answers to qualitative questions: for example, the exponent of the n vector model depend only on ε and n. Three-dimensional series calculations give more wishy-washy results in some cases.

(5) The ε expansion shows that the fixed point hypothesis survives a quantitative calculation. The fixed point hypothesis, without calculation, is only an assumption; without calculation the derivations of scaling from the fixed point hypothesis is no more than a restatement of the scaling assumption.

Martin: The ε-expansion is certainly asymptotic and not convergent but the expansion parameter is not so bad in 3 dimensions. For $n = 0$, the parameter is $(\varepsilon/8)$, and for $n = 1$ it is $(\varepsilon/6)$; for large n it is $(\varepsilon/2)$ which is much worse, i.e. the parameter is

$$\varepsilon/2((n+2)/(n+8)).$$

Even for $\varepsilon = 2$, it doesn't give such bad answers for small $n = 0$ or 1.

Fisher: To answer Wilkins' question there may be room for optimism regarding the ε-expansion for the exponents. Thus even if the series is only asymptotic it may well be accurately summable by Padé approximate or other techniques. Indeed Dr Nickel has recently told me of his estimates of the main fourth order and fifth order contributions. The values of the coefficients suggest that the series might have the appropriate (Stiltjies) character such that rigorous upper and lower bounds can be constructed from them. (This is actually the situation which arises in perturbation theory of the anharmonic oscillator, as shown by Simon, Bender & Wu; the ε-expansion is somewhat analogous in character.)

de Gennes: We may have an amusing application of the renormalization group ideas with the liquid-gas transition of adsorbed monolayers. Measurement on the coexistence curve for rare gases adsorbed in graphite have been carried out in particular by Dural and coworkers. The data have been analyzed recently in terms of an exponent β by M. Bienfait. Naively one would expect a very small β, as for the 2-dimensional Ising model: $\beta_{\text{Ising}} = 0.125$. Experimentally β appears much larger (at least $\beta \cong \frac{1}{3}$). This paradox might be interpreted in various ways:

(1) very unpleasant: if we had in fact more than one layer in the experiments, we would soon recover the three-dimensional behavior. I am

told that this is not really plausible in the Dural set up.

(2) more interesting: each argon atom adsorbed on the graphite distorts the solid surface. This distortion reacts on other neighboring atoms. This "mattress effect" gives (I think) an attraction decreasing like $1/R^3$. Renormalization group argument for long range forces (Fisher et al., Susuki, Sorma) show that with $1/R^3$ interactions the dimension $d=2$ is the critical dimension at which mean field becomes applicable. Thus we might interpret the apparent β as resulting from a crossover between normal Ising behavior (β small) rather far from T_c and mean field behavior ($\beta=\frac{1}{2}$) close to T_c.

Halperin: The ε and $1/n$ expansions serve many purposes beyond the approximate calculation of exponents—one can test relations between various quantities. One thing we have learned, for example, is that the conventional theory of critical slowing down fails for systems where it was previously expected to hold.

Srinivasan: Is the Wilson procedure applicable if there are cubic terms present in the Hamiltonian; for example, in the context of a structural transformation?

Halperin: You are referring to the unsymmetric case such a occurs in the gas-liquid transition. Consider a model in which H is given by

$$H\{\varrho\} = \int \{(\nabla\varrho)^2 + a\varrho + b\varrho^2 + c\varrho^3 + d\varrho^4\} d^d x$$

We define the order parameter as $\psi = \varrho - \varrho_0$ where ϱ_0 is an arbitrary constant. We can always choose this constant so that the ψ^3 term is missing when we express H in terms of ψ. If one chooses the chemical potential (or pressure) and temperature correctly, the coefficient of ψ will also vanish, and the coefficient of ψ^2 will take on its critical value. H then has the same form as in Wilson's work, and clearly the exponents must be the same. In general, H will also contain terms like ψ^5, ψ^7 etc. which have not been eliminated. At least near $d=4$ however, it is clear that these terms are "irrelevant", and disappear when the renormalization group is repeatedly applied. At $d=3$, it is possible to check the stability of the Ising fixed point against odd perturbations of this type within the framework of the approximate recursion relations. I believe this has been done by Grover & Swift.

P. W. Anderson: Where does one go from here?

Fisher: To partially answer this question, it is worth stressing the capability for detailed calculation which we now have. One can study the effects of all sorts of perturbations (dipolar interactions, cubic anisotropics, etc.) which may be present in the original Hamiltonian. Furthermore, one can also compute the detailed scaling functions, the amplitudes that enter, and the successive corrections to scaling. In particular it is important to do this, in order to check that the various "deviations" from scaling and universality that are seen in experiment and in Ising model calculations, can be explained. If this cannot be done one will have to worry seriously about the whole formulation!

P. W. Anderson: Are efforts being made to solve the integral equation directly instead of the ε-expansion?

Wilson: The approximate recursion formula (see the Wilson–Kogut lecture notes) seems not to be generalizable to a quantitative, improvable approximation, which is what one needs at present. However, Leo Kadanoff (unpublished), Niemeyer & Van Leeuwen (preprint from Delft), and people at Cornell are now studying renormalization group equations in two dimensions which are quantitative and improvable if no formal difficulties (e.g., long range forces) arise. The idea is to define an effective Hamiltonian depending on Ising-like spins (± 1 only) which correspond in the Kadanoff sense to block spins. For example, Niemeyer & Van Leeuwen define block spins on a triangular lattice to be the sign of the sum of three spins on a triangle, the lattice being divided into non-overlapping triangles for this purpose. Hopefully the calculation of the block spin Hamiltonian can be carried out keeping only a few of the most local terms (nearest and second nearest neighbour interactions, for example) and also studying a finite block of spins containing only a few triangles, rather than an infinite lattice.

The Kondo Problem

Solution of the Spin-1/2 Kondo Hamiltonian

K. G. Wilson

Laboratory of Nuclear Studies, Cornell University, Ithaca, N.Y. 14850, USA

This is a preliminary summary of an extensive calculation. The Hamiltonian to be solved [1] is, in dimensionless form,

$$H_K = \int_{-1}^{1} k a_k^+ a_k \, dk - JA^+ \boldsymbol{\sigma} A \cdot \boldsymbol{\tau} \tag{1}$$

where

$$A = \int_{-1}^{1} a_k \, dk \tag{2}$$

and $a_{k\mu}$ is an electron destruction operator of momentum k and spin μ (the spin index μ is suppressed); the components τ_i of $\boldsymbol{\tau}$ are the spin matrices for a single spin $\frac{1}{2}$ impurity, the σ_i are the spin matrices for the electrons, k is the momentum relative to the Fermi momentum ($k \propto q - q_F$) and normalized to the maximum conduction band momentum. Only s wave conduction band electrons are considered. J is the exchange coupling constant. The calculation outlined below gives the zero temperature susceptibility χ_{imp} of the impurity and the near-zero temperature behaviour of the specific heat C_{imp} of the impurity, for small negative J (antiferromagnetic coupling). The results are as follows (in the notation of [1]):

$$\chi_{imp}\big|_{T=0} = \frac{0.103 \pm 0.003}{\tilde{D}(J)} \frac{1}{\sqrt{|2J\varrho|}} \times \exp\left\{\frac{1}{|2J\varrho|} - 1.58\,|2J\varrho| + O(J^2\varrho^2)\right\} \tag{3}$$

$$\lim_{T\to 0}\left\{\frac{C_{imp}(T)}{T\chi_{imp}(T)}\right\} = \frac{2\pi^2}{3}(1 \pm 0.03)\,k^2 + O(e^{-1/(2|J\varrho|)}) \tag{4}$$

where $\tilde{D}(J)$ is closely related to the band edge energy, and has a regular expansion in J (the definition of $\tilde{D}$ will be given following eq. (25)), ϱ is the density of states at the Fermi surface, and k is Boltzmann's constant. The factor $2\pi^2/3$ is not an analytic result; this factor happens to fit a numerical result within the numerical accuracy of about 3%. For comparison, in the limit $J \to \infty$ [2] the ratio $C/\chi T$ is $4\pi^2/3$, and the Anderson-Yuval theory [3–5] gives $C/\chi T$ as $\pi^2/3$ [6]. In the formula for χ, the J dependence had previously been determined [7], except for the $1.58|2J\varrho|$ term. The coefficient 0.103 ± 0.003 is new.

The essence of the calculation is a numerical solution within a renormalization group framework of a "hopping" Hamiltonian

$$H = \sum_{n=0}^{\infty} \Lambda^{-n/2}(f_n^+ f_{n+1} + f_{n+1}^+ f_n) - \tilde{J} f_0^+ \boldsymbol{\sigma} f_0 \cdot \boldsymbol{\tau} \tag{5}$$

where Λ is a parameter of order 2 to 3 (see below), $\tilde{J}$ is proportional to J, and the $f_n(f_{n\mu})$ are discrete orthogonal electron destruction operators. Three steps are involved in obtaining the hopping Hamiltonian from H_K. The first step is a discretization of H_K. The continuum operators a_k are replaced by a set of discrete operators a_n and b_n, a_n corresponding to momentum $k = \Lambda^{-n}$ and b_n correponding to momentum $k = -\Lambda^{-n}$. The constant Λ can be varied: $\Lambda = 1$ is the continuum limit; numerical calculations were done for $\Lambda = 2$, 2.5, and 3. The calculations for these three values of Λ are in agreement within 3%; other arguments (see later) also suggest that the results for $\Lambda = 2$ to 3 are a good approximation to the continuum case $\Lambda = 1$. The numerical computations become very difficult for Λ smaller than 2. The discretization is done on a logarithmic scale (the discrete values Λ^{-n} are uniformly distributed in the variable ln k). The motivation for this distribution is that each order of magnitude range of k contributes equally to the logarithmic divergences found in perturbation solutions [1] of the Kondo Hamiltonian at $T=0$.

The result of the discretization is a Hamiltonian

$$H_a = \sum_{n=0}^{\infty} \Lambda^{-n}(a_n^+ a_n - b_n^+ b_n) - JA^+ \boldsymbol{\sigma} A \cdot \boldsymbol{\tau} \tag{6}$$

where

$$A = \sum_{n=0}^{\infty} \Lambda^{-n/2}(a_n + b_n) \tag{7}$$

and the a_n and b_n satisfy the usual discrete anticommutation rules, $\{a_n, a_m^+\} = \delta_{nm}$, etc. (Possible overall scale factors in H_a have been ignored and will not affect the final result.)

The second step is a conversion to a new orthogonal basis of operators $f_{0\mu}, f_{1\mu}, \ldots$ from the above basis $a_{n\mu}$ and $b_{n\mu}$. The first operator f_0 is chosen to be proportional to A itself, with f_0 normalized to satisfy $\{f_0, f_0^+\} = 1$. It can be proven that one can then construct a sequence of orthogonal operators f_1, f_2, etc. such that the operator H_a has the form

$$H_a = \sum_{n=0}^{\infty} \Lambda^{-n/2} \varepsilon_n (f_n^+ f_{n+1} + f_{n+1}^+ f_n) - \tilde{J} f_0^+ \boldsymbol{\sigma} f_0 \cdot \boldsymbol{\tau} \tag{8}$$

where $\tilde{J}$ is proportional to J, and the ε_n are a uniquely determined set of constants. A less trivial theorem can also be proven, namely that $\varepsilon_n \to 1$ for $n \to \infty$.

The second step is an exact transformation, unlike the first step which is an approximation unless $\Lambda = 1$.

The third step is to replace all the ε_n by 1. One then has the hopping Hamiltonian (5). This step is done only for convenience. The low temperature properties of the Kondo Hamiltonian are governed only by the large n terms where $\varepsilon_n = 1$ anyways; the calculations reported below are specifically designed to be independent of the details of the low n terms.

The solution of the hopping Hamiltonian will now be described. It is convenient to define H_N:

$$H_N = \Lambda^{(N-1)/2} \times \left\{ \sum_{n=0}^{N-1} \Lambda^{-n/2}(f_n^+ f_{n+1} + f_{n+1}^+ f_n) - \tilde{J} f_0^+ \boldsymbol{\sigma} f_0 \cdot \boldsymbol{\tau} \right\} \tag{9}$$

Then the original H is

$$\lim_{N\to\infty} \Lambda^{-(N-1)/2} H_N.$$

The purpose of the factor $\Lambda^{(N-1)/2}$ is so that the low-lying excited states of H_N have energies ~ 1 above the ground state, for any N. The H_N satisfy a recursion formula:

$$H_{N+1} = \sqrt{\Lambda}\, H_N + f_N^+ f_{N+1} + f_{N+1}^+ f_N \tag{10}$$

One can define a formal procedure for obtaining the eigenstates and the eigenvalues of H_{N+1}, given the eigenstates and eigenvalues of H_N. Let the eigenstates of H_N be a set of states $|l\rangle$, $0 \leqslant l \leqslant L_N$. Let the eigenvalues be E_l. Assume the matrix elements $\langle l|f_{N\mu}|l'\rangle$ are known for all l, l' and μ. One first forms a basis for H_{N+1} consisting of the states $|l\rangle, f_{N+1,1/2}^+|l\rangle, f_{N+1,-1/2}^+|l\rangle$ and $f_{N+1,1/2}^+ f_{N+1,-1/2}^+|l\rangle$, for all l. There are $4 \times (L_N + 1)$ such states. The matrix elements of H_{N+1} are easily constructed in this basis. Since the basis is finite one can in principle diagonalize H_{N+1}, computing its eigenvalues E_l' and eigenvectors $|l'\rangle$; one can then compute matrix elements $\langle l'|f_{N+1\mu}^+|l''\rangle$ between eigenstates. Hence one can proceed to construct and diagonalize H_{N+2}, etc.

As a practical matter diagonalizing H_{N+1} is prohibitively expensive for large N because it is too large a matrix. For $N = 50$ for example, H_N has 2^{101} eigenstates. The numerical approximation consisted in dropping all but the first 100–300 eigenstates of H_N before introducing f_{N+1} and diagonalizing H_{N+1}. For a fast calculation only about 130 energy levels of H_N were kept; for a more thorough calculation about 276 energy levels were kept. Direct comparison of the two calculations indicated that errors of at most 1% to 10^{-2}% were caused by this truncation, at least for the lowest lying energy levels (E_1, E_2, etc.). Even the accumulated errors of many iterations (in practice 50 to 80 iterations were performed) did not exceed 1 to 2% (for $N = 2$) and were much less than 1% for $\Lambda = 2.5$ and 3. The reason for this small error is twofold. The first is that in the $f_{N+1}^+ \otimes |l\rangle$ basis H_N is already diagonal and the off diagonal terms due to $f_N^+ f_{N+1} + f_{N+1}^+ f_N$ are fairly small. Hence the low-lying levels of H_{N+1} involved mainly the low-lying levels of H_N. One consequence of this is that matrix elements of f_{N+1} tend to be large only for eigenstates of H_{N+1} with nearby energies. But this means that f_N tends to connect only eigenstates of H_N with nearby energies. This is the second effect reducing the importance of the energy levels thrown away in the numerical treatment.

Calculations were performed for $\Lambda = 2, 2.5$, and 3; values of $\tilde{J}$ from -0.053 to -0.09, and for 130 or 276 energy levels kept. Due to conservation of electron number and spin the full matrix H_{N+1} could be reduced to submatrices no larger than about 50×50. One iteration required at most 30 sec on an IBM360–65 (less on a CDC 6600). The calculations involved 50 to 80 iterations each.

As background for the results of numerical calculations we need the results of two special cases: $\tilde{J} = 0$ and $\tilde{J} = \infty$. When $\tilde{J} = 0$ one has a free electron Hamiltonian which can be solved by finding the single electron eigenstates. This is much simpler than calculating the many-electron states for nonzero $\tilde{J}$ but still must be done numerically. The result, for large N, is a simple limit cycle, with one set of levels for all odd large N and another set of levels for all even large N. For example, for $\Lambda = 2$ the single particle energy levels are:

N odd: 0.6555, 1.976, 4, 8, ..., 2^l, ...

N even: 0, 1.297, 2.827, $4\sqrt{2}$, $8\sqrt{2}$, ..., $2^{l-1}\sqrt{2}$, ...

When $\tilde{J} = -\infty$ the $-\tilde{J} f_0^+ \sigma f_0 \cdot \tau$ term by itself has a singlet ground state. All excited states of this term have infinite energy and can be neglected. This means in effect that f_0 and the impurity can be ignored, leaving an effective Hamiltonian

$$H_N(\tilde{J} = -\infty) = \Lambda^{(N-1)/2} \sum_{n=1}^{N-1} \Lambda^{-n/2}(f_n^+ f_{n+1} + f_{n+1}^+ f_n) \tag{11}$$

This differs from the $\tilde{J} = 0$ Hamiltonian only by having the sum start at $n = 1$ instead of $n = 0$. As a result the solution for large N is the same limit cycle but with the phase reversed: now the single electron energy levels are 0.6555, 1.976, etc., for N even and large and 0, 1.297, etc. for N odd and large.

If one looks only at even N iterations, one sees a fixed point instead of a limit cycle; there is a $\tilde{J} = 0$ fixed point and a $\tilde{J} = \infty$ fixed point. By "$\tilde{J} = \infty$ fixed point" we mean a fixed point Hamiltonian generated from the initial Hamiltonian with $\tilde{J} = -\infty$. As usual in the modern renormalization group approach the fixed point Hamiltonian is a single point in a many dimensional space (the variables in this case being at least the single particle energies and single particle matrix elements of f_N) and is distinct from any of the initial Hamiltonians, in particular is distinct from the initial $\tilde{J} = -\infty$ Hamiltonian. That is, the complete set of single particle energies of the original H for $\tilde{J} = -\infty$ are not identical with the single particle energies of H_N for $N \to \infty$, with or without the rescaling factor $\Lambda^{-(N-1)/2}$. What is true is that for very large N, the single particle energies of H_N which are $\gg 1$ and $\ll \Lambda^{(N-1)/2}$, when multiplied by $\Lambda^{-(N-1)/2}$, give correctly the corresponding range of single particle levels of H for $\tilde{J} = -\infty$.

The result of the numerical calculation for small negative $\tilde{J}$ is that there is a crossover from near the $\tilde{J} = 0$ fixed point for small N to the $\tilde{J} = \infty$ fixed point for large N. For example, for $\Lambda = 2$ and $\tilde{J} = -0.053$, the following results were obtained for two excited state energies:

N	20	22	44	46	66	78
E_1	0.0945	0.1007	0.296	0.3418	0.6537	0.6555
E_2	0.1269	0.1353	0.4177	0.4929	1.3037	1.3109

For $\tilde{J} = 0$ these energies would have been both 0. As N increases the energies increase, reaching almost exactly the $\tilde{J} = -\infty$ results by $N = 78$. The state with energy E_1 is a two particle state (both with 0 energy) for $\tilde{J} = 0$, while for $\tilde{J} = \infty$ it is a single particle state. Conversely the state with energy E_2 starts out as a single particle state and ends up as a two particle state. In the middle of the crossover region ($N = 42$ or 44) the electrons are strongly coupled and the levels cannot be fitted with additive single particle energies.

When one starts with different values of $\tilde{J}$, but with the same Λ (say $\Lambda = 2$) one gets the same crossover behaviour but translated in N. This is true at large N; for small N there are transient effects depending on $\tilde{J}$. Thus if one compares E_1 versus N for two different values of $\tilde{J}$ one obtains what is shown in fig. 1.

For large N the two sets of energies fit on the same smooth curve – to the extent that one can define a smooth curve through the discrete points. By adjusting $\tilde{J}_B$ one can move the circles on top of the crosses and see an exact matching. (In

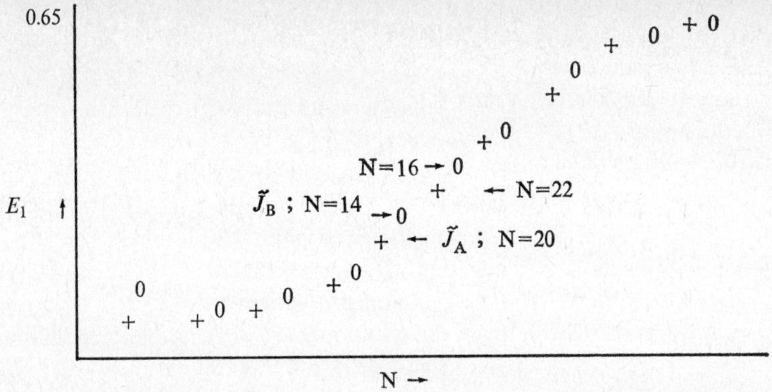

Fig. 1. Excited state energy E_1 vs N for two values of $\tilde{J}$.

practice there have only been calculations with the circles close to the crosses, not on top of them; some interpolation is always required between two crosses.) It has been verified that the translation in N required to relate the $\tilde{J}_B$ results to the $\tilde{J}_A$ results is independent of which energy level (E_1, E_2, or whatever) one plots.

To calculate the zero temperature results for the susceptibility and specific heat it turns out to be necessary to know how the energy levels of H_N approach their limiting values for $N \to \infty$. That is, for the $\Lambda = 2$ case one must have a first approximation for $E_{1N} - E_{1\infty} = E_{1N} - 0.6555$, $E_{2N} - E_{2\infty} = E_{2N} - 1.311$, etc. One can analyze this problem analytically, with the conclusion that these energy level differences can all be calculated from a two parameter pseudopotential. Namely one takes the Hamiltonian

$$H_N = \Lambda^{(N-1)/2}\left\{\sum_{n=1}^{N-1} \Lambda^{-n/2} \times (f_n^+ f_{n+1} + f_{n+1}^+ f_n) + \lambda(f_1^+ f_2 + f_2^+ f_1) + \eta(f_1^+ f_1 - 1)^2\right\} \quad (12)$$

and calculates all the energy levels of H_N, for large N, to first order in λ and η. The terms of first order in λ and η (disregarding the zero order terms) will then fit all the differences like $E_{1N} - E_{1\infty}$. Even though for large but finite N, the zero order eigenvalues of H_N are *not* identical with the fixed point eigenvalues $E_{1\infty}$, the pseudopotential rule is still to extract only the terms of order λ and η. Typical results for energy levels are (N even only):

$$E_{1N} - E_{1\infty} = \tilde{\lambda}\Lambda^{-N/2} \quad (13)$$

$$E_{2N} - E_{2\infty} = 2\tilde{\lambda}\Lambda^{-N/2} - \tilde{\eta}\Lambda^{-N/2} \quad (14)$$

$$E_{3N} - E_{3\infty} = 2\tilde{\lambda}\Lambda^{-N/2} + \tilde{\eta}\Lambda^{-N/2} \quad (15)$$

etc., where $\tilde{\lambda}$ is proportional to λ and $\tilde{\eta}$ is proportional to η. The values of $\tilde{\lambda}$ and $\tilde{\eta}$ can be fit from numerical results for E_{1N} and E_{2N}; for example, for $\tilde{J} = -0.053$ and $\Lambda = 2$, using the N = 64 results one finds $\tilde{\lambda} = 2^{32} \times (-3.668 \times 10^{-3})$ and $\tilde{\eta} = 2^{32} \times 7.2496 \times 10^{-3}$. One can then calculate $E_{3N} - E_{3\infty}$, etc., and these calculations agree with the numerical results for $E_{3N} - E_{3\infty}$, etc. Eqs (13)–(15) also agree with the numerical calculations for N = 64, 66 etc.

It is impractical to do numerical calculations for $|\tilde{J}|$ much less than 0.05 because too many iterations are required to reach the crossover region. However, we know that H_{20} for $\tilde{J} = -0.053$ will be say $\tilde{H}_{80}$ for a smaller value of $\tilde{J}$, and to complete the calculation for the smaller value one needs only $\tilde{H}_1$ to $\tilde{H}_{79}$, all of which will have eigenvalues close to the $\tilde{J} = 0$ case since H_{20} for $\tilde{J} = -0.053$ has eigenvalues near the $\tilde{J} = 0$ ones. Therefore these eigenvalues can be calculated by perturbation theory. Specifically the eigenvalue E_{1N} for N odd has been studied, which would be 0.6555 for $\tilde{J} = 0$ and large enough N (at $\Lambda = 2$). Let

$$Z_N = 0.6555 - E_{1N} \quad (16)$$

for odd N. A perturbation theoretic analysis gives a recursion formula (cited for $\Lambda = 2$) (the coefficients involve complicated sums over the single particle levels for $\tilde{J} = 0$; these were calculated by computer)

$$Z_{N+2} = Z_N + 0.71436\, Z_N^2 + 0.14212\, Z_N^3 - 2.92\, Z_N^4 + O(Z_N^5) \quad (17)$$

provided N is large (otherwise the coefficients depend on N). Inverse iteration of this formula allows one to extrapolate backwards the values of Z_N starting from the numerical result (for $\tilde{J}=-0.053$) $Z_{25}=0.10960$; for example $Z_{-137}=0.01524$. (Iteration -137 for $\tilde{J}=-0.053$ corresponds to $N \gg 1$ for sufficiently small $\tilde{J}$.) Asymptotically for large negative N one finds from eq. (17) that Z_N satisfies

$$\frac{N}{2}+\Psi(Z_N)=25.53 \tag{18}$$

where (for $\Lambda=2$)

$$\Psi(Z)=(0.71436\,Z)^{-1}-0.72137\ln(0.71436\,Z)-7.719(0.71436\,Z)+O(Z^2) \tag{19}$$

Now consider the specific heat of the impurity. The definition used in the calculation of χ is (g factors and moment factors have been dropped)

$$\chi_{\text{imp}}(T)=\frac{1}{kT}\lim_{N\to\infty}\left\{\frac{\langle S_z^2\exp[-\beta\Lambda^{-(N-1)/2}H_N]\rangle}{\langle\exp[-\beta\Lambda^{-(N-1)/2}H_N]\rangle}-\left.\frac{\langle S_z^2\exp[-\beta\Lambda^{-(N-1)/2}H_N]\rangle}{\langle\exp[-\beta\Lambda^{-(N-1)/2}H_N]\rangle}\right|_{\text{no impurity}}\right\} \tag{20}$$

where S_z is the z component of the total spin (electrons plus impurity). The limit $N\to\infty$ is a thermodynamic limit; hopefully this is a correct definition of the thermodynamic limit. The susceptibility has been calculated analytically using perturbation theory for large T (meaning $T \gg T_K$, T_K being the Kondo temperature [1] but $T \ll 1$) and using the pseudopotential for $T \ll T_K$. The discretization results in a rather complicated T dependence, namely a slight wobble as T varies through a factor Λ. This wobble is a 1 or 2% effect for $\Lambda=3$, and a fraction of a percent for $\Lambda=2$ or 2.5. It turns out that the easiest calculation to perform is to obtain χ_{imp} for a sequence of values $T=T_0\Lambda^{-N/2}$ and the results below will reflect this. For weak coupling one finds the following: for $\Lambda=2$, $\tilde{J}=-0.053$, $T=2\times2^{-N/2}$, and N odd and not too large ($\lesssim 20$), one has

$$\chi_{\text{imp}}=\frac{1}{kT}\left\{\tfrac{1}{4}-0.25765\,Z_N+0.31471\,Z_N^2-1.13\,Z_N^3+O(Z_N^4)\right\} \tag{21}$$

For a different value of T, say $T=2\sqrt{2}\times2^{-N/2}$, the coefficients 0.314 and -1.13 are replaced by new coefficients 0.407 and -1.43. The necessary consistency conditions between the Z_N dependence of χ_{imp}, the T dependence of the coefficients, and the N dependence of Z_N, are satisfied.

From the fixed point Hamiltonian for $N\to\infty$ one gets χ_{imp} and C_{imp} both to be identically 0. One gets nonzero values for χ_{imp} and C_{imp} only from the pseudopotential terms. The calculations give, for $T=2\times2^{-N/2}$ again, N even and $\gtrsim 50$,

$$\chi_{\text{imp}}=\frac{1}{kT}\{-0.834\,\tilde{\lambda}+0.422\,\tilde{\eta}\}2^{-N/2} \tag{22}$$

$$C_{\text{imp}}=k\{-10.97\,\tilde{\lambda}\}2^{-N/2} \tag{23}$$

Using the values for $\tilde{\lambda}$ and $\tilde{\eta}$ reported earlier, one finds the ratio

$$\frac{C_{\text{imp}}}{T\chi_{\text{imp}}}\to\frac{2\pi^2}{3}k^2 \tag{24}$$

for $T\to0$, within the numerical errors of the calculation. To be precise, the values of the coefficient of k^2 are 6.57, 6.64, and 6.67 for $\Lambda=2$, 2.5, and 3 respectively, compared with $2\pi^2/3=6.58$.

In order to determine the zero temperature limit of χ_{imp} as a function of J one makes use of the scaling law [3] for χ_{imp}, namely that χ_{imp} depends only on $T/T_K(J)$ rather than on T and J separately, where $T_K(J)$ is any one of the many definitions of the Kondo temperature and kT must be much less than the band edge. This scaling law follows in the present analysis from the uniqueness up to a translation in N of the sequence of Hamiltonians H_N once N is large. This means in practice that the formulae given previously for χ_{imp} for $T=2\times2^{-N/2}$ are valid for any J provided one uses $T=T_0(J)\times2^{-N/2}$ instead, where $T_0(J)$ is proportional to the Kondo temperature $T_K(J)$. To determine $T_0(J)$ it is convenient to match χ_{imp} in eq. (21) to previous perturbation calculations of χ_{imp} for the continuum Kondo Hamiltonian. Using the formulae of [1], one can write

$$\chi_{\text{imp}}=\frac{1}{kT}\tfrac{1}{4}\{1+2J\varrho+4J^2\varrho^2\ln(kT/\tilde{D})+\text{order}\,(J^3\ln^2 kT/\tilde{D},\,J^3\ln kT/\tilde{D})+O(J^4)\} \tag{25}$$

One defines $\tilde{D}(J)$ to be $c_1 D + c_2 J\varrho + O(J^2)$ where c_1 and c_2 are constants chosen so that no constant (non-logarithmic) terms appear in χ_{imp} either in order J^2 or J^3. Here D is the band edge and ϱ is the density of states at the Fermi surface, as in [1]. Terms of order T, T^2, etc. have been dropped in eq. (25). To determine $T_0(J)$ one matches eq. (25) to eq. (21). The T dependence is found to match (apart from the wobble in the discrete approximation, mentioned earlier). Furthermore the T dependence of eqs (25) and (21) has the same extrapolation to $kT \sim \tilde{D}$, which simplifies the calculation of $T_0(J)$. Namely one matches eqs (25) and (21) for $kT = \tilde{D}$:

$$\tfrac{1}{4}(1+2J\varrho) = 0.25 - 0.258\, Z_N + 0.314\, Z_N^2 - 1.13\, Z_N^3 \quad (26)$$

for

$$kT_0(J)\, 2^{-N/2} = \tilde{D} \quad (27)$$

This equation can be solved for Z_N, and N itself can then be determined from eq. (18). Then $T_0(J)$ is determined from eq. (27). Then using eq. (22), one obtains the formula (3) for χ. To be precise, the results for different values of Λ for the coefficient in front of the exponential in eq. (3) are 0.100, 0.103, and 0.104–0.108 for $\Lambda = 2$, 2.5, and 3 respectively. The range of values for $\Lambda = 3$ are due to the wobble of the discrete theory: the near-zero temperature susceptibility varies as T changes from Λ^{-N} to $\Lambda^{-(N+1)}$ even for very large N, with 0.104–0.108 giving the range of variation.

In the calculation described above the value of N obtained by solving eq. (18) will not be an integer, in general. Eq. (18) defines an interpolation for Z_N between integer values of N which in turn allows one to calculate χ from eq. (21) even if N is not an integer. The results for χ obtained this way agree with the results obtained from the temperature dependence of the coefficients in eq. (25), except for the wobble which, as noted earlier, is a few percent for $\Lambda = 3$ and negligible for $\Lambda = 2.5$ and 2.

The agreement of the calculations for $\Lambda = 2$, 2.5, and 3 is rather good. A formal analysis suggests that this is not accidental, and there is a good chance that an accurate calculation for $\Lambda = 2$ would give results agreeing with the continuum limit ($\Lambda = 1$) to one part in 10^4 or less. The analysis is the following. The discretization is an operator generalization of replacing an integral by a sum. Using the logarithmic distribution means one is approximating integrals over $\ln k$ by a uniformly spaced sum with spacing $\ln \Lambda$. The range of the variable $\ln k$ is $-\infty < \ln k < 0$ (considering positive k, for example), but for $kT \ll D$, the upper limit is irrelevant (see below) and in effect the range is $-\infty < \ln k < \infty$.

In perturbation calculations one sees explicitly that the discretization replaces integrals over k by sums; however, in order to replace the upper limit 0 by ∞ one must exercise caution due to renormalization effects. The present calculation for the susceptibility is in effect a calculation for the susceptibility at low temperatures T given the susceptibility at one particular value of T from a continuum theory; when this calculation is studied in perturbation theory it is found that all upper limits 0 can be replaced by ∞. A general result is that any calculation which avoids any explicit reference to $\tilde{J}$ or results of the first few iterations in N can be represented by integrals with upper limit ∞ (to any order in perturbation theory).

There is a theorem [8] that the error in replacing $\int_{-\infty}^{\infty} f(z)\,dz$ by $\sum_{n=-\infty}^{\infty} af(na)$ is exponential in $1/a$, namely the error behaves as $\exp\{-2\pi r/a\}$ where r is the distance from the real axis of the nearest singularity of $f(z)$ in the complex z plane. This exponential convergence is obvious in the perturbation theoretic calculations of the susceptibility; for instance, for $\Lambda = 2$ the temperature dependence of χ_{imp}, to third order, agrees to within 1 part in 10^5 with the continuum limit. For $\Lambda = 2.5$ the error is of order 0.1% and for $\Lambda = 3$ there is a 1–2% error. The calculation of the zero temperature behaviour is more complicated than a calculation of integrals, but the existence of exponential convergence to all orders in perturbation theory suggests it is a more general result. In particular, this suggests that the zero temperature results for $\Lambda = 2$ and $\Lambda = 2.5$ should agree to a fraction of a percent. Unfortunately the error in the $\Lambda = 2$ calculations due to truncating the states of H_N is of order a few percent so the present numbers agree only to a few percent.

The model solved here has been a bottleneck for Kondo theory for some time. Hopefully the availability of a solution will enable progress to resume, especially in attacking more realistic models.

One should be able to compute many properties of the Kondo Hamiltonian at temperatures $T \ll T_K$ from the pseudopotential. A continuum form of the effective Hamiltonian is

$$H = \int_{-1}^{1} k a_k^+ a_k \, dk + J' A^+ A + g(\bar{A}^+ B + B^+ \bar{A}) + \eta (B^+ B - 1)^2 \qquad (28)$$

where

$$\bar{A} = \int_{-1/2}^{1/2} a_k \, dk$$

and

$$B = \frac{\sqrt{3}}{\sqrt{2}} \int_{-1}^{1} k a_k \, dk; \{a_k, a_{k'}^+\} = \delta(k - k')$$

The Hamiltonian is to be solved for $J' \to \infty$, keeping zero *and* first order terms in g and h. The constants g and h can be determined by fitting to the zero temperature results for χ_{imp} and C_{imp} reported above.

The ratio $C/\chi T$ for $T \to 0$ reported above is valid only for small J. As J becomes large, the crossover region moves to small N and the small N transients begin to affect the parameters $\tilde{\lambda}$ and $\tilde{\eta}$ which determines the ratio. A rough analysis suggests that the corrections to $C/\chi T$ for small J due to these transients are of order $\exp\{-1/|2J\varrho|\}$.

The calculations described here grew out of general renormalization group work described elsewhere [9] and from specific work on fixed source meson-nuclear theories [10].

I am grateful to many people for advice and encouragement, including V. Emery, A. Luther, J. Wilkins, K. Murata, and others. R. Hoffman provided the diagonalization routine; some of the calculations were performed at Brookhaven, courtesy of G. Baker.

References

1. Kondo, J, Solid state phys 1969, 23, 184.
2. This result is generally known and apparently unpublished.
3. Anderson, P W & Yuval, G, Phys rev 1970, B1, 1522.
4. Anderson, P W, Yuval, G & Hamann, D, Phys rev 1970, B1, 4464.
5. Andersson, P W, Yuval, G & Hamann, D, Solid state comm 1970, 8, 1033.
6. Anderson, P W. Private communication.
7. Abrikosov, A A & Migdal, A A, J low temp phys 1970, 3, 519.
8. Fisher, M E & Barber, M N, Arch rat mech analysis. 1972. Appendix.
9. Wilson, K & Kogut, J, Phys repts. Review to be published. (Includes further references on the Kondo problem.)
10. Wilson, K G, Phys rev 1970, D2, 1438.

Discussion

Nozières: Although your theory does not work as such in the case of strong coupling, its general structure may be compared to that of a $1/J$ expansion. For $J \to \infty$, the state f_0 (the central site in a Wannier function representation) is locked in a singlet state and does not enter. The t-matrix for the electrons corresponds to a phase shift $\delta = \pi/2$, and the extra density of states $\delta\varrho$ at the Fermi level is $2/\pi D$; it is not clear how this limit fits into your description. If one starts from $J = \infty$, and scales the index N to infinity, I do not see why the extra density of states is related to the approach to the fixed point, rather than to the fixed point itself.

If now J is large, but finite, one may expand in powers of the hopping integral—or equivalently in powers of the $\epsilon_0 f_0^+ f_1$ term. That gives rise to a self energy on site 1, $\Sigma_{11}(z)$, beginning at order $1/J$, and to an interaction on site, $(f_1^+ f_1 - 1)^2 \lambda(z)$, beginning at order $1/J^3$. It turns out that the corrections to $\delta\varrho$ is of order $1/J^2$, the net $\delta\varrho$ remaining finite. The interaction term acts like a Landau Fermi liquid enhancement and acts to reduce the ratio of specific heat to spin susceptibility, $C_v/\chi T$. Again, it is not clear how (i) one can neglect the energy dependence of Σ and λ when studying the low lying states; (ii) how Σ_{11} can be transformed into an "effective term" coupling sites 1 and $2(f_1^+ f_2)$ of the type that you use.

In order to elucidate these questions, it would be interesting to apply the renormalization group machinery starting from an already large J, and to compare its detailed structure with elementary perturbation theory.

Wilson: (1) My technique is less accurate for large J because there are band edge effects and the convergence is not exponential in $1/\ln \Lambda$. But this is not a serious problem, and I agree that a large J calculation would be interesting.

(2) In conventional language the specific heat near zero temperature is related to $d\delta(E)/dE$ near the Fermi surface. For $J=\infty$ $\delta(E)$ still has an energy dependence. However there is a rescaling factor $\Lambda^{N-\frac{1}{2}}$ involved in the definition of H_N, which stretches out the energy dependence of the phase shift $\delta_N(E)$: in the limit $N\to\infty$, $\delta_N(E)$ becomes flat and $d\delta_N(E)/dE$ is 0.

(3) The complex effective Hamiltonian discussed by Nozières is needed to give the complete high energy behaviour of the system. The two term pseudopotential gives only the first order corrections in T for T small, with higher order terms in T being neglected.

P. W. Anderson: It seems to me that a very intuitively appealing way to understand the nature of the final effective Hamiltonian is to see what could be expected if the scaling were stopped at a finite but very large J. Then the "pseudo-Wannier function" or localized function at the spin site will be so strongly coupled as to be out of the problem. But the first neighbor Wannier function f_1 will experience two first-order effects in $1/J$: (*a*) when it has a single electron on it ($n_{1\uparrow}n_{1\downarrow}-1=0$) it will be favorably affected because it can polarize the central singlet, and this will be equivalent to a small U or l.s.f. term; (*b*) its wave function will be renormalized and so its hopping term to the next function, f_2, will be modified (reduced). These are the two terms Wilson has. In an asymmetric case the third, potential term will also be present.

Toulouse: Can you say something about the time dependence of the local spin correlation function?

Wilson: No, in the sense that I have not studied it. However, one may be able to extract some time dependent quantities from the effective low temperature Hamiltonian. For local spin properties one will need a formula for the impurity spin within the effective Hamiltonian picture, i.e. an operator which gives the matrix elements of τ between low energy states. This can be done but is not presently included in the computer calculations.

P. W. Anderson: Without being very concerned about whether the results of Anderson, Yuval & Hamann, of Anderson, and of Fowler were on the level of speculation, conjecture, plausibility argument or demonstration (which is very much a matter of taste) it is clear that these authors gave arguments why (in the present language) the fixed point which was finally approached was of $J\to\infty$ type. This (with the Toulouse conjecture) gives one a theory which, in terms of Wilson's remark of the first day, is a "$\pm 100-\varepsilon\%$ theory", i.e. it tells you the form of the result but not its magnitude to a factor of 5 or 50. Wilson has replaced this with a 3% theory—an achievement which must not be underestimated (for one thing, it goes far beyond the level of plausibility argument), but I think it should be stated in those terms for comparison with the remark on the critical point problem.

For the benefit of our Proceedings, I have reproduced the best argument, to my mind, from AYH, which is not the Toulouse limit argument. (This was brought out in private discussions). This is the Ising-Kondo correspondence, which is exact in the limit of small J. In this limit one may then scale (by orthodox methods, if one likes) the Ising problem up to a reasonably finite J (which corresponds to fairly weak short- and long-range ferromagnetic interactions); one is still in the region of validity of the simple scaling laws. It is now obvious that one is at a point well above the Ising model's critical point, and it is almost inconceivable to us that further scaling will lead one to a non-trivial fixed point. For one thing, one can limit exponents on the correlation functions by Griffiths' inequalities.

Review of the Application of the Renormalization Group Method to Logarithmic Problems

A. Zawadowski[1]

Institut Laue-Langevin, Grenoble, France[1]

Summary

The applicability of the renormalization group method to typical logarithmic problems such as X-ray absorption, the Kondo effect and one-dimensional superconductivity is discussed. It is concluded that if the problem scales into a strong coupling regime, then the orthodox version of the renormalization group method does not distinguish between strong but finite coupling and infinitely strong coupling.

The renormalization group method was originally developed in quantum field theory [1, 2] and it has been applied to logarithmic problems such as infrared and ultraviolet divergencies in quantum electrodynamics. Not much later, Bonch-Bruevich [3] applied this method to solid state physics by treating the static screening of an external potential and plasma in electron gas. It is quite surprising that the renormalization group method was not used again for a long time until Larkin & Kmel'nitskii [4] attacked the problem of phase transitions in uniaxial ferro-electrics with it. They summed up the "parquet diagrams". More recently there have been two classes of problems to which the renormalization group has been applied:

(i) Dynamical problems exceeding the simple "parquet" approximation in the Kondo and X-ray absorption problem [5–9], critical behavior in one-dimensional problems at zero temperature [10, 11] and localized spin fluctuations in dilute alloys [12].

(ii) Critical behavior of thermodynamical systems, critical exponents in phase transitions and the Kondo problem [13] which are the subject of other papers in this volume (pp 14, 16 and 68).

The common nature of these problems (except the localized spin fluctuations [11]) is that the perturbation theory typically provides logarithmic contributions and the summation of these terms results in scaling behavior. The essential difference between the two classes is that dynamical energy variables are handled in the first (i) rather than the second group (ii).

Brief Outline of the Method

In a typical logarithmic problem, a change of a cut-off parameter, D (the width of the range of interaction) from D to D' has an effect on different functions such as vertices, Green's function, etc. in a way which can be simply formulated. Namely, the function is multiplied by a constant z and the value of the coupling g must be replaced by another value g' keeping the analytic form of the function. A typical example is the following:

$$\frac{1}{1-g\ln\dfrac{\omega}{D}} = z\,\frac{1}{1-g'\ln\dfrac{\omega}{D'}}$$

where

$$z = \frac{1}{1-g\ln\dfrac{D'}{D}} \quad \text{and} \quad g' = \frac{g}{1-g\ln\dfrac{D'}{D}}$$

Moreover, logarithmic expressions of this type can be expanded with respect to the coupling g if the energy ω is near the cut-off, i.e. $\omega \sim D'$.

---

[1] Permanent address: Central Research Institute for Physics, 1525 Budapest, POB 49, Hungary.

Thus, one may proceed to calculate such an expression by using first perturbation theory with an artificial set of parameters D' and g', and afterwards, by scaling, extend the results obtained to the physical value of the coupling and cut-off, g and D respectively. This scaling program is the essential feature of the renormalization group method.

The formal multiplication of different quantities such as vertices, Green's functions and coupling constants by a set of arbitrary constants, e.g. z_1, z_2, z_3, does not alter the mathematical structure of the problem and these transformations form a continuous group called the renormalization group [1]. Let us consider briefly the example of two interacting Fermion fields described in terms of Green's functions G and $\mathcal{G}$, and coupled by a four-legged vertex with coupling constant g. The renormalization group transformation is

$$G \to z_1 G, \mathcal{G} \to z_2 \mathcal{G}, \Gamma \to z_3^{-1}\Gamma, g \to z_3 z_1^{-1} z_2^{-1} g$$

In general, the physical set of quantities is scaled into an unphysical set in which the functional forms of the quantities are modified. The logarithmic problems are a special case where the functional forms are not affected; only the parameters (cut-off and coupling strength) are scaled. In the general case, in order to describe the new functional form, a new energy-like parameter, λ should be introduced [2], which measures how far the scaling takes one from the physical set. Thus an arbitrary function can be written as $A(\omega/\lambda, ..., D/\lambda, g)$, which for the physical set has the more simple form $A(\omega/D, ...; g)$. In logarithmic problems the parameter λ is not necessarily introduced. For the sake of simplicity, we will not introduce it. Furthermore, the following notations are used: $\bar{d} = G/G^{(0)}$ and $d = \mathcal{G}/\mathcal{G}^{(0)}$, where $G^{(0)}$ and $\mathcal{G}^{(0)}$ represent unrenormalized Green's function.

Applying the renormalization group transformation, one gets the following set of equations

$$\bar{d}\left(\frac{\omega}{D'}; g'\right) = z_1 \bar{d}\left(\frac{\omega}{D}; g\right)$$

$$d\left(\frac{\omega}{D'}; g'\right) = z_2 d\left(\frac{\omega}{D}; g\right)$$

and

$$\Gamma\left(\frac{\omega}{D'}; g'\right) = z_3^{-1}\Gamma\left(\frac{\omega}{D}; g\right)$$

$$g' = z_3 z_1^{-1} z_2^{-1} g \tag{1}$$

This transformation can be established in the form of a Lie equation [2] for an arbitrary quantity A

$$\frac{\partial}{\partial \ln \frac{\omega}{D}} \ln A\left(\frac{\omega}{D}; g\right) = \frac{\partial}{\partial \xi}\left[\ln A\left(\xi; g\left(\frac{\omega}{D}; g\right)\right)\right]_{\xi=1} \tag{2}$$

where the infinitesimal generator forming the left-hand side of this equation is expressed in terms of an effective coupling constant $g(\omega/D; g)$ whose strength depends on the variable ω/D. It is defined by transformation (1) taking $\omega/D' = \xi = 1$, thus getting

$$z_1^{-1} = \frac{\bar{d}\left(\frac{\omega}{D}; g\right)}{\bar{d}\left(\frac{\omega}{D'}; g'\right)} \quad \text{or} \quad \frac{\bar{d}\left(\frac{D'}{D}; g\right)}{\bar{d}(1; g')} \tag{3}$$

and

$$g'\left(\frac{\omega}{D}; g\right) = \frac{\bar{d}\left(\frac{\omega}{D}; g\right)}{\bar{d}(1; g')} \frac{d\left(\frac{\omega}{D}; g\right)}{d(1; g')} \frac{\Gamma\left(\frac{\omega}{D}; g\right)}{\Gamma(1; g')} g \tag{4a}$$

or

$$g'\left(\frac{D'}{D}; g\right) = \frac{\bar{d}\left(\frac{D'}{D}; g\right)}{\bar{d}(1; g')} \frac{d\left(\frac{D'}{D}; g\right)}{d(1; g')} \frac{\Gamma\left(\frac{D'}{D}; g\right)}{\Gamma(1; g')} g \tag{4b}$$

where $g' = g'(\omega/D, g)$ notation is used. By solving this equation, one can calculate g'. The main idea of the renormalization group method is to calculate the left hand-side of eq. (2) by perturbation theory in terms of the effective coupling $g'(\omega/D', g)$ but at $\omega/D = \xi = 1$ and then to scale this result to $\omega/D \neq 1$ and to the bare value of the coupling g. However, it must be assumed that $g(\omega/D, g) \ll 1$ in order to ensure the applicability of perturbation expansion. It is interesting to note that the quantity $g\bar{d}d\Gamma$ is invariant under transformation (1), thus

$$g'\bar{d}(1, g')\, d(1; g')\, \Gamma(1; g') = g\bar{d}\left(\frac{D'}{D}; g\right) d\left(\frac{D'}{D}; g\right) \Gamma\left(\frac{D'}{D}; g\right) \tag{5}$$

In the case where one finds that

$$\tilde{d}(1;g') = d(1;g') = \Gamma(1;g') = 1 \tag{6}$$

g' itself is invariant under this transformation and then it is called an invariant coupling.

In order to make use of this method, one should calculate the infinitesimal generator occurring in eq. (2) by perturbation theory. If the normalization given by eq. (6) is automatically fulfilled, then the procedure is straightforward. This is the situation in logarithmic problems if one neglects imaginary parts, thus retaining logarithmic terms only. In the more general case, two methods have been suggested. Bogoliubov & Shirkov [2] have introduced the extra variable λ in such a way that the normalization (6) is generally valid. They had to introduce a particular subtraction procedure in the infinitesimal generator. Recently, Sólyom [8] has pointed out, that one can directly calculate the multiplicative constants z_i ($i = 1, 2, 3$) and in that way g' by using perturbation theory and solving eqs (3) and (4*a*) in a self-consistent way. By using this method, the spurious variable does not occur and one should not worry about normalization (6). He has shown that using his method the constants z_i and g' depend only on the variable D'/D for logarithmic problems, and are independent of the choice of variables in the vertex function Γ. These calculations have been performed to the lowest orders (first and second) of perturbation theory for the Kondo effect and one-dimensional metallic systems. This method does not work for non-logarithmic problems like localized spin fluctuations [12].

The first step must be to calculate $g(D'/D; g)$ which satisfies the Lie eq. (2).

The general form of this equation is

$$\frac{\partial}{\partial \mathrm{x}} g(\mathrm{x}) = -\mathrm{F}(g(\mathrm{x})) \tag{7}$$

where $\mathrm{x} = \ln D/\omega$ and the infinitesimal generator is given in a general form by the function $\mathrm{F}(g)$. Furthermore, the boundary condition is $g(0) = g$ in terms of the bare coupling. In the region of interest $0 < \mathrm{x} < \infty$, this equation can be solved as

$$\mathrm{x} = -\int_g^{g(\mathrm{x})} \mathrm{d}g' \frac{1}{\mathrm{F}(g')} \tag{8}$$

The value of the coupling constant g_0, where $\mathrm{x} = +\infty$, is called the fixed point using the expression introduced in critical phenomena. Such a point corresponds to a singular point in the scaling trajectories if more than one coupling is considered (9). One can distinguish between the following two cases:

(i) g_0 = finite, and $\mathrm{F}(g_0) = 0$.

Starting with weak coupling, the problem is scaled into the smallest zero of the function F assuming that $(\partial \mathrm{F}/\partial g)_{g=g_0} > 0$. In that case, the scaling trajectories run into the singular point and one can say that the singular point is stable. If $(\partial \mathrm{F}/\partial g)_{g=g_0} < 0$, then the trajectories start from the singular point, thus it is unstable.

(ii) $g_0 = \infty$, thus $\mathrm{F}(g)$ does not have any positive zero.

One should notice that using perturbation theory in case (ii), the asymptotic form of $\mathrm{F}(g)$ is g^{n} for $g \to \infty$ with $\mathrm{n} \geq 1$, so that x remains finite at $g = \infty$, thus g diverges in the scaling region $0 < \mathrm{x} < \infty$. This very likely indicates some inconsistency in the approximation used to calculate F. In general there is the serious difficulty that F is available only for $g < 1$ from perturbation theory.

The Kondo Problem

The Kondo problem can be formulated in terms of electrons and pseudofermions describing the localized spin with Green's function G and $\mathcal{G}$. The renormalization of the electrons is not essential, thus $z_1 \equiv 1$. In the Hamiltonian one can introduce a bare vertex Γ_K diagonal in the spin variables with coupling K and an exchange term with Γ_J and J having the spin structure $\boldsymbol{S} \cdot \boldsymbol{\sigma}$ where $\boldsymbol{S}$ and $\boldsymbol{\sigma}$ are the spins of the impurity and the electron, respectively. In the Kondo problem, the diagonal term does not occur in the Hamiltonian; therefore the limit $K \to 0$ should be taken in the final results, however, the coupling between these two scattering channels is of importance. Furthermore, calculating the generators F in the two lowest orders of perturbation theory, it turns out that the Lie equations for the invariant couplings $J(\mathrm{x})$ and $K(\mathrm{x})$ are not coupled [5], thus $K(x) \equiv 0$ can be taken. This result holds even when imaginary parts are considered as well [8].

The X-ray absorption problem is a special case where the spin-dependent channel does not occur and using first and second-order perturbation theory for the generator [5] one finds that the effective coupling is not enhanced, $K(x) = K$.

Starting with the lowest approximation in the generator of the Kondo problem one obtains the Lie eq. [5, 6].

$$\frac{\partial}{\partial x} \ln J(x) = -J(x) \tag{9}$$

with solution $J(x) = J(1 - Jx)^{-1}$ and $J(x)$ diverges as $x \to 1/J$ which is a spurious divergency. This result corresponds to the summation of the parquet diagrams.

In the second approximation for the generator, vertex and self-energy corrections enter as well, and one obtains the following Lie equation with fixed point $J_0 = 2$ [5, 6].

$$\frac{\partial}{\partial x} \ln J(x) = -(J(x) - \tfrac{1}{2} J^2(x)) \tag{10}$$

It is interesting to note that this result is in exact agreement with a theory worked out by Nozières [14] in which differential equations of "parquet" type are solved including self-energy corrections and related vertex corrections which do not occur in a simple "parquet" approximation.

The only conslusions which can be drawn from these calculations are that the Kondo effect is a strong coupling problem and therefore the generator cannot be calculated using perturbation theory; furthermore, the logarithmic behavior is changed into power laws (scaling) with critical exponents independent of the bare coupling in the small energy region $x \gg 1$ [5], e.g. in the approximation corresponding to equation (11) one obtains

$$d\left(\frac{\omega}{D}\right) - \left(\frac{\omega}{D}\right)^{2S(S+1)} \tag{11}$$

where the exponent depends only on the value of the impurity spin.

It is easy to see that only those terms of perturbation theory which are linear in logarithmic terms contribute to the generator. However, it has been shown by Nozières [15] and his collaborators that all skeleton diagrams of "non-parquet" type provide such terms. Thus the generator is an infinite series unless there is some cancellation of higher order terms as in the X-ray absorption problem. Furthermore, it is important to note that eqs (9) and (10) do not alter by considering imaginary parts as well or by changing the energy variables of the vertices as has been pointed out by Sólyom [8].

Thus, we are not able to determine whether the fixed point is finite, as in eq. (10), or infinite, as it has been suggested by Anderson and his collaborators [13] considering thermodynamical scaling, and by Wilson [p. 68] on the basis of numerical calculations. Furthermore, we cannot decide whether the fixed points of the scaling of dynamical quantities [5, 6] at $\omega < D$ (e.g. vertex ...) and of the thermodynamical scaling of free energy ([13 and p. 68]) coincide or not.

Moreover, Anderson [7] has suggested a simple way of scaling to derive dynamical scaling. He obtained a spurious divergency discussed before in analysing the solution of eq. (9). Recently, Sólyom & Zawadowski [9] have reformulated this method to obtain the second order scaling given by eq. (10). They have scaled the scattering amplitude of an electron with Fermi momentum, which is normalized according to the initial and final states $|i\rangle$ and $|f\rangle$, respectively

$$T'(\omega) = \frac{\left\langle f \left| \sum_{n=0}^{\infty} H_1 \left(\frac{1}{\omega - H_0} H_1 \right)^n \right| i \right\rangle}{\left(\left\langle f \left| \sum_{n=0}^{\infty} \left(\frac{1}{\omega - H_0} H_1 \right)^n \right| f \right\rangle \left\langle i \left| \sum_{n=0}^{\infty} \left(\frac{1}{\omega - H_0} H_1 \right)^n \right| i \right\rangle \right)^{\frac{1}{2}}}, \tag{12}$$

where H_0 and H_1 are the Hamiltonian of the electron gas and exchange interaction. The idea of scaling is to replace the effect of a change of the cut-off ΔD by altering the coupling constant J to $J + \Delta J$; thus one gets

$$\frac{\partial T'(\omega)}{\partial D} \Delta D = \frac{\partial T'(\omega)}{\partial J} \Delta J \tag{13}$$

By calculating these derivatives up to the third order one can reproduce the scaling given by eq. (10).

Imaginary Parts

The crucial test for any theory dealing with the Kondo problem is to get the unitarity limit for the electron scattering amplitude at the Fermi energy, $\omega=0$. In order to do that, one should calculate $\Gamma_K(\omega=0)$ which must be imaginary; however, this problem is far from being solved. In order to understand the problem, we consider a quantity like $\mathcal{G}(\omega)$, which can be calculated on the negative energy axis, where it must be real at zero temperature. One obtains for $|\omega| < D$ that

$$\mathcal{G}(-\omega)=\frac{1}{\omega}\left(\frac{-\omega}{D}\right)^{\alpha(g_0)} \tag{14}$$

where the exponent α depends on the fixed point g_0 and must be positive to fulfill the sum-rule for its spectral function. The analytic continuation on the upper half energy plane gives

$$\mathcal{G}(\omega)=\frac{1}{\omega}\left(\frac{\omega}{D}\right)^{\alpha(g_0)} \mathrm{e}^{-i\pi\alpha(g_0)}$$

One can see that the factor $\mathrm{e}^{-i\pi\alpha(g_0)}$ is an infinite series in the coupling even if we keep only the first terms in the power series expansion of α. In general, we should sum up such series, or in other words we should calculate the contribution of all diagrams which do not contain logarithmic terms. This seems to be an unsolvable problem.

Returning to the problem of $\Gamma_0(\omega=0)$, one should notice that analytic continuation does not work, because Γ_0 has a cut along the positive and negative axis as well. Thus it looks very difficult to express $\Gamma_0(\omega)$ in terms of the invariant coupling $J(x)$. If this could be done, then one could get the value of the coupling at the fixed point by assuming the unitary limit at $\omega=0$.

One Dimensional Metallic Systems at Zero Temperature

Menyhárd & Sólyom [10, 11] have applied a similar theory to one-dimensional metallic systems, where the electrons are directly interacting in two scattering channels with respect to the spin variables. The situation is exactly like the Kondo problem. The first-order scaling contains the spurious divergence which would correspond to a phase transition prohibited by exact theorems. The next order approximation provides a finite value of the coupling of the order or unity at the fixed point and critical behavior at zero energy for several quantities. The critical exponents are determined by the coupling at the fixed point. Depending on the signs and ratio of the two bare couplings one or two of the calculated susceptibilities responsible for superconducting, antiferromagnetic and density-wave-fluctuations show critical behavior. Thus one can say that the phase transition "occurs" at zero temperatures. The difficulties of this theory are very similar to those discussed in the case of the Kondo problem.

Conclusion

The renormalization group method is very effective to demonstrate scaling behavior in different logarithmic problems. In general, it seems to be very difficult to reliably calculate critical exponents because this would require the summation of the contributions of all diagrams which are linear in logarithmic expressions or do not contain logarithmic expressions at all. This problem is closely related to the calculation of imaginary parts and the unitarity condition. However, the renormalization group method is very efficient in those cases where the effective coupling is not enhanced, in other words, when the weak coupling limit scales into a weak coupling problem. This situation is realized where there are e.g. two scattering channels and a strong cancellation between them. Typical examples are X-ray absorption [15] and localized spin fluctuation problems [12]; in the latter the electron electron and electron-hole correlations exhibit a strong cancellation.

The author would like to express his gratitude to N. Menyhárd and J. Sólyom for letting him use their results before publication and for valuable discussions. The author is also grateful to P. Nozières and G. Iche for their helpful comments.

References

1. Gell Mann, M & Low, F, Phys rev 1954, 95, 1300.
2. Bogoliubov, N N & Shirkov, D V, Introduction to the theory of quantized fields. Interscience, New York, London, 1959.

3. Bonch-Bruevich, V L & Tyablikov, S V, The Green function method in statistical mechanics. North-Holland, Amsterdam, 1962.
4. Larkin, A I & Khmel'nitskii, D E, Sov phys JETP 1969, 29, 1123.
5. Fowler, M & Zawadowski, A, Solid state comm 1971, 9, 471.
6. Abrikosov, A A & Migdal, A A, J low temp phys 1970, 3, 519.
7. Anderson, P W, J phys C 1970, 3, 2436.
8. Sólyom, J. To be published.
9. Sólyom, J & Zawadowski, A, J phys F, metal phys (1973). In press.
10. Menyhárd, N & Sólyom, J low temp phys 1973, 12, 529.
11. Sólyom, J low temp phys 1973, 12, 547.
12. Iche, G, J low temp phys 1973, 11, 215.
13. Anderson, P W, Yuval, G & Hamann, D R, Phys rev B 1970, 1, 446.
14. Nozières, P. Private communication.
15. Roulet, B, Gavoret, F & Nozières, P, Phys rev 1969, 178, 1072 and Nozières, P, Gavoret, F & Roulet, B, Phys rev 1969, 178, 1084.

Superfluid ^3He

The Low Temperature Phases of Liquid ^{3}He[1,2]

R. C. Richardson and D. M. Lee

Laboratory of Atomic and Solid State Physics, Cornell University, Ithaca, N.Y. 14850, USA

Summary

A brief summary is given of the recent experiments in liquid ^{3}He at very low temperatures. Several new phases of the liquid have been observed in various experiments that are discussed.

In the surprisingly brief interval of a little more than a year there have been quite a number of experiments [1–16] measuring various properties of liquid ^{3}He at temperatures below 3 mK. We know as a result of this work that there are several new phases of the liquid at these temperatures with properties that differ markedly from those of the Fermi liquid characterizing ^{3}He at higher temperatures. A summary of the measurements is presented in that which follows. It is clear that the efforts in this area, to date, are but a small beginning in a field that is developing quite rapidly.

Phase Diagrams

We begin our summary with a set of "maps" which mark the boundaries of the new phases. Fig. 1 shows the intersection of the liquid phase boundaries with the surface traced by the melting curve in various magnetic fields. Points in pressure, temperature and magnetic field above the melting surface lie in the solid phase of ^{3}He. The experiments which have produced the information shown in the figure have been performed using the technique of compressional cooling [17] so that the liquid is always in coexistence with the solid phase. The pressures and magnetic fields shown in fig. 1 are known rather accurately; however, the temperatures below 2.6 mK are only approximate and have been estimated from the assumed melting properties of solid ^{3}He. A satisfactory thermometer for measurements in ^{3}He below 2 mK has not yet been developed.

Upon cooling from the higher temperatures it is observed that the first phase boundary in low magnetic fields is at 2.65 mK [1, 2, 5]. The boundary has been identified in measurements of the cooling rate in compression experiments [1–3, 5–8, 16] nuclear magnetic susceptibility [2, 4], heat capacity [6, 10, 13], sound attenuation [7, 9, 14, 15], and viscous damping of a vibrating wire [11]. As the magnetic field is increased the transition, which is quite narrow in low magnetic fields, broadens and then divides into two distinct branches [2, 7, 15]. The thermal separation between the phases increases linearly with magnetic field. In a field of 10 kOe the separation is approx. 65 μK. In fig. 1 the two branches of the transition are labeled A_1 and A_2. There is apparently no latent heat or thermal hysteresis associated with these transitions.

As the liquid is cooled to still lower temperatures a second transition is encountered at approx. 2 mK [1–5, 8, 13, 16]. There is a small latent heat associated with this transition and the thermal hysteresis can be quite large, more than several tenths of a millikelvin under some circumstances. The latent heat of the transition is of order 20 ergs/mole [13]. The phase line shown in fig. 1, labeled B, was determined in observations of pressure changes upon warming in a compression cell. On warming, the transition occurs at a reproducible temperature. The temperature of the "B transition" is suppressed quadratically as the field is increased [2]. The

[1] Report 2002 issued by the Materials Science Center.
[2] Read by R. C. Richardson.

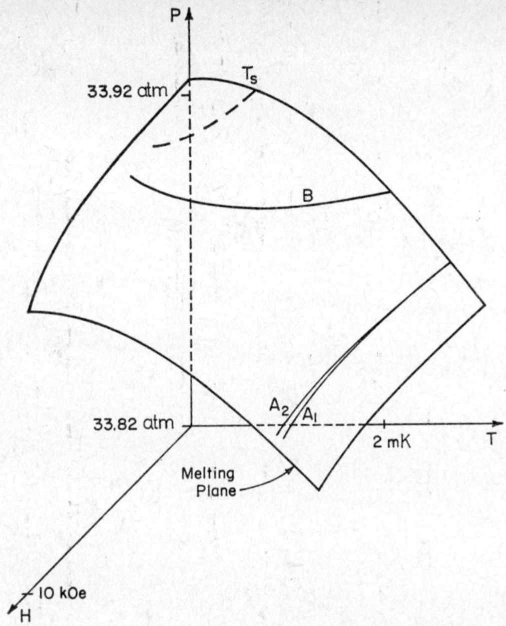

Fig. 1. The melting surface. The temperatures in the sketch are estimated from extrapolations of the melting curve in pressure, temperature and magnetic field from the temperature 2.65 mK and zero magnetic field.

critical field, extrapolated from measurements in fields up to 3 kOe, is 5.5 kOe. In the B phase some rather striking transient phenomena have been observed in nuclear magnetic resonance (NMR) experiments [4] and a large change has been observed in the resistance which the liquid offers to the motion of a vibrating wire [11].

The lowest temperature feature which has been observed on the melting curve is labeled T_s in fig. 1 and probably corresponds to the point at which the solid undergoes its spin ordering transition [13]. The pressure at which the solid transition occurs can be easily identified and is 0.055 atm above the A transition in zero magnetic field. Because the theory of the solid transition is now quite uncertain, it is not possible to make an accurate calculation of the temperature from the melting pressure at a point so far away from the last reliable direct temperature measurements, near the A transition. However, the upper limit of the solid transition temperature, established by extrapolating the melting curve with constant slope from the B transition, is 1 mK. Since it is anticipated that the melting curve slope should decrease substantially as the solid entropy drops near its spin ordering transition, we estimate that the pressure of the transition corresponds to a temperature near 0.5 mK.

The approximate phase diagram in the *P–T* plane in zero magnetic field is shown in fig. 2. The phase boundary labeled A is based upon measurements of heat capacity [6] and sound attenuation [9, 12] in experiments in which the cooling was achieved through the adiabatic demagnetization of CMN.

The temperature of the transition in this work was determined through measurements of the magnetic susceptibility of the CMN and through the extrapolation of Fermi liquid parameters (a linear dependence of the heat capacity and quadratic dependence of the sound attenuation coefficient upon temperature) [9, 12]. The temperature of the A transition decreases with pressure in the liquid so that the lowest pressure at which the A transition has been observed is 9.5 atm where the transition occurs at a temperature of about 2 mK, the lowest temperature to which ^{3}He has been cooled in the demagnetization of CMN.

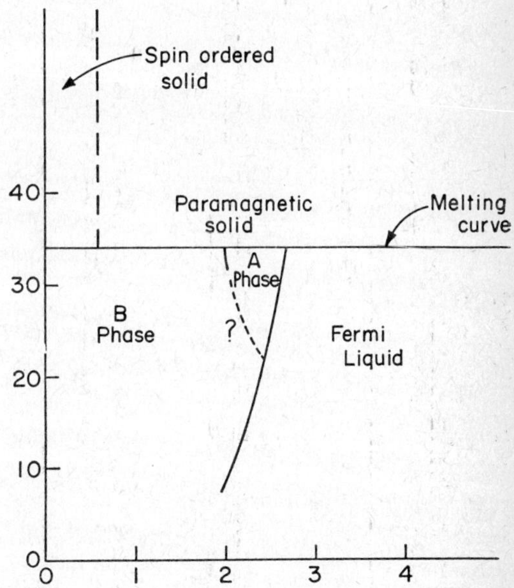

Fig. 2. *Abscissa*: *T* mK; *ordinate: P* atm.
P–T diagram of low temperature phases in liquid ^{3}He. The curves are representations of the data reported in refs [6, 9, 12]. The dashed line labeled B is speculative (see text).

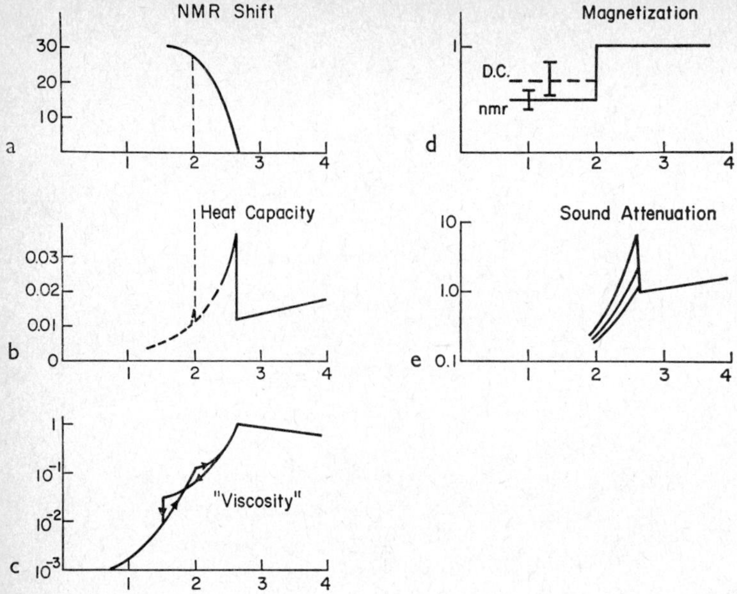

Fig. 3. *Abscissa:* T mK; *ordinate:* (*a*) $[H_0^2 - H_1^2]^{\frac{1}{2}}$ Gauss; (*b*) C/nR; (*c*) $(\eta \varrho_n)_T/(\eta \varrho_n)_A$; (*d*) χ_l/χ_0; (*e*) α cm^{-1} (log).

Experiments in new phases of liquid ^{3}He. (See text for references and explanation of experiments. In most of the range shown the temperature scales are estimates.)

Note added in proof: Recent precise measurements of the static nuclear magnetization in liquid ^{3}He by D. N. Paulson, R. T. Johnson & J. C. Wheatley (Phys rev lett 1973, 31, 746) reveal that the lines drawn in fig. 3*d* are incorrect. When the magnetization is measured at a pressure of 21.5 atm, so that the liquid passes directly from the normal Fermi liquid phase into the B phase, it is observed that the magnetization decreases monotonically with temperature, falling to a value of 0.5 times the Fermi liquid magnetization at $T/T_c = 0.85$.

The dashed line labeled B in fig. 2 is somewhat speculative and is based upon measurements of thermal conductivity of the liquid which have been recently reported [12]. It is observed that the temperature gradient across a tube of liquid ^{3}He, with a constant heat flux through it, decreases sharply at the A transition for liquid with pressure less than about 20 atm. For liquid at higher pressures it is observed that the temperature gradient decreases less rapidly at the A transition and that there is a second point which moves to lower temperatures with increasing liquid pressure at which a small but sharp decrease in the temperature gradient is also observed. At pressures near the melting pressure the temperature at which the second feature is observed corresponds to the B temperature in experiments performed on the melting curve. No latent heat has been detected in connection with the thermal conductivity measurements (or heat capacity measurements at pressures below the melting pressure) so that the phase line sketched in the figure is quite tentative. If the phase diagram of fig. 2 is correct, there is a tri-critical point in the liquid system in low magnetic fields and the A phase does not exist in low magnetic fields at pressures less than about 20 atm.

Summary of Experiments in Liquid ^{3}He

The experiments performed to date are represented schematically in fig. 3. Once again, the temperature scales are only approximate. For the precise data the reader is referred to the original publications in which the work is described.

Magnetic Measurements

As the liquid is cooled from high temperatures through the A phase transition the susceptibility changes little [2, 4], if any, but the field necessary to observe the resonance in a fixed frequency NMR spectrometer begins to shift. The sign of the shift is such that the liquid is apparently in a larger local field than the applied magnetic field. Measurements have been made between applied fields of 30 and 850 Oersteds.

The amount of the shift at a given temperature is strongly dependent upon the applied field but the parameter $(H_0^2 - H_1^2)^{\frac{1}{2}}$, plotted in fig. 3, is independent of applied field over the entire range of the measurements. H_0 is the field at which the signal appears in solid ^3He or in the Fermi liquid, and H_1 is the field in which the A phase liquid signal appears. The strength of nuclear magnetic field that a ^3He atom sees from one of its neighbors is less than $\frac{1}{2}$ Oersted. The existence of such a large additional "effective field" in the ^3He requires an extraordinary amount of correlation in the way in which the magnetic dipoles are arranged.

As the ^3He is cooled into the B phase the field shift disappears and the NMR signal is found back at the place where it is measured in the Fermi liquid (and in the solid) [2, 4]. On the other hand, the amplitude of the resonance decreases by approximately a factor of two. In static magnetization measurements on the melting liquid, a sharp drop in magnetization is observed when the B transition occurs [13]. The measured magnitude of the drop corresponds to a decrease in the liquid magnetization of slightly less than $\frac{1}{3}$. However, in the latter measurement the uncertainties in the demagnetization coefficient do not preclude agreement with the NMR result. Both types of measurements are indicated in the figure.

A curious transient phenomenon has also been observed in the liquid in the B phase [2, 4]. If a small heat pulse is applied to the liquid near the NMR coil, the resonant absorption signal disappears for several seconds and then recovers in a wave that fills the coil from the bottom to the top. In this work the bottom of the cell was probably the coldest so that it is not clear whether the recovery is determined by the thermal gradient or by gravity. It has also been observed in NMR experiments probing the spatial variation of the absorption along the axis of the signal coil that certain regions of the liquid may have apparent susceptibilities much less than the surrounding liquid. The regions with essentially zero absorption signal at the spectrometer frequency persist until the liquid is warmed into the A phase and then may not reappear when the sample is cooled back into the B phase again.

Heat Capacity Measurements

The change in heat capacity of the liquid at the A transition has been measured at pressures between 16 atm and the melting pressure [6]. The measurements are represented schematically in fig. 3. The solid line represents experiments performed in a demagnetization cell in which the heat capacity was determined from observations of the changes in the magnetic susceptibility of CMN in contact with liquid ^3He, after the application of known quantities of heat to the liquid. The dashed line corresponds to the liquid heat capacity deduced from the analysis of the pressure changes in a compressional cooling cell due to the formation of solid ^3He following the application of a heat pulse [13]. The temperature scale of the latter measurements is at this time only "approximate" because of the lack of knowledge of the melting pressure as a function of temperature.

In the demagnetization experiments it was found that the transition occurs over a temperature interval that is less than 3 μK. The shape of the transition is strongly suggestive of heat capacity change observed in superconductors. Immediately below the transition the heat capacity is a factor of 2.9 larger than the heat capacity of the Fermi liquid at the transition temperature. The precise functional form of the heat capacity with temperature has not been determined because of the uncertainties in the temperature scale of the CMN susceptibility. The data may be equally well fitted to a power law with the heat capacity C varying as T^4, or to an exponential law with C varying as $\exp(-3.6[(T/T_c)-1])$.

The heat capacity obtained from the analysis of changes in the melting pressure after heat pulses is in good agreement with the demagnetization experiments and extends to quite low temperatures, probably below 1 mK. At the B transition there is a latent heat of approx. 20 ergs/mole and apparently a sharp increase in the heat capacity of the liquid in a narrow temperature interval (less than 10 μK) below the transition. The latter may be an effect due to thermal inhomogeneity in the compression chamber but simultaneous measurements of the magnetization change of the liquid which are sensitive to more than 90% of the liquid volume support the interpretation that heat capacity increases before any of the liquid is warmed into the A phase.

Sound Attenuation Measurements

Some of the data from sound attenuation measurements performed in a demagnetization cell are also represented in fig. 3 [9]. The data shown are based upon measurements at a pressure near the melting pressure. The 3 lines correspond with measurements made at frequencies of 25 MHz, 15 MHz and 5 MHz. At the A transition the sound attenuation increases sharply as the temperature is lowered, passes through a very narrow peak and then decreases quite rapidly with further temperature decrease. The height of the peak increases as the frequency of the sound is increased and the temperature at which the peak occurs decreases as the frequency increases. In the same experiments it was observed that the velocity of the sound decreased after the attenuation passed through the peak. The velocity decreases by 0.5% in a narrow temperature interval below the peak in the sound attenuation and continues to decrease slowly with further decrease in temperature. It is speculated that the sound velocity is asymptotically approaching the velocity of first sound which is 1% less than the zero sound velocity that characterized the vibration mode of the Fermi liquid at these frequencies at temperatures just above the A transition.

In measurements of the sound attenuation with a liquid pressure of 19.4 atm it is observed that the width of the attenuation peak is much narrower than that shown in fig. 3 which was obtained at 32.7 atm. The qualitative difference in the temperature dependence of the sound attenuation may perhaps be associated with the possibility that the liquid passes directly into the B phase at pressures below 20 atm.

The attenuation of sound has also been measured in a compressional cooling cell [7]. The effects observed in low magnetic fields are qualitatively the same as those found in the demagnetization experiments. A peak is observed after the liquid is cooled through the A transition. In a compression cell, however, the quantitative analysis of the shape of the peak is marred by the spurious signal from solid that also forms in the chamber. The technique of compressional cooling, on the other hand, permits studies in magnetic fields of at least 14 kOe. As the magnetic field is increased it is found that the attenuation peak broadens and then separates into two separate peaks. Simultaneous observations of the time rate of change of cell pressure indicate that the heat capacity of the liquid increases sharply at the two temperatures at which increases in the sound attenuation are observed. It is therefore concluded that the liquid passes through two separate "A transitions" that are easily resolved in magnetic fields greater than 4 kOe.

Viscous Damping of a Vibrating Wire

Finally we wish to outline the results of an experiment [11] that has produced information that may qualitatively be interpreted as evidence for superfluidity in the liquid phases below the A transition. A representation of the data is also presented in fig. 3. Here, great liberties have been taken with the form of the original data reported for the experiment. The work was performed in a compressional cooling cell. The experiment consisted of observing the changes in the amplitude of resonant vibration of a wire stretched across the cell. A large magnetic field was applied perpendicular to the axis of the wire. A current is induced in such a wire as it vibrates in the field. In this work it was observed that starting above the A transition, as the temperature was lowered the amplitude of the resonant vibrations first decreased because the Fermi liquid viscosity increases, and then at the A transition the amplitude began to increase. At the B transition, the vibration amplitude increased very rapidly. At the lowest temperatures of their observations the amplitude of the resonance signal was more than 30 times greater than the amplitude at the A transition.

In fig. 3 we have plotted the logarithm of the inverse square of the wire vibration amplitude, normalized to be unity at the A transition, versus the assumed temperature scale for the pressures reported. The application of a crude two-fluid model to the experiment suggests that the inverse square of the amplitude of the resonant vibrations is proportional to the product of the viscosity and the normal fluid density fraction. We have optimistically labeled the ordinate of the sketch with the characters, $\eta\varrho_n$, associated with the two-fluid model. In the experiment it was observed that a hysteresis in the vibration amplitude occurred in the thermal cycling of the cell. In cooling the B transition occurs at a lower temperature than on warming. The hysteresis loop is indicated in the figures with arrows pointed in the sense of the temperature change during the observation of the data. Apart from the basically remarkable general nature of the results the data has the curious feature that the signal size is found to be different in the A phase upon warming from the value found in cooling. This is the only reported experiment in which a property of the A phase seems to be multivalued.

Conclusions

In the report that we have given here of the experiments in liquid ^3He in the new phases, theoretical interpretations based upon pairing models of the transitions have been avoided. There are several good reasons for this. The first is that many of the principals of the theoretical investigations are present at this symposium and may wish to apply their own interpretations to the data. The second is that from the point of view of an experimentalist, at least, it seems to spoil part of the excitement in the work to categorize the results in terms of a broad general theory before more information is collected. We have been strongly influenced in the way we think about these data by: the suggestion of Leggett [18] that pairing in an odd angular momentum state is responsible for the dipole spin correlations that produce the NMR shift in the A phase; Ambegaokar & Mermin [19] for the suggestion that the two A phases correspond to the separate transitions of the Fermi spheres with spins parallel and anti-parallel to the field; and Anderson & Brinkman [20, 21] for the general discussion of how the spin-fluctuations cause the selection first of the paired states with $m_z = \pm 1$ for the transition to the A phases which they call an anisotropic superfluid and the later selection at lower temperatures of the state $m_z = 0$ to produce the B phase called the isotropic superfluid. Eventually, these names may replace "A" and "B" but we would like to hold out for the possibility that the transitions result from a mechanism that has not occurred to anyone yet. Perhaps (or more accurately, we hope) the idea that will correctly explain all of the data will even occur to one of us first.

We wish to recognize the stimulating recent conversations that we have had with John Wheatley and Richard Johnson. Their valuable experimental contributions to understanding the new phases of the liquid are cited throughout this work. We owe a special debt of gratitude to our colleagues in these experiments at Cornell: to Doug Osheroff, who was the first to see the transitions and was the first to really know that they occur in liquid ^3He; to Bill Halperin, who was the most careful critic of the original work and whose subsequent measurements have added a great amount to the knowledge of the new liquid phases; and to Dewey Lawson, Steve Goldstein, Willy Gully, Bob Buhrman, Michel Bernier, Bill Truscott and Eric Smith, all of whom have participated in the experiments and discussions that we have had. We are also indebted to John Reppy, John Wilkins, Joe Serene, Vinay Ambegaokar, David Mermin, Phil Anderson, Bill Brinkman and Tony Leggett for criticism and theoretical insight about the transitions in liquid ^3He.

This work was supported by NSF under grant #GP-29682 and grant #GH-35692, and by NSF under grant #GH-33637 through the facilities of the Materials Science Center at Cornell University, Report #2002.

References

1. Osheroff, D D, Richardson, R C & Lee, D M, Phys rev lett 1972, 28, 885.
2. Osheroff, D D, Ph.D. thesis. Cornell University, 1973. Unpublished.
3. Halperin, W P, Buhrman, R A, Webb, W W & Richardson, R C, Proc 13th int conf on low temperature physics. To be published.
4. Osheroff, D D, Gully, W J, Richardson, R C & Lee, D M, Phys rev lett 1972, 29, 920.
5. Johnson, R T, Paulson, D N, Pierce, C B & Wheatley, J C, Phys rev lett 1973, 30, 207.
6. Webb, R A, Greytak, T J, Johnson, R T & Wheatley, J C, Phys rev lett 1973, 30, 210.
7. Lawson, D T, Gully, W J, Goldstein, S, Richardson, R C & Lee, D M, Phys rev lett 1973, 30, 541.
8. Anufriev, Yu D, Alvesalo, T A, Collan, H K, Opheim, N T & Wennerström, P, Phys lett 1973, 43 A, 175.
9. Paulson, D N, Johnson, P T & Wheatley, J C, Phys rev lett 1973, 30, 829.
10. Dundon, J M, Stolfa, D L & Goodkind, J M, Phys rev lett 1973, 30, 843.
11. Alvesalo, T A, Anufriev, Yu D, Collan, H K, Lounasmaa, A V & Wennerström, P, Phys rev lett 1973, 30, 962.
12. Johnson, R T, Presentation at Washington APS meeting, April 1973. EC2, Bull Am phys soc 1973, 18, 614.

13. Halperin, W P, Buhrman, R A & Richardson, R C, Bull Am phys soc 1973, 18, 642 (and to be published).
14. Gully, W J, Lawson, D T, Goldstein, S, Richardson, R C & Lee, D M, Bull Am phys soc 1973, 18, 642.
15. Gully, W J, Osheroff, D D, Lawson, D T, Richardson, R C & Lee, D M. To be published.
16. Landau, J B, Technion Institute, Haifa, Israel. Private communication.
17. See e.g. Johnson, R T & Wheatley, J C, J low temp phys 1970, 2, 423.
18. Leggett, A J, Phys rev lett 1972, 29, 1227; Phys rev lett 1973, 31, 352.
19. Ambegaokar, V & Mermin, N D, Phys rev lett 1973, 30, 81.
20. Anderson, P W, Phys rev lett 1973, 30, 368.
21. Anderson, P W & Brinkman, W F, Phys rev lett 1973, 30, 1108.

Discussion

Kohn: Wheatley has not observed a latent heat along the line separating the A and B phases but considers a value of 1 erg/cm^3 possible near the "B"-point.

Richardson: This is consistent with the result we have obtained in melting curve studies at Cornell in which we measure a latent heat of 0.6 erg/cm^3 for the transition at the melting pressure.

Richardson: During the time of this symposium I received a telegraph communication from John Wheatley related to recent results by the group in La Jolla. He reports that they observe a 26% reduction in the normal paramagnetic susceptibility of the liquid ^{3}He as the liquid passes from the A into the B phase at a pressure of 26 bar. The temperature at which the susceptibility change is observed is the same as that of the AB transition observed in thermal resistance measurements.

Kohn: I think it would be interesting if you would mention Wheatley's estimates of the critical velocity of superfluid ^{3}He.

Richardson: His estimate, based upon the onset temperature of a non linear regime of the resistance in heat flow experiments, is that the critical velocity in the A phase is 0.1 cm/sec. On the other hand the measurements in the B phase indicate very much larger critical velocities.

Experimental Evidence for Superfluidity in the Newly Found Phases of ^{3}He[1]

T. A. Alvesalo, Yu. D. Anufriyev, H. K. Collan, O. V. Lounasmaa, and P. Wennerström

Department of Technical Physics, Helsinki University of Technology, 02150 Otaniemi, Finland

Summary

The relative bulk viscosity η of liquid ^{3}He has been measured along the melting curve from 15 mK to about 1 mK with a vibrating string viscometer. As the temperature is reduced, η first increases as expected until the pressure anomaly "A" at $T_A = 2.6$ mK is reached. The viscosity then starts to decrease, has a discontinuous drop at the pressure anomaly "B", and then continues to diminish rapidly. The lowest measured value of η, approximately at 1 mK, is more than three orders of magnitude smaller than expected on the basis of the Fermi liquid theory and experimental data above T_A; this is regarded as strong evidence for superfluidity. If the two-fluid model is applied to liquid ^{3}He, the results can be interpreted as showing that the density of the normal component rapidly diminishes below the A transition, with a corresponding increase in the density of the superfluid component.

After the original discovery [1], a little more than a year ago, considerable amount of experimental and theoretical work (pp 84, 97) has been done in trying to explain the nature and the origin of the pressure anomalies "A" and "B" of ^{3}He below 3 mK. We have measured the relative bulk viscosity η of liquid ^{3}He from 15 mK to well below these points [2]. Our results show several interesting features and strongly infer that two superfluid transitions occur in the liquid below 3 mK. The data also confirm earlier results suggesting that the A and B transitions both are properties of the liquid phase.

The ^{3}He specimen was refrigerated inside a Pomeranchuk cell which had first been precooled to 15 mK by means of a dilution refrigerator. The cell, illustrated in fig. 1, is equipped with a capacitive pressure gauge and a vibrating string viscometer. The string, 2.5 cm long, 0.25 mm thick, and made of superconducting NbTi wire is under a tension which gives the system a natural resonant frequency of 1 245 Hz and an intrinsic Q of about 2 500. During measurements an 0.1 T magnetic field was applied perpendicular to the string and an a.c. excitation current of 2 mA was passed through it.

Runs were made in the conventional way [3] by reducing the ^{3}He volume at a steady rate with the aid of ^{4}He in the inner compartment (cf fig. 1). During an experiment the ^{3}He pressure was measured with a capacitance bridge, employing phase-sensitive detection. The viscosity was monitored by sweeping the frequency of the viscometer a.c. excitation current semi-continuously through the resonance. Since the signal amplitude, frequency shift at the resonance, and the width of the resonance line are all related to the viscosity η and density ϱ of the liquid [4], a considerable amount of information can be deduced from this type of measurement.

Fig. 2 illustrates the data obtained during one run. It is seen that, as cooling proceeds after the melting curve has been reached at X, the signal amplitude at resonance decreases continuously and the resonant frequency is reduced. Immediately after the A transition this behavior is reversed. The amplitude starts to increase and the resonant frequency shifts upwards; this shows that damping of the viscometer string is decreasing. Exactly simultaneously with the pressure drop at B the vibration amplitude jumps discontinuously. After this has happened the amplitude increases very quickly and then remains constant as the temperature no longer

[1] Read by T. A. Alvesalo.

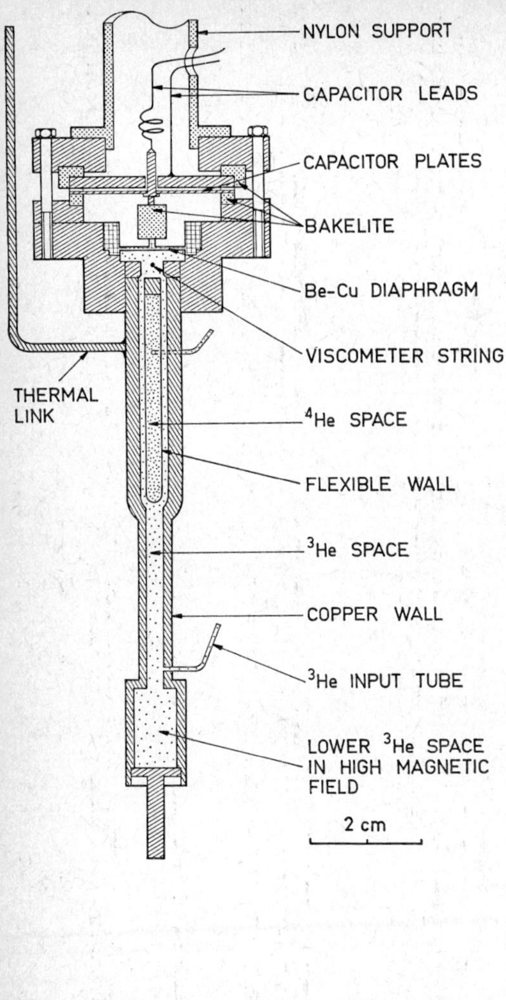

Fig. 1. The Pomeranchuk cell, connected to a dilution refrigerator through a thermal link for precooling. When pressure is applied to the ^{4}He space the walls of the inner cell bulge and thereby reduce the volume of the ^{3}He space. Direct compression of ^{3}He is impossible owing to an unavoidable solid plug in the ^{3}He input tube; the plug is first formed at 0.32 K, i.e. close to the minimum of the melting curve. The ^{3}He cell is equipped with a capacitive pressure gauge, consisting of a Be–Cu diaphragm and capacitor plates, and with a vibrating string viscometer (the string is perpendicular to the plane of paper). The superconducting solenoid, providing the 0.1 T magnetic field necessary for operating the viscometer, is not shown.

It was expected that frictional heating might promote the formation of solid ^{3}He around the string which, in turn, would have prevented the viscometer from working. For this reason a superconducting solenoid (not shown in the figure), capable of producing a sharply profiled field up to 4.5 T, was fitted around the lower ^{3}He space. Because a magnetic field depresses the melting curve of ^{3}He it was hoped in this way to restrict the formation of solid to the lower ^{3}He space only, at least at the beginning of a compressional cooling experiment. It turned out, however, that a magnetic field did not have the expected effect; the lower ^{3}He space has subsequently been removed.

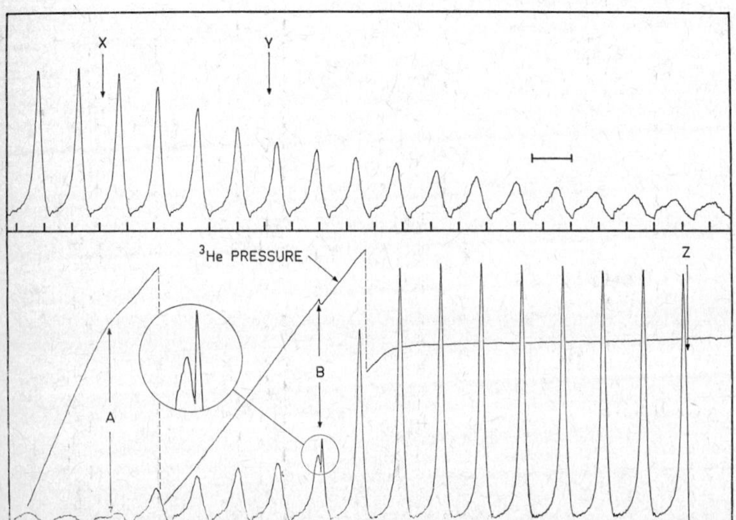

Fig. 2. *Abscissa:* time (20 s/division).

An actual chart recorder tracing of the viscometer signal amplitude as a function of time. In the lower part of the figure the output from the capacitance bridge, measuring the pressure in the ^{3}He cell, is also shown. The melting curve is reached and Pomeranchuk cooling starts operating at point X. At Y the compression speed is reduced. The A and B transitions (A is manifested by a kink in the pressure vs time curve and B by a sudden drop in pressure), and point Z, at which the string sticks, are also marked. The inset illustrates the discontinuous increase of the viscometer signal at B. The horizontal bar in the upper part of the figure shows the frequency sweep range, from 1 177 to 1 245 Hz.

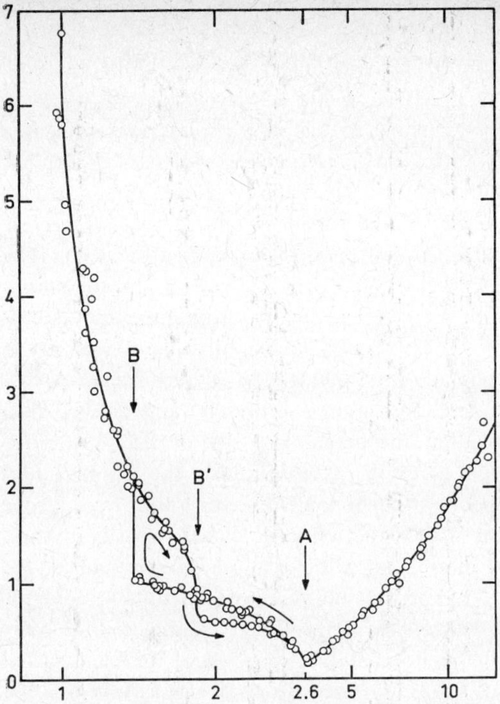

Fig. 3. *Abscissa:* temp (mK; provisional scale); *ordinate:* signal amplitude (a.u.).

Normalized viscometer signal amplitudes at resonance from 6 runs as a function of temperature. Note that the provisional temperature scale is different on each side of the A transition, assumed at $T_A = 2.6$ mK. The double hysteresis loop, corresponding to cooling and subsequent warming through the transition region, is marked with bent arrows. The B′ transition, at $T_{B'} = 1.9$ mK, is manifested by a flat portion on the pressure vs time curve during warming [1].

decreases. Finally, at point Z the signal sudddenly, in about 0.1 s, completely disappears, presumably because solid ^{3}He impedes further vibrations of the wire.

So far we have made about two dozen runs of the type illustrated in fig. 2. The specific run shown was selected as an example because both the A and B transitions happen to occur near a resonance. It is clear from this figure that η changes continuously through the A transition and discontinuously through B. During some experiments the viscometer string stuck soon after the A point was passed and in a number of early experiments solid ^{3}He started to form inside the viscometer chamber already before the A point had been reached. The importance of a low starting temperature, in the vicinity of 15 mK, for the operation of our string viscometer inside the Pomeranchuk cell was clear from these experiments. Measurements of η could be carried out up to the highest attainable pressure only if during the last 4 h of precooling and during the compression itself the dilution refrigerator was operated in the single-cycle mode.

Results of several runs are shown in fig. 3; the viscometer signal amplitude at resonance has been plotted as a function of temperature. Because the melting pressure P_M of ^{3}He, employed as the thermometer in our experiments, is inaccurately known as a function of T below 10 mK, the temperature scale must be regarded provisional, especially at the low temperature end. We have assumed that on the melting curve the A transition occurs at $T_A = 2.6$ mK and that below 2.6 mK dP_M/dT is constant at -30 bar/K.

From fig. 3 it is again observed that as the temperature decreases from 15 mK, the viscometer signal amplitude first decreases, reaches a minimum at T_A, and then starts to increase. After a fast initial growth the signal continues to increase more slowly while the temperature is further reduced, until at T_B the vibration amplitude jumps discontinuously and then increases rapidly to a value at which the signal somewhat later suddenly disappears. The viscometer signal amplitude at the lowest experimental temperature, probably near 1 mK, was almost 4 times the amplitude at 10 mK and was increasing with decreasing T very quickly indeed.

The behavior of the signal amplitude as a function of temperature suggests that along the melting curve below 2.6 mK liquid ^{3}He has two different phases: phase A between 2.6 and 1.9 mK, and phase B below 1.9 mK. From fig. 3 it is also clear that the A-phase can be supercooled well below 1.9 mK before the transition from A to B occurs. A natural explanation for this behavior is that A→B is a first order phase transition. Because of supercooling the location of point B on the temperature scale varied considerably for different experiments.

During many of the runs we stopped compression of the ^{3}He cell soon after point B was passed, then reduced the pressure slowly, after which a new pressurization was started. By this procedure it was possible to traverse the hysteresis loop A–B–B′–A (cf fig 3) many times

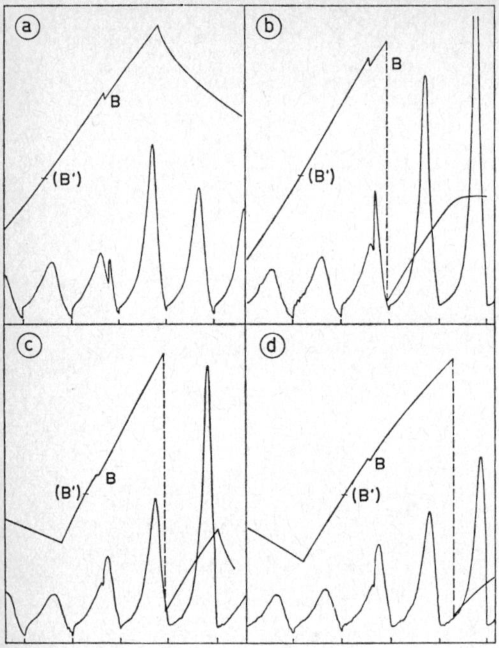

Fig. 4. *Abscissa:* time (32 s/sweep); *ordinate:* pressure and signal amplitude.

Examples of discontinuities in the vibration amplitude at B.

before the viscometer string eventually stuck. If the cell was allowed to warm less and less above $T_{B'}$, then the temperature at which the B transition occurred was each time, for a new pressurization cycle, closer to $T_{B'}$ than before whereby the hysteresis B–B′–B finally almost vanished and B appeared as a "mirror image" of B′. The jump of the signal amplitude at T_B simultaneously decreased but never vanished. This is evident from fig. 4 illustrating several successive passages through the pressure anomaly B. In fig. 5 we show an experimental sequence during which the double hysteresis loop of fig. 3 was traversed several times. Within the limits of our resolution it seems that with increasing temperature η changes rather steeply but continuously at $T_{B'}$.

An interesting feature of these experiments was that during decompression, between $T_{B'}$ and T_A, the measured values of the signal amplitude consistently fell below the values recorded during the preceding (or subsequent) compression, and did not coincide with the upper curve until the temperature was rather close to T_A. We conclude that some kind of thermal relaxation within the liquid, and not just simple supercooling, is responsible for the hysteresis B′–A–B′.

According to the simplest interpretation of a vibrating string experiment [4] assuming that the liquid is an ordinary fluid of a constant density ϱ, the signal amplitude varies inversely as the square root of the bulk viscosity η. The minimum signal amplitude at T_A and the amplitude at the lowest temperature differ by a factor of 34. The lowest measured value of viscosity is thus about 1 000 times smaller than at T_A. Calculating more accurately and taking into account the actual viscosity of ^{3}He, one finds $\eta_A/\eta_{1mK}=530$. According to the Fermi liquid theory, predicting a $1/T^2$ temperature dependence for η, the expected value of the viscosity at 1 mK is more than a factor of six larger than at T_A. The experimental value of η is thus a factor of 4 000 smaller at 1 mK than expected on the basis of the Fermi liquid theory and the measured η at T_A. If the viscosity at 10 mK is taken as the standard for comparison, the measured viscosity at 1 mK is about a factor of 1 500 smaller than expected.

Allowing for possible inaccuracies in the data and in the temperature scale used the agreement between the two sets of calculations is adequate. Our results thus show that the bulk viscosity of liquid ^{3}He near 1 mK is more than 3 orders of magnitude smaller than expected on the basis of the Fermi liquid theory and experimental measurements of η above T_A. This must be regarded as strong evidence for superfluidity.

If the two-fluid model is adopted for ^{3}He, i.e. if it is assumed that liquid ^{3}He below T_A is made of two components, normal fluid (density ϱ_n) and superfluid (density ϱ_s), damping of a vibrating string in the previous approximation is proportional to $(\varrho_n\eta_n)^{\frac{1}{2}}$. When the fraction of the normal fluid diminishes as the temperature is reduced below T_A, the simplest approximation states that η_n should stay constant or even increase [5] ($\eta_s=0$ by definition). This type of behavior is well known from measurements on liquid ^{4}He below $T_\lambda=2.17$ K; it has been found that η_n, from T_λ down to $\frac{1}{2}T_\lambda$, is constant within a factor of two. The present experiments thus suggest, within the framework of the two-fluid model, that the reduction of the bulk

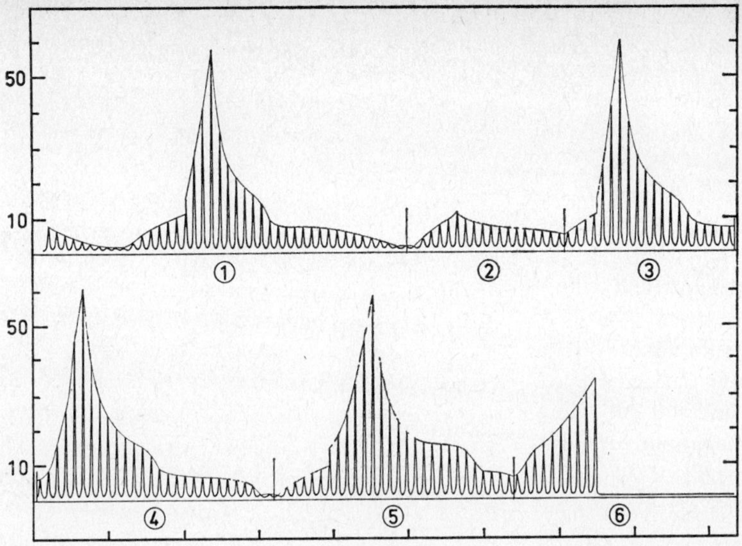

Fig. 5. *Abscissa:* time (4 min/division); *ordinate:* signal amplitude.

Study of the hysteretic properties of the transitions. The "envelope curve" is drawn as a visual aid.

Fig. 6. *Abscissa:* reduced temp.; *ordinate:* reduced bulk viscosity.

The reduced bulk viscosities of liquid ^{3}He and liquid ^{4}He (viscosity divided by its value at the respective transition temperatures, $T_A = 2.6$ mK for ^{3}He and $T_\lambda =$ 2.17 K for ^{4}He), plotted on a logarithmic scale as a function of the reduced temperature ($= T/T_A$ for ^{3}He and T/T_λ for ^{4}He). Notice the rapid decrease of the viscosity of ^{3}He immediately below T_A.

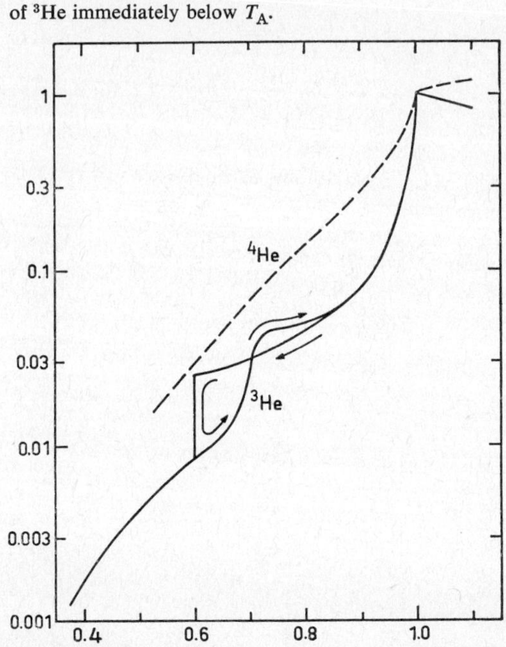

viscosity η of liquid ^{3}He below 2.6 mK is caused by a reduction of ϱ_n and, therefore, by a corresponding increase in ϱ_s.

It is illustrative to try to make a quantitative comparison between the present results on ^{3}He and the known viscosity of ^{4}He below T_λ. As noted earlier, the bulk viscosity approximately corresponds to $\varrho_n \eta_n$ in a two-fluid picture. In fig. 6 we have plotted the reduced bulk viscosities of liquid ^{3}He and liquid ^{4}He on a reduced temperature scale. It is observed that the overall behavior is somewhat similar, except for the extra features observed for ^{3}He because of the B transition. However, the absolute viscosity of liquid ^{3}He is, at a given reduced temperature, four orders of magnitude larger than the viscosity of ^{4}He owing to differences in the bulk viscosities of ^{3}He and ^{4}He at T_A and T_λ, respectively. One should also note that the quantitative interpretation of the ^{3}He data below 2.6 mK is not quite unambiguous but depends on the ratio between the diameter d of the vibrating wire and the viscous penetration depth l. In fact, l is not known below 2.6 mK. For simplicity we have assumed that $d/l > 1$. With the current knowledge of the physical properties of the new phases of ^{3}He, we have not been able to carry out a more detailed analysis of our data below T_A.

The present measurements, together with other experimental data (p. 84) and several theoretical predictions (pp 103, 109, 112, 116), support the existence of an anisotropic superfluid ^{3}He below 2.6 mK due to a pairing type of condensation. Our viscosity measurements seem to give the most direct evidence for this hypothesis. The observed continuous viscosity, with a discontinuity in $d\eta/dT$ at T_A (cf fig. 6), and the rapid decrease of η below T_B seem to fit the suggestion [6] that between T_A and T_B the liquid is a superfluid in which the energy gap disappears in certain directions and that the transition at T_B leads to a superfluid with a uniform energy gap.

By a careful analysis of our viscosity data, taking into account changes in the signal amplitude, resonance frequency, and the line width, it may be possible [4] to determine ϱ_n and η_n separately, at least approximately. This analysis is in progress.

References

1. Osheroff, D D, Richardson, R C & Lee, D M, Phys rev lett 1972, 28, 885.
2. Alvesalo, T A, Anufriyev, Yu D, Collan, H K, Lounasmaa, O V & Wennerström, P, Phys rev lett 1973, 30, 962.
3. Lounasmaa, O V, Experimental principles and methods below 1 Kelvin. Academic Press, London, 1974.
4. Black, M A, Hall, H E & Thompson, K, J phys C 1971, 4, 129.
5. Seiden, J, Compt rend acad sci (Paris) 1973, 277, B-115.
6. Anderson, P W, Phys rev lett 1973, 30, 368.

Key words

Liquid ^{3}He, superfluidity, Pomeranchuk cooling, ultra-low temperatures

Discussion

Anderson: A remark which should perhaps more aptly be made by several other participants in this conference is that the behavior of the apparent viscosity looks characteristically superconducting (i.e. BCS) and not superfluid, e.g. it looks very like sound attenuation in superconductors, with the characteristic vertical drop at T_c due to coherence factors. In general, it may be dangerous to use the two-fluid theory as more than a guide to intuition because of pairing and coherence effects. In general BCS is not hard to calculate with and it is usually not very necessary to use two fluids.

The Order Parameter in Liquid ^{3}He

N. D. Mermin and V. Ambegaokar[1]

Laboratory of Atomic and Solid State Physics, Cornell University, Ithaca, N.Y. 14850, USA

The extraordinary low temperature phases of liquid ^{3}He were discovered while studying magnetic properties of the solid phase along the melting curve, using the technique of compressional cooling [1]. Subsequent measurements along the melting curve [2, 3] have provided substantial clues about the nature of the liquid ordering, despite some obvious difficulties:

(i) Determining absolute temperature dependences requires a knowledge of the temperature and magnetic field dependence of the melting curve itself, which is particularly uncertain due to the likelihood of incipient magnetic ordering in the solid; (ii) other lines in the P–T plane (e.g. isobars) that would yield less ambiguous information are inaccessible; (iii) the coexisting solid phase can make it difficult to disentangle from the data the liquid phase contributions.

However, melting curve measurements have the advantage of using a cooling method that is not incompatible with the application of high magnetic fields. If, as seems increasingly likely, liquid ^{3}He has become in the anomalous phase (or phases) a magnetic superfluid, then there arises a class of field dependent phenomena hitherto unknown in superfluids or superconductors which are, quite apart from their intrinsic interest, of great value in determining the detailed features of the phases.

The following melting curve data have proved especially informative in reducing the considerable range of possibilities one must face in considering ab initio the pairing problem for general pair orbital angular momentum L:

(1) In zero magnetic field there is a transition to a phase A at a temperature $T_A = 2.7$ mK, followed by a second transition to a phase B at a T_B of about 2.0 mK.

(2) T_B is depressed by an applied field; roughly $T_B(H) = T_B(0) - [H(\text{kilogauss})]^2 \times 0.06$ mK, with measurements going up to about 4 kG. (The coefficient is particularly subject to the uncertainty in the slope of the melting curve in this temperature range.)

(3) No such strong quadratic depression of the A transition is observed. (The pressure at which the A transition first occurs does have a quadratic field dependence a few percent of that observed at B, but uncertainty in the field and temperature dependence of the melting curve prevents translating this into a variation in temperature.) However the A transition is observed to split into two occurring at temperatures separated by a small linear term in the field: $T_{A_1}(H) - T_{A_2}(H) = [H(\text{kG})] \times 6\ \mu$K.

(4) The susceptibility in the A phase (deduced from the strength of the NMR signal) appears to be unaltered from the normal state susceptibility (and therefore temperature independent).

(5) The susceptibility of the B phase appears to be roughly half of the normal state susceptibility and also shows no temperature dependence.

(6) The NMR signal (in an r.f. field perpendicular to the static field) is shifted away from the Larmor value as T drops below T_A. The shift grows with declining temperature, but the signal returns to the Larmor frequency as the B phase appears.

(7) The attenuation of ultrasound exhibits a peak as the temperature is lowered below T_A. The width of the peak is larger for higher frequencies. At 10 MHz the width is approx. 25 μK.

---

[1] Read by V. Ambegaokar.

Of the mechanisms which could account for these phenomena, a condensation into a paired state is strongly suggested. It is hard to understand the splitting of the A transition in a magnetic field on any other basis, for reasons which will be given in the next paragraph. Assuming that the A and B phases are characterized by pairing of the BCS type, the phenomena (1)–(7) provide important clues about the structure of the pairs near the transition temperature. The transition temperature itself is determined by the linearized gap equation, solutions to which can be classified according to their orbital angular momentum L. The field dependence of the transition temperature will be quite different depending on whether the L at which a solution is first possible is even (singlet pairing) or odd (triplet pairing). In the singlet case paired atoms are necessarily of opposite spin, so that pairing competes with the field induced paramagnetic imbalance in spin populations, and the effect of a field is to lower (quadratically) the transition temperature. In the triplet case pairing can occur within each separate spin population (if only pairs with $S_z = \pm 1$ are formed) thereby maintaining the net spin alignment of the normal state. In this case the primary effect of the field is a slight alteration in the transition temperature for each population, due to the opposite linear shifts in field of the effective Fermi energies. Thus the signature of triplet pairing is a linear splitting of the transition, without appreciable depression; the singlet transition temperature is unsplit and depressed quadratically in field.

This excludes singlet pairing as a model of the A phase. (It also excludes itinerant antiferromagnetism, for such ordering could not set in one spin population at a time without (in the A_1 phase) setting up a spatially inhomogeneous mass density, which would be very expensive of compressional energy.) It is also tempting to conclude that the B transition is a shift to even L pairing, brought on, for example, by changes in the p and d wave coupling parameters as the pressure varies along the melting curve, or more subtle considerations of free energy balance. There is, however, another possibility within pairing theory: triplet pairing in which the spin populations are coherently mixed (i.e. in which there is a component of $S_z = 0$ pairing in addition to $S_z = \pm 1$) is depressed by a magnetic field for the same reasons the singlet transition is depressed. This type of triplet pairing is distinguished from singlet pairing by the magnetic susceptibility, which drops to zero with declining temperature for singlet pairing, but only to some finite fraction of the normal state value in the triplet case. This is currently the only direct evidence in favor of a triplet pairing model for the B phase as well as the A phase.

One is therefore led (with rather less confidence for the B phase than the A) to examine the theory of triplet pairing. Such pairing is characterized by a 2×2 order parameter:

$$\psi_{\mu\nu} = \langle a_\mu(\boldsymbol{k}) a_\nu(-\boldsymbol{k})\rangle = [\mathrm{i}\boldsymbol{d}(\hat{\boldsymbol{k}}) \cdot \boldsymbol{\sigma}\sigma_y]_{\mu\nu}. \qquad (1)$$

(The second form has the required symmetry under interchange of spins if $\boldsymbol{d}(-\hat{\boldsymbol{k}}) = -\boldsymbol{d}(\hat{\boldsymbol{k}})$ and implies that $\boldsymbol{d}(\hat{\boldsymbol{k}})$ is a vector under rotations in spin space.) The properties of ψ under rotations are summarized in the identities:

$$(\hat{\boldsymbol{n}} \cdot \boldsymbol{S})\,\psi[\boldsymbol{d}(\hat{\boldsymbol{k}})] = \psi[\mathrm{i}\hat{\boldsymbol{n}} \times \boldsymbol{d}(\hat{\boldsymbol{k}})],$$

$$(\hat{\boldsymbol{n}} \cdot \boldsymbol{L})\,\psi[\boldsymbol{d}(\hat{\boldsymbol{k}})] = \psi\left[\hat{\boldsymbol{n}} \cdot \left(\frac{\boldsymbol{k}}{\mathrm{i}} \times \frac{\partial}{\partial \boldsymbol{k}}\right) \boldsymbol{d}(\hat{\boldsymbol{k}})\right]. \qquad (2)$$

Analysis to date has concentrated on the case of p-wave pairing, though there is as yet no direct evidence supporting $L = 1$ over higher odd L. Neither, however, in spite of some earlier suggestions to the contrary [4], is there any evidence that $L = 1$ is incompatible with what has been observed. Since microscopic calculations of the pairing interaction have been notoriously unreliable, until more detailed experimental probes (and the accompanying theories) of the order parameter are developed, theoretical attention will probably remain focused on p-wave pairing, if only because of the quite formidable mathematical problems posed by a comparably thorough analysis of the f-wave case.

When $L = 1$ the vector $\boldsymbol{d}$ is a linear function of the components of $\hat{\boldsymbol{k}}$ (one assumes, as usual, that pairing is confined to a spherical shell of radius k_{F}, and that the significant dependence of $\boldsymbol{d}$ is on the direction $\hat{\boldsymbol{k}}$):

$$d_\mu(\hat{\boldsymbol{k}}) = \sum A_{\mu\nu}\,\hat{k}_\nu. \qquad (3)$$

If the very weak nuclear dipolar interactions are ignored, then the Hamiltonian is separately invariant under rotations in real space and spin space. Together with gauge invariance this implies that in the absence of a magnetic field, to fourth order in ψ, the free energy f must have the general form [5]:

$$f = \alpha \mathrm{Tr}AA^\dagger + \beta_1 |\mathrm{Tr}A\tilde{A}|^2 + \beta_2(\mathrm{Tr}AA^\dagger)^2 + \beta_3 \mathrm{Tr}(A^\dagger A)(A^\dagger A)^* + \beta_4 \mathrm{Tr}(AA^\dagger)^2 + \beta_5 \mathrm{Tr}(AA^\dagger)(AA^\dagger)^*. \tag{4}$$

The degeneracy of the family of A's that minimize the free energy (4) is reduced by minimizing the dipolar energy V_D. To leading order in ψ this can be shown to have the form:

$$V_\mathrm{D} \propto \int \frac{\mathrm{d}\Omega}{4\pi}(3|\boldsymbol{d}\cdot\hat{\boldsymbol{k}}|^2 - |\boldsymbol{d}|^2), \tag{5}$$

which, for p-wave pairing, gives

$$V_\mathrm{D} \propto |\mathrm{Tr}A|^2 + \mathrm{Tr}A^*A - \tfrac{2}{3}\mathrm{Tr}AA^\dagger. \tag{6}$$

In contrast to the theory of s-wave pairing, in which there is a single fourth order invariant, p-wave pairing is characterized by 5 independent fourth order terms. Consequently the order parameter can be sensitive in important structural features (instead of simply in its numerical amplitude) to the microscopic model from which the β_i are calculated. A full analysis of (4) has yet to be given, and current models for the order parameter are based almost entirely on that subset of possible order parameters satisfying the condition that the matrix ψ be proportional to a unitary matrix. This is equivalent to requiring that $\boldsymbol{d}^* \times \boldsymbol{d} = 0$. Since $(\psi, \hat{\boldsymbol{n}}\cdot\boldsymbol{S}\psi) = -\mathrm{i}\hat{\boldsymbol{n}}\cdot(\boldsymbol{d}^*\times\boldsymbol{d})$, it amounts to neglecting the possibility that there might be a correlation between a non-vanishing pair spin polarization with position on the Fermi surface.

In the simplest weak coupling BCS model this assumption can be justified. There the fourth order term in f is simply proportional to

$$\int \frac{\mathrm{d}\Omega}{4\pi}\mathrm{Tr}(\psi^\dagger\psi)^2 = \int \frac{\mathrm{d}\Omega}{4\pi}[2|\boldsymbol{d}|^4 - (\boldsymbol{d}\times\boldsymbol{d}^*)^2] = \frac{4}{15}[-\tfrac{1}{2}|\mathrm{Tr}A\tilde{A}|^2 + (\mathrm{Tr}AA^\dagger)^2 + \mathrm{Tr}(A^\dagger A)(A^\dagger A)^* + \mathrm{Tr}(AA^\dagger)^2 - \mathrm{Tr}(AA^\dagger)(AA^\dagger)^*] \tag{7}$$

In other words, in the weak coupling model one has the following relations connecting the coefficients β in (4): $-\beta_5 = \beta_4 = \beta_3 = \beta_2 = -2\beta_1$. From the first line of (7) one readily verifies that for $\boldsymbol{d}$ linear in $\boldsymbol{k}$ the lowest f is achieved when $\int(\boldsymbol{d}^*\times\boldsymbol{d})^2\mathrm{d}\Omega/4\pi$ vanishes. There is no general reason to carry this feature of the weak coupling model over to the general case, and the neglect of possible subtle spin polarization effects that it entails reflects only the present lack of a more thorough analysis rather than any compelling argument.

If the "unitarity" assumption is made, however, a complete analysis of (4) is possible. The results are summarized in table 1. Parameter space divides into 3 regions, specified in the first 2 columns. The third column gives a form of the best equilibrium order parameter, taking into account the reduction in degeneracy by the dipole interaction (6) and the vanishingly small magnetic field along the z-axis. The fourth column gives the energy gap, $\Delta^2 \propto |\boldsymbol{d}|^2$. The fifth contains the zero field susceptibility under the assumption that the quadratic term in $\boldsymbol{H}$ in the free energy has the form:

$$\boldsymbol{H}\cdot AA^\dagger\cdot\boldsymbol{H} \tag{8}$$

with a positive coefficient (to represent the depairing effect of an applied field). There is also a linear term (of undetermined sign)

$$\mathrm{i}\varepsilon_{\mu\nu\lambda}H_\mu(AA^\dagger)_{\nu\lambda} \tag{9}$$

responsible for the linear splitting of the transition [6]. The final column specifies whether the Leggett sum rule [4] for the resonance shift has a non-zero value.

The first row of fig. 1 includes the region of parameter space given by conventional weak coupling BCS theory. The solution, first investigated by Balian & Werthamer [7], is given in the form determined by the dipolar and Zeeman energies: $\boldsymbol{d}(\hat{\boldsymbol{k}})$ is $R\hat{\boldsymbol{k}}$, where R is a rotation through arc cos $(-1/4)$ about the direction of the (vanishingly small) field. The reduced susceptibility and lack of a transverse resonance shift exclude this most obvious model of the A phase, but are consistent with the properties of the B phase. (For a time it was asserted that the Leggett sum rule gave a transverse shift in the B phase, casting suspicion on either the Balian-

Table 1.

Unitary order parameters

$\beta_1+\beta_3$	$\frac{2}{3}(\beta_3+\beta_4+\beta_5)$	d	Gap	χ	NMR shift (sum rule) Trans.	NMR shift (sum rule) Long.
>0 <0	$>\beta_1+\beta_3$ or >0	$\hat{x}\left(\frac{-\hat{k}_x+\sqrt{15}\,\hat{k}_y}{4}\right)+\hat{y}\left(\frac{-\sqrt{15}\,\hat{k}_x-\hat{k}_y}{4}\right)+\hat{z}\hat{k}_z$	isotropic	reduced	no	yes
>0	$<\beta_1+\beta_3$	$\hat{x}(\hat{k}_y+i\hat{k}_z)$	$\hat{k}_y^2+\hat{k}_z^2$	normal	same	
<0	<0	$\hat{x}(\hat{k}_y\cos\gamma+\hat{k}_z\sin\gamma)$	$(\hat{k}_y\cos\gamma+\hat{k}_z\sin\gamma)^2$	normal	$\sin^2\gamma$	$\cos^2\gamma$

Werthamer model [7] or the relevance of the sum rule; however a correct evaluation of Leggett's formula yields zero shift.)

The second and third rows of table 1 describe regions of parameter space in which gaps of the form first discussed by Anderson & Morel [8], in an early attempt to solve the conventional BCS gap equation. Both types of order parameter describe Cooper pairs in a zero eigenstate of the component of the total spin along a fixed axis in the $x-y$ plane. Their gaps are anisotropic and their susceptibilities are not reduced from the normal state values (the vector $\boldsymbol{d}$ can lie perpendicular to the field thereby causing the term in H^2 to vanish). Both types of order parameter, according to the sum rule, should have a transverse resonance shift, but the size of the shift in the third type of phase depends on the remaining degeneracy in the order parameter. Thus both possibilities are consistent with the properties of the A phase, though the stability of the resonance shift can be viewed as evidence in favor of the second row. This is the form of the gap indicated by the microscopic calculations of Anderson & Brinkman [9] who have suggested a way in which the paired state may be determined by the almost ferromagnetic nature of liquid ^{3}He. By associating a part of the interaction between atoms in the liquid with the exchange of paramagnons, and seeking a BCS type of self-consistent solution, Brinkman & Anderson [10] find a free-energy of the form (4) with however the coefficients related by the equations: $\beta_1=-S$, $\beta_2=S(2+\delta)$, $\beta_3=2S$, $\beta_4=S(2-\delta)$, $\beta_5=-S(2+\delta)$. Here δ is a measure of the spin fluctuation effects and S describes the remaining interactions which are assumed to be describable by a weak coupling model, corresponding to eq. (7). It can be shown that the state described in the second row of table 1 is more stable than the isotropic state if $\delta>\frac{1}{4}$. In loose physical terms the state of lowest free energy is obtained by balancing the pairing energy which is most negative when both spin populations are coherently mixed and the spin-fluctuation energy which favors states of large susceptibility. These ideas of Anderson & Morel offer a possible resolution of the difficulties which an apparently stable anisotropic state pose for a simple theory based on eq. (4).

A theory of ultrasonic propagation in the paired state has been given by Wölfe [11], who identifies the attenuation peak with the breaking up of pairs, the binding energies of which are less than $\hbar\omega$ where ω is the frequency of the sound. This analysis also rules out the isotropic Balian–Werthamer state below T_δ since for such a solution the requirement $\omega>2\Delta(T)$ restricts one to a region near T_c an order of magnitude smaller than that observed. A qualitative agreement with experiment is obtained by assuming a state of the Anderson–Brinkman form, averaging the preferred axis over all directions.

Thus the melting curve data, the structure of the Ginzburg–Landau free energy (subject to the unitarity assumption), and the Anderson–Brinkman calculation are all compatible with a model of p-wave pairing for both A and B phases; the A phase having an Anderson–Morel type of order, stabilized by spin-fluctuations, and the B, Balian–Werthamer. To confirm this picture considerable further theoretical and experimental work is required:

(1) A more elaborate calculation of the kind sketched by Anderson & Brinkman is needed. In particular, it does not seem excluded that a careful treatment of real processes may lead to solutions more subtle than the simple BCS type (in which the order parameter is directly related to the energy gap). It would be useful to know how to make correspondence between results based on the simple spin-fluctuation Hamiltonian and a theory of the Landau type, in which both the density of states and the susceptibility are separately enhanced. Even within the spin-fluctuation model in which the interaction is taken to be a short range repulsion, it would be useful to know whether or not vertex corrections introduce terms of the same size as those calculated.

(2) To complement the above, a more systematic study of the minima to the variational problem posed by (4) is needed. There are regions of parameter space in which the unitarity assumption can be demonstrated to fail; the relation of such solutions to experiment has yet to be explored, and the plausibility of the parameters falling in these regions in a microscopic theory has yet to be conclusively excluded.

(3) The slopes of the phase boundaries in a T–H plane of the linearly split A-transition are currently unknown. Their values, along with the associated specific heat discontinuities, reduce the allowed region in β-space, and would supply a stringent test to be met by a microscopic calculation.

(4) A theory of the NMR shift is needed. It is clear from the sum rule that the absorption spectrum for parallel and perpendicular r.f. fields provides the kind of detailed information necessary to diagnose the order parameter. Leggett [12] has recently produced a more detailed theory of the resonance, applicable to non-linear phenomena, in which detailed features of the gap can be further probed, but his analysis is based on an adiabatic approximation the validity of which is difficult to assess. Anderson's phenomenological theory of the resonance [13] is equally informative, but in disagreement with both the theory of Leggett and (at least for the longitudinal resonance) the sum rule.

(5) Experiments (and accompanying theories) that probe in detail the direction dependence of the order parameter are needed; thus the p-wave Balian-Werthamer model of the B phase is unambiguously distinguished from other possibilities by a completely isotropic energy gap, and at the moment there seems no striking ways to distinguish $L=1$ pairing from pairing with higher odd L for either phase other than through this dependence. Sound attenuation anomalies and the NMR properties of the fluid will certainly reflect this direction dependence, but much theoretical effort will be required to extract the information.

Work supported in part by the US Office of Naval Research under contract N0014-67-A-0077-0010, technical report #36; by NSF, grant #GH-36457; and by NSF grant #GH-33637 through the Materials Science Center at Cornell University, MSC Report #2022.

References

1. Osheroff, D D, Richardson, R C & Lee, D M, Phys rev lett 1972, 28, 885.
2. Osheroff, D D, Gully, W J, Richardson, R C & Lee, D M, Phys rev lett 1972, 29, 920.
3. Lawson, D T, Gully, W J, Goldstein, S, Richardson, R C & Lee, D M, Phys rev lett 1973, 30, 541.
4. Leggett, A J, Phys rev lett 1972, 29, 1227.
5. Mermin, N D & Stare, G, Phys rev lett 1973, 30, 35.
6. Ambegaokar, V & Mermin, N D, Phys rev lett 1973, 30, 81.
7. Balian, R & Werthamer, N R, Phys rev 1963, 131, 1553.
8. Anderson, P W & Morel, P, Phys rev 1961, 123, 1911.
9. Anderson, P W & Brinkman, W F, Phys rev lett 1973, 30, 1108.
10. Brinkman, W F & Anderson, P W. Private communication.

11. Wölfe, P, Phys rev lett 1973, 30, 1169.
12. Leggett, A J, Phys rev lett 1973, 31, 352.
13. Anderson, P W, Phys rev lett 1973, 30, 368.

Discussion

Luther: Where are the Goldstone particles?

Ambegaokar: These are the various collective modes which appear in the random phase approximation (see Maki & Ebisawa, preprint).

Anderson: There are a great many of them because of the complexity of the order parameter. Orbit waves and the nuclear resonances ($\varrho = 0$ of spin waves) have already been discussed. De Gennes will discuss this general topic.

Aligned Orbital Angular Momentum in the Low Temperature Phase of ^{3}He

P. W. Anderson

Bell Laboratories, Murray Hill, N. J. 07974, USA, and Cavendish Laboratory,[1] Cambridge, UK

What I would like to do here is two things: (1) to give a bit of background, from a more physical point of view than Ambegaokar has, to the problem of the nature of these anisotropic superfluid states; (2) to say a little more about the resonance problem.

Very soon after BCS came out several groups independently suggested that ^{3}He might undergo a BCS transition, but to an anisotropic state [1, 2].

^{3}He seemed a very likely case because of the form of the interaction. He atoms have a strong repulsive core and a relatively weak long-range van der Waals' attraction. The old theorem that the lowest bound state is always an s state does not hold in the case of BCS theory, because the particles are already forced to have some relative orbital angular momentum by the exclusion principle. Thus it was relatively easy to show that the crucial question is simply whether or not the quasi-particle scattering amplitude is attractive for total momentum zero and some finite angular momentum L at the Fermi surface. For some sufficiently high L, the hard core will become ineffective and the long-range attraction will take over.

In fact, the problem of dealing with the hard cores was essentially equivalent to the Bethe-Goldstone equation technique used in nuclear matter theory, which gives an effective repulsive part of the interaction nearly independent of the wave vector $\boldsymbol{q}$, which averages out for higher angular momentum. All agreed that $L=2$ would furnish an attractive interaction, with $L=1$, and 3 coming in second. Transition temperatures in the range of 10^{-2} degrees were predicted.

I do not know whether my collaborators would agree, but even at the time I was extremely skeptical of the value of making any such quantitative estimate of T_c. The interactions in He are very strong relative to those between electrons, and so all kinds of unexpected renormalization effects could not be ruled out. That indeed turned out to be the case, as we shall see. But that the effective interactions should remain repulsive for all L seemed very unlikely, and so did the possibility of attractive interactions for $L=0$; so I think it is fair to say that anisotropic superfluidity was predicted.

To me, at least, a far more interesting problem than T_c, which as I say could not be predicted with any confidence, was the properties of such a state once it had occurred. The present theory of these properties is based on the work Morel did with me for a Ph.D. thesis on this question [3, 4], as improved upon by the monumental paper of Balian & Werthamer [5] which corrected our treatment of the spins and improved many aspects of the formalism.

The equation determining the transition temperature is the condition for determining the stability of the normal state against the appearance of a small pair amplitude $\langle c_{k\sigma} c_{-k\sigma'} \rangle = (X_k)^{\sigma\sigma'}$, and is therefore linear. This leads to two simplifications. (1) Different total angular momenta L separate: a simplification which remains a valid approximation so long as $T_c \ll E_F$; (2) it is the same for all functions belonging to a given L, $X_k = X^0 Y^L(\hat{k})$.

This second point is *not* the case below T_c, where the equation determining the pair function

---

[1] Work at the Cavendish Laboratory supported in part by the Air Force Office of Scientific Research Office of Aerospace Research, US Air Force, under grant No. 73-2449.

is nonlinear and therefore only some special spherical harmonics furnish valid solutions to it, and different solutions exist, which then must be compared on the basis of their total free energies.

To demonstrate these complications let me present a few equations from the Balian–Werthamer treatment. We start from a pair interaction potential in which we have already separated out the L'th component, and replaced $V(\boldsymbol{k})$ by a constant pseudopotential at the Fermi surface.

$$V=\tfrac{1}{2}V_L\sum_{kk',\sigma\sigma}\mathrm{P}_L(\hat{\boldsymbol{k}}\cdot\hat{\boldsymbol{k}}')\,a^+_{k\sigma}a^+_{-k\sigma'},a_{-k'\sigma'}a_{k'\sigma}.$$

We assume there may be a pair amplitude

$$X^{\sigma\sigma'}(\boldsymbol{k})=\langle a_{-k\sigma'}a_{k\sigma}\rangle$$

which will introduce an anomalous self-energy or energy gap

$$\Delta^{\sigma\sigma'}(\boldsymbol{k})=V_L\sum_{k'}\mathrm{P}_L(\hat{\boldsymbol{k}}\cdot\hat{\boldsymbol{k}}')\,X^{\sigma\sigma'}_{k'}.$$

This will then lead to a 4×4 secular equation for the quasiparticle energies E_k

$$H_k=\begin{pmatrix}\varepsilon_k & 0 & & \\ 0 & \varepsilon_k & \multicolumn{2}{c}{\Delta^+_k} \\ & & -\varepsilon_k & 0 \\ \multicolumn{2}{c}{\Delta_k} & 0 & -\varepsilon_k\end{pmatrix}$$

The energies occur in $\pm$ pairs, satisfying the 2×2 matrix equation obtained by squaring H_k,

$$E_k^2=\varepsilon_k^2+\Delta\Delta^+$$

which will give Kramers degenerate pairs only if $\Delta\Delta^+$ is proportional to the unit matrix (the "unitary" case).

Finally, we determine X_k by the equation (for the unitary case)

$$X_k=\frac{\Delta_k}{E_k}\tanh\tfrac{1}{2}\beta E_k$$

(a slight generalization in the nonunitary case) which gives us a complete self-consistent set of equations.

The anticommutation relations of the a's show us that when Δ is odd in $\boldsymbol{k}$, it must be a symmetric matrix with three independent components—this is the spin triplet case. If it is even in $\boldsymbol{k}$ (L = even) then $S=0$ and Δ is antisymmetric. Thus we have two possibilities

$$L\text{ odd: }\Delta_L=\begin{pmatrix}d_u(\boldsymbol{k})+\mathrm{i}d_v(\boldsymbol{k}) & d_w(\boldsymbol{k})\\ d_w & d_u-\mathrm{i}d_v\end{pmatrix}$$

$$L\text{ even: }\Delta=\begin{pmatrix}0 & d\\ -d & 0\end{pmatrix}$$

We have written the first Δ_L in such a way that if all three components of the vector d_u, d_v, d_w are real, Δ_L is "unitary". d_u, d_v, d_w, which are functions in $\hat{\boldsymbol{k}}$ space (which we may call orbital space), transform like a vector under rotations in *spin* space: they are essentially the amplitudes of the pair $\boldsymbol{k}$, $-\boldsymbol{k}$ along three orthogonal directions. In the absence of magnetic forces, the energy is wholly independent of the relative orientation of the triad *uvw* describing spin space relative to any coordinate system (x, y, z) we may choose in orbital space.

Morel and I soon arrived at a very useful heuristic way of understanding the search for solutions. There is in fact little or no difference between the problem near T_c where an expansion of the gap equation in powers of the function d may be made, or at absolute zero. Those states are relatively unstable in which the gap is small in a large fraction of $\boldsymbol{k}$-space, essentially on the basis that any gapless electron will improve its energy by participating in the condensation if it can. Near T_c, one tries to minimize $\overline{|\Delta|^4}/(\overline{|\Delta|^2})^2$; at lower temperatures, a different, logarithmic function of Δ^2, with similar properties.

Thus what we saw immediately, in the $L=2$ case, or any case of a single spin component, was that it was always going to be advantageous to use *complex* d's; complex gap functions, since a single real function (such as k_z for $L=1$, or $3k_z^2-k^2$ for $L=2$) has great areas of small gap, which can easily be filled in by adding $\mathrm{i}k_y$ in the one case, $\mathrm{i}(k_x^2-k_y^2)$ in the other. But this immediately leads us to a state which has fascinating dynamical properties, at least microscopically and sometimes macroscopically. In any BCS theory the existence of anomalous Green's functions $X_k^{\sigma\sigma'}$ implies the existence of a bump in the density–density correlation functions $\langle n_\sigma(0) n_\sigma(\boldsymbol{r})\rangle$. We found in fact that

$$\langle nn\rangle \propto |I(\hat{\boldsymbol{r}})|^2$$

where $I(\hat{\boldsymbol{r}})$ is proportional to the appropriate spherical harmonic in $\hat{\boldsymbol{r}}$ space: if

$$d_i(\boldsymbol{k}) \propto f_L(\hat{\boldsymbol{k}})$$

$$I(\boldsymbol{r}) \propto f_L(\hat{\boldsymbol{r}}).$$

If $f_L(\hat{\boldsymbol{r}})$ is *complex*, there is in addition to the correlation $\langle n_\sigma n_{\sigma'}\rangle$, a correlation $\langle n_\sigma(0) j_{\sigma'}(\boldsymbol{r})\rangle$, where if

$$I = |I| e^{i\gamma},$$

$$\langle nj\rangle \propto (\nabla\gamma)\,|I(\boldsymbol{r})|^2.$$

This means essentially that for complex gap functions the $\langle nn\rangle$ correlation bump is *rotating* This can easily be shown to imply a *finite total orbital angular momentum* if f_L correponds to a function with such a momentum; an orbital ferromagnet. In such a substance, for instance, if one had a space-varying $f_L(\boldsymbol{r})$, one would observe orbital supercurrents according to the relation

$$\boldsymbol{j} = \nabla \times \boldsymbol{L},$$

which is included in the results de Gennes will discuss. As de Gennes points out, this implies a strong interaction of supercurrent flow with L domains, which is undoubtedly the source of the heat current anomalies observed by Wheatley in the A phase [6].

Unfortunately, with the appearance of Balian–Werthamer the one case in which we expected an orbital ferromagnet evaporated, because as they showed, for $L = 1$ the lowest state of the simple gap equation uses the three independent real $L = 1$ orbital functions for d_u, d_v, and d_w. This is a unique case; for all $L \geqslant 2$ one always gains by using complex d and the "unitary" assumption is wrong. For instance, the best solution for $L = 3$ is

$$d_u = 3A(5k_z^3 - 3k_z) + iBk_z(k_x^2 - k_y^2)$$

and for $u \to v \to w$, $x \to y \to z$. But this does not have a net orbital momentum, nor does the corresponding solution for $L = 2$.

It is an amusing—but I suspect not general—result that all of the best solutions so far found for $L = 1$, 2, and 3 have the identical point-group symmetry when one considers total (spin and orbital) space—they belong to Γ_0 of the cubic point group. That is, they are transformed into themselves, not into equivalent states, by $x \to y \to z$ permutations. This has not been obvious of B–W but is in fact true, though one may have to rotate spin space to make it true. B–W is *not*, in fact, isotropic, therefore, since most rotations do not carry it into a multiple of itself.

Now, finally, we come to the exciting point: the discovery by Osheroff et al. [7] that ³He exhibited not just one but two transitions in the Fermi liquid region, and that the upper of these undoubtedly went into an anisotropic state, because of the appearance of a nuclear resonance shift. The experimental data, as they came out, increasingly favored the suggestion of Leggett that the A phase was, first, odd L, and second, not a three-component, real, Balian–Werthamer state, but what was then called an Anderson–Morel state, in which only one or two of the three spin components d_u, d_v, d_w appeared, and they were indeed complex of the form

$$d_u \propto k_x + ik_y.$$

What were these data?

First, the nuclear spin susceptibility. It is in fact very easy to understand the susceptibility from the effective single-particle Hamiltonian I have already written down. In this Hamiltonian we insert the perturbation due to the magnetic field:

$$H_k \to \begin{pmatrix} \varepsilon_k - \boldsymbol{\sigma}\cdot\boldsymbol{H} & \Delta_k^+ \\ \Delta_k & -\varepsilon_k + \boldsymbol{\sigma}\cdot\boldsymbol{H} \end{pmatrix}$$

In the original AM state for odd L in which spin-up and spin-down pairs are independent, $\sigma_z H_z$ simply shifts the energies ε_k for $\sigma = +$ and $-$ independently, and the susceptibility is just the same as the normal one. In BCS, on the other hand, or any L even state, Δ connects the "wrong" spins and the energy which appears in the quasi-particle energy expression is unshifted because it is the sum of $\varepsilon_{k\uparrow}$ and $\varepsilon_{k\downarrow}$. Thus no response to the external field occurs: at $T = 0$, $\chi \to 0$. The simple way to think of this is to ask: can Δ and $\boldsymbol{\sigma}\cdot\boldsymbol{H}$ be diagonalized simultaneously by a spin axis rotation or maximally not? The math is a little funny because Δ transforms by $U\Delta\tilde{U}$ while $\boldsymbol{\sigma}\cdot\boldsymbol{H}$ goes by $U\boldsymbol{\sigma}\cdot\boldsymbol{H}U^{+}$, but the idea is obvious. In general, the rule is: the susceptibility is a tensor and, the values in the direction u, v, or w corresponding to the energy gap vector $\boldsymbol{d}$ components decrease. This decrease should be especially striking because of the "Stoner enhancement" of a factor of four or so which will be lost when χ decreases.

Thus when χ did not change, and also because of the nuclear resonance, we had to assume an odd-L anisotropic state. To see how this could come about we have to make two points, which will be explained much better by Brinkman. (A) In the presence of the strong hard-core interactions which are responsible for the susceptibility enhancement, T_c for $S=1$ will be enhanced, for $S=0$ decreased. A simple physical way to see this is to make three observations: (1) hard cores are much more effective for singlets than triplets, because of the exclusion principle; for very small hard cores the exchange diagram completely cancels the interaction for parallel spins. (2) We have already noted that superfluidity for $L=1$ increases the parallel spin correlations locally. (3) The relatively low compressibility means that total density fluctuations are small; thus a parallel spin correlation which is positive implies a decrease in antiparallel spin correlation, which is reflected in relatively large T_c because the repulsive interactions are reduced in this way.

The change in susceptibility caused by the pairing, however, makes the state *less* stable than it would be expected to be in weak coupling theory. This tendency to "cut its own throat" was, however, (B) shown by Brinkman and myself not to be operative for some of the anisotropic states, and in particular, as might be expected, for any state in which only one component of d is finite. For instance, a convenient example is the "ABM" state

$$\Delta=\begin{pmatrix}0 & d_w\\ d_w & 0\end{pmatrix},$$

(i.e. spin component $\alpha\beta+\beta\alpha$) for which $\chi_{ww}\to 0$ as $T\to 0$ but χ_{uu} and χ_{vv} remain normal. Bill Brinkman will discuss the fact that we can produce quantitative agreement with a number of data, particularly the specific heat jumps, using these ideas.

While spin fluctuation theory is suspect in its application to phenomena at high energy and large q, the application to the ABM–BW differences involves only wavelengths of order the coherence length (100 Å or more) and ω's of order Δ, i.e. a few millidegrees. In this regime a Landau Fermi-liquid version of SF theory is valid, as Pethick has shown, and it is not surprising that our results are quantitative.

Finally, I would like to discuss a few ideas on the dynamics of these systems, with application to the problem of nuclear spin resonance.

The fascination of the problem lies in the very complicated nature of the order parameter; actually, it contains, in the case $L=1$, nine complex numbers or 18 parameters. If we specialize to the ABM state, this can be replaced (aside from magnitudes of Δ) by two directions in space, that of $\boldsymbol{w}$ and that of $\boldsymbol{L}$, and the phase of the order parameter (which as de Gennes remarks is equivalent to a rotation about $\boldsymbol{L}$). The condensation energy itself is completely invariant under rotations of $\boldsymbol{w}$ and of $\boldsymbol{L}$ independently, but there is a tiny dipolar energy which couples the spins to the orbits via the spin-dependent density–density correlation function. It is easy to see that this will make $\boldsymbol{L}$ prefer to be along $\boldsymbol{w}$,

$$F_{dd}=-V_d\cos^2(\hat{\boldsymbol{L}}\cdot\hat{\boldsymbol{w}}).$$

A second semiphenomenological energy we have already discussed: the anisotropic susceptibility. If there were an external magnetic field and we were to pin down the gap function to a direction $\boldsymbol{w}$, the free energy must satisfy

$$\begin{aligned}F&=\tfrac{1}{2}\boldsymbol{H}\cdot\boldsymbol{\chi}\cdot\boldsymbol{H}\\&=\tfrac{1}{2}(\boldsymbol{S}\cdot\boldsymbol{\chi}^{-1}\cdot\boldsymbol{S})\\&=\tfrac{1}{2}\chi_N^{-1}(S_u^2+S_v^2)+\frac{1}{2\chi^{-1}}(S_w^2).\end{aligned}$$

Finally, there are forces which may interact with $\boldsymbol{L}$, but these will be very weak; nonetheless one could expect a definite effect of supercurrent, heat current, and boundaries on $\boldsymbol{L}$.

The gap function itself, one may guess from experience in superconductivity, probably relaxes according to some quasiparticle relaxation time, which may be slow at low temperatures but is probably not very much so in the A phase, relative to NMR frequencies. Just above T_c it is about 10^{-6} s, if we equate it with the τ which appears in zero sound. But there will be two quantities which must relax very slowly, because of conservation laws: $\boldsymbol{L}$, because it is a macroscopic angular momentum, and $\boldsymbol{S}$, for the same reason and also because of the weakness of spin–spin forces. These quantities can, to a good approximation, only change by the application of torques, which can only come from the angle-dependent free energies we have already introduced as well as from the external field $\boldsymbol{H}$;

$$\frac{d\boldsymbol{L}}{dt} = \text{torque} = \frac{\partial F}{\partial \hat{\theta}_L}$$

$$\frac{d\boldsymbol{S}}{dt} = \text{torque} = \boldsymbol{S} \times \boldsymbol{H} + \frac{\partial F}{\partial \hat{\boldsymbol{\theta}}_S}$$

The spin-quantization direction w, however, has no intrinsic conservation law and is just a property of the gap function. If ω is low compared to its relaxation rate, we may assume that $\boldsymbol{w}$ takes its equilibrium direction instantaneously, which will in general be a compromise between V_d and $\chi^{-1}S^2$. At high frequencies, it follows S, which gives Leggett's formula [8]

$$\omega^2 = \omega_0^2 = \frac{V_d}{\chi},$$

but at low fields and frequencies the coefficient gets multiplied by $H^2/(H^2 + H_c^2)$, where H_c is of order a few gauss.

At low enough T, there is likely to be a collective mode of $\boldsymbol{w}$ with a frequency in the general range of Leggett's frequency shift, but as far as I know not necessarily precisely at it. It is likely that Leggett's theory is qualitatively, and perhaps even quantitatively, right in this regime.

If I were to make a best guess at the correct theory at this time, I would suggest that in the A phase, which we identify with the ABM state, the equation of motion of the vectors $\boldsymbol{T}$ (in Leggett's notation) or of $\boldsymbol{w}$ (in mine) should contain both a relaxation term which reduces to my result at low frequencies, and the kinematic terms of Leggett's theory. That is, I propose the following set of equations:

$$\frac{d\boldsymbol{S}}{dt} = \boldsymbol{S} \times \boldsymbol{H} + (\text{torque})$$

$$\frac{d\boldsymbol{w}}{dt} = \boldsymbol{w} \times \left(\boldsymbol{H} - \frac{\boldsymbol{S}}{\chi}\right) - \frac{1}{\tau}(\boldsymbol{w} - \boldsymbol{w}_0)$$

where "torque" is the derivative appearing in my L equation essentially $\delta F/\delta \hat{\boldsymbol{w}}$. My theory does not give longitudinal resonances. I am not sure if this is correct but it probably is if $\Delta\omega < 1/\tau$.

Since in the BW case χ is isotropic, we must assume that $\hat{\boldsymbol{w}}$ is not affected by the external magnetic fields, a.c. or d.c., and retains its equilibrium orientation. Then there will be no resonance shift. To make this fact come out in a nonphenomenological way is very hard and I do not think has been done yet.

It is to be expected that many phenomena will interact with the "orbital ferromagnetic moment" $\boldsymbol{L}$, and because of the above ideas all of these will affect NMR strongly. Thus the NMR studies probably offer a unique opportunity for detailed confirmation of the nature of the state, and discovery of further new physical phenomena, such as the longitudinal resonance suggested by Leggett.

I would like to close with a reminder of the extraordinary good fortune we have had to find that the state actually exhibits these wild possibilities of which we caught a glimpse 13 years ago.

References

1. Pitaevskii, L P, Sovj phys JETP 1960, 37, 1267.
2. Brueckner, K A, Soda, T, Anderson, P W & Morel, P, Phys rev 1960, 118, 1442.
3. Anderson, P W & Morel, P, Phys rev lett 1960, 5, 136.
4. — Phys rev 1961, 123, 1911.
5. Balian, R & Werthamer, N R, Phys rev 1963, 131, 1553.
6. Wheatley, J, Talk at US–USSR symposium, Berkeley, Calif., 1973.
7. Osheroff, D D, Gully, W J, Richardson, R C & Lee, D M, Phys rev lett 1972, 29, 920.
8. Leggett, A J, Phys rev lett 1972, 29, 1227.

Discussion

Doniach: Is the energy gap anisotropic? In this case the sound attenuation should depend on the direction of sound relative to an applied magnetic field. This would presumably be a good test of anisotropy of the ground state.

Anderson: Yes, and I understand it will be done. But the Wölfle & Patton results differ in detail on this point.

Fisher: Is there at present any direct experimental evidence of anisotropy?

Anderson: What do you mean by "direct"? The nuclear resonance shift seems direct to me.

Martin: If the spin fluctuations are so effective in making the A state favorable to the B, can we understand why the B-transition takes place when the temperature is slightly lowered?

Anderson: We indeed do not understand why. This remains the major unsolved problem in the formalism.

NMR in an Anisotropic Superfluid (^3He)

A. J. Leggett

Cornell University, Ithaca, N.Y. 14850, USA and University of Sussex, Falmer, Brighton, BN1 9QH, UK

In this contribution I do not aim to discuss this topic in any great detail (see [1–3]) but just to comment on a few of the important physical ideas involved.

To remind you of the basic experimental facts known at present [4] about NMR in liquid $^3He - A$ and B: in the normal phase the resonance line is a very narrow line at the free Larmor frequency $\omega = \omega_L \equiv \gamma\varkappa_0$. As we enter the A phase, the line shifts to progressively higher frequencies, apparently according to the formula

$$\omega^2 = \omega_L^2 + \omega_0^2(T) \tag{1}$$

where $\omega_0^2(T)$ is proportional to $T_A - T$ near the temperature T_A of the A transition. However, it should be emphasized at this point that the values of ω_L used to date are such that ω_L is always greater than $\omega_0(T)$ at all T; thus, we cannot necessarily conclude that (1) is a general formula. As we go into the B phase, the resonance frequency drops discontinuously back to ω_L. The susceptibility appears to have approximately at least the normal value in the A phase and to drop discontinuously at B to about half the normal value.

What general conclusions can we draw about the A phase from the existence of the NMR shift? First, the effect must be due to the dipole forces, since these are the only significant forces in the problem which do not conserve total spin. If we assume that all the absorption is concentrated in the single observed (shifted) line, then sum-rule arguments show [1, 2] that (crudely speaking) the expectation value of the dipole energy, $\langle H_D \rangle$, must be large—in fact of order $g_D \equiv \mu_N^2/a^3$ rather than of order g_D^2 ($\mu_N \equiv$ nuclear magneton, a = typical atomic spacing). This in turn implies that the A phase must possess the property of "spontaneously broken spin-orbit symmetry" [5]; the system is so ordered that all pairs of particles i, j with the same direction of total spin $\boldsymbol{S}_{ij}$ must have also the same direction of relative orbital angular momentum $\boldsymbol{L}_{ij}$. It is the "strong" (hard-core and van der Waals) forces which produce this effect, but because of their invariance properties they cannot fix the direction of $\boldsymbol{L}_{ij}$ relative to $\boldsymbol{S}_{ij}$: this task is performed by the dipole forces, despite their weakness (see [5] for details).

A specific microscopic mechanism which will give rise to this kind of ordering is the BCS-like anisotropic superfluid phase which was predicted many years ago [6–8]. It is clear from the above that the Cooper pairs must form with $S = 1$ and hence odd L. It turns out that if the theory developed in [3] is correct, the NMR characteristics should help us to decide exactly what form the order parameter has.

To describe the NMR behaviour we can work with the following semi-classical variables: the total spin $\boldsymbol{S}$, and a vector $\boldsymbol{T}(\boldsymbol{n})$ defined (apart from constants) by

$$\boldsymbol{T}(\boldsymbol{n}) \equiv \sum_{|\boldsymbol{p}|} \sum_{\alpha\beta} a_{-p\alpha}(\sigma_2\,\boldsymbol{\sigma})_{\alpha\beta}\, a_{p\beta} \tag{2}$$

$\boldsymbol{T}(\boldsymbol{n})$ describes the Cooper pairs which form on the Fermi surface at the points ($\boldsymbol{n}$, $-\boldsymbol{n}$). Its magnitude gives the pair condensation amplitude: its direction also has a simple interpretation, at least in the so-called "unitary-gap" case [9] in which $\boldsymbol{T}(\boldsymbol{n}) \times \boldsymbol{T}^*(\boldsymbol{n}) = 0$. In this case the Cooper pairs at point $\boldsymbol{n}$ are always in the eigenstate $S = 1$, $\boldsymbol{S}\cdot\boldsymbol{a} = 0$ for some axis $\boldsymbol{a}$ (in general a function of $\boldsymbol{n}$), and $\boldsymbol{T}(\boldsymbol{n})$ lies along this axis. $\boldsymbol{T}(\boldsymbol{n})$ is an operator: its expectation value in equilibrium is closely related to the vector $\boldsymbol{d}(\boldsymbol{n})$ defined by Balian & Werthamer [9]. If $\boldsymbol{T}$ is in a common direction for all $\boldsymbol{n}$ (as in the state proposed by

Anderson & Brinkman [10], then the direction of $\boldsymbol{T}$ is Anderson's vector $\boldsymbol{w}$.

In particular, notice that if $\boldsymbol{T}$ lies in the xy-plane then the Cooper pair wave function is (as far as the spins are concerned) of the general form

$$\psi \sim c_{\uparrow} |S=1, S_z=1\rangle + c_{\downarrow} |S=1, S_z=-1\rangle$$

with no $S_z=0$ components (the so-called equal-spin-pairing or ESP state); the direction of $\boldsymbol{T}$ in the xy-plane gives the phase relation between $c_{\uparrow}$ and $c_{\downarrow}$. In particular, if $\boldsymbol{T}$ rotates around the z-axis, then the phase of the "up" pairs changes relative to that of the "down" pairs.

To construct equations of motion for $\boldsymbol{S}$ and $\boldsymbol{T}(\boldsymbol{n})$ we make the following assumptions. First, in the dipole Hamiltonian we keep only the term involving Cooper pairs. Secondly, in the rest of the Hamiltonian we make a "Born-Oppenheimer" type of approximation, arguing that for motion on the time-scale of typical NMR frequencies the system has plenty of time to follow adiabatically, so that we can write the energy as a function only of the conserved or nearly conserved quantities, i.e. of $\boldsymbol{S}$ and $\boldsymbol{T}(\boldsymbol{n})$. We therefore write [3]

$$H = \gamma^2 \chi^{-1} \boldsymbol{S}^2 - \gamma \boldsymbol{S} \cdot \boldsymbol{\mathcal{H}} + H_{\mathrm{D}}\{\boldsymbol{T}, \boldsymbol{T}^*\} \quad (\boldsymbol{\mathcal{H}} \equiv \text{external field}) \tag{3}$$

Moreover, we have the usual commutation relations between the components of $\boldsymbol{S}$, and from the definition (2) also

$$[S_{\mathrm{i}}, T_{\mathrm{j}}(\boldsymbol{n})] = i\hbar \varepsilon_{\mathrm{ijk}} T_{\mathrm{k}}(\boldsymbol{n}) \tag{4}$$

This kinematical relation between $\boldsymbol{S}$ and $\boldsymbol{T}$ plays an essential role in the problem. It is the analogue of the statement for superconductors that the particle number and condensate phase are conjugate variables.

After neglect of terms of order $g_{\mathrm{D}}/\varepsilon_{\mathrm{F}}$ relative to those kept, we obtain from (3) and (4) the equations of motion of $\boldsymbol{S}$ and $\boldsymbol{T}$. That for $\boldsymbol{T}$ is:

$$\frac{\mathrm{d}\boldsymbol{T}(\boldsymbol{n})}{\mathrm{d}t} = \boldsymbol{T}(\boldsymbol{n}) \times \boldsymbol{H}, \quad \boldsymbol{H} \equiv -\frac{\partial E}{\partial \boldsymbol{S}} = \gamma \boldsymbol{\mathcal{H}} - \gamma^2 \chi^{-1} \boldsymbol{S} \tag{5}$$

Thus, $\boldsymbol{T}(\boldsymbol{n})$ is constant (actually, apart from a trivial phase factor $\mathrm{e}^{2i\mu t/\hbar}$) in equilibrium, since then $\boldsymbol{H} \equiv 0$. Eq. (5) is actually just the analogue of the Josephson phase equation for superconductors. To see this, consider the case when $\boldsymbol{S}=0$, $\boldsymbol{\mathcal{H}} = \mathcal{H}_z \hat{\boldsymbol{z}} \neq 0$. Then (5) says that $\boldsymbol{T}$ precesses around the z-axis, which, as we pointed out above, means that the phase of the up-spin pairs changes in time relative to that of the down-spin pairs. This is exactly what we would expect, since the two chemical potentials are unequal if the spin polarization has not come to equilibrium with the field.

From eqs (3)–(5) we can derive the absorption spectrum [3]. If the r.f. field is polarized perpendicular to the external field (the usual "transverse" case) then we get in general two resonances, with frequencies given by

$$\omega^2 = \tfrac{1}{2}[\omega_{\mathrm{L}}^2 + \Omega_{\mathrm{x}}^2 + \Omega_{\mathrm{y}}^2 \pm \{(\omega_{\mathrm{L}}^2 + \Omega_{\mathrm{x}}^2 + \Omega_{\mathrm{y}}^2)^2 - 4\Omega_{\mathrm{x}}^2 \Omega_{\mathrm{y}}^2\}^{\frac{1}{2}}] \tag{6}$$

while in the "longitudinal" case (r.f. parallel to external field) we get a single finite-frequency resonance at $\omega^2 = \Omega_{\mathrm{z}}^2$. Here the Ω_{i}^2 are the principal values of a tensor function of the equilibrium values of $\boldsymbol{T}(\boldsymbol{n})$, and are proportional to g_{D} (see [3]). The ratios of the three Ω_{i}^2 are characteristic of the particular configuration of the order parameter occurring: in particular, for the state proposed by Anderson & Brinkman [10] $\Omega_{\mathrm{y}}^2 = 0$ and hence eq. (6) reduces exactly to the form (1). In other cases it reduces to (1) only for large ω_{L}.

The "longitudinal" resonance has nothing to do with precession; it describes a coherent tunnelling of pairs from up-spin states to down-spin states and back, the role of the tunnelling matrix element being played by the dipole force. The flow of particles into the up-spin states increases the chemical potential of these states and tends to drive them back again. In fact this excitation can be regarded as a coupled fluctuation of the pair of conjugate variables constituted by the population difference between the "up" and "down" bands and the difference in phase of the Cooper pairs in these bands.

For the Balian–Werthamer (BW) state it turns out (if one makes plausible assumptions [3] about the behaviour of such a state in weak fields) that $\Omega_{\mathrm{x}}^2 = \Omega_{\mathrm{y}}^2 = 0$ but $\Omega_{\mathrm{z}}^2 \neq 0$. Hence there is no effect on the transverse NMR but there is a finite-frequency longitudinal resonance. Thus the existing NMR results do not exclude the possibility that ^{3}He − B is in a BW state, and this can be tested by longitudinal NMR experiments.

One may argue that the above treatment, and in particular eq. (3), is not strictly correct, because there may also be terms in the energy of the general form $(\boldsymbol{S}\cdot\boldsymbol{T})^2$. However, taking such terms into account does not affect the qualitative results, though it may in certain circumstances change the detailed form of eq. (6) when T is far from T_c (see [3]).

In conclusion, if the above theory is correct, NMR should prove one of the most sensitive indicators of the nature of the low-temperature phases of liquid ^{3}He.

References

1. Leggett, A J, Phys rev lett 1972, 29, 1227.
2. — Ibid 1973, 30, 411 (E).
3. — Ibid 1973, 31, 352.
4. Osheroff, D D, Gully, W J, Richardson, R C & Lee, D M, Phys rev lett 1972, 29, 920.
5. Leggett, A J, J phys C. In press.
6. Pitaevskii, L P, JETP 1959, 37, 1794. (Translation: Soviet phys JETP 1960, 37, 1267.)
7. Brueckner, K A, Soda, T, Anderson, P W & Morel, P, Phys rev 1960, 118, 1442.
8. Emery, V J & Sessler, A M, Phys rev 1960, 119, 43.
9. Balian, R & Werthamer, N R, Phys rev 1963, 131, 1553.
10. Anderson, P W & Brinkman, W F, Phys rev lett 1973, 30, 1108.

Discussion

Anderson: It was a surprise to me to discover that Tony Leggett had never seen the equations of antiferromagnetic resonance, since as it turns out his theory, in the ABM case of interest, is virtually identical in all its consequences with the simplest case of antiferromagnetic resonance (including the longitudinal resonance). To give the precise correspondence may be helpful in visualizing the effects. We write a Hamiltonian

$$H = \boldsymbol{H}\cdot\boldsymbol{S} + J\boldsymbol{S}_1\cdot\boldsymbol{S}_2 - H_A(S_{1x}^2 + S_{2x}^2)$$

The x-axis is the preferred axis for $\boldsymbol{S}_1$ and $\boldsymbol{S}_2$, and thus for the antiferromagnetic polarization $\boldsymbol{S}_1 - \boldsymbol{S}_2 = \boldsymbol{D}$. χ is large $\perp$ to $\boldsymbol{D}$, small along $\boldsymbol{D}$, so $\boldsymbol{D} \Leftrightarrow w$, $k\chi$, t; $\hat{\boldsymbol{x}} \Leftrightarrow \boldsymbol{L}$ (i.e. the orbital wavefunction plays the role of the crystal axes. ^{3}He corresponds to a freely-rotating antiferromagnetic crystal which may set $\boldsymbol{x}$ at will).

The equations of motion are (if we define $\vec{H}_A = \hat{H}_A\hat{x}$)

$$\frac{\mathrm{d}\boldsymbol{S}}{\mathrm{d}t} = \boldsymbol{S}\times\boldsymbol{H} + \boldsymbol{D}\times\boldsymbol{H}_A$$

$$\frac{\mathrm{d}\boldsymbol{D}}{\mathrm{d}t} = \boldsymbol{D}\times(\boldsymbol{H} - J\boldsymbol{S}) + \boldsymbol{S}\times\boldsymbol{H}_A$$

the first corresponding exactly to Leggett's equation for $\boldsymbol{S}$, the second to his equation for $\boldsymbol{T}$ if we realize that $\boldsymbol{S}\times\boldsymbol{H}_A$ is always small and $J = 1/\chi$.

It is interesting that also in the antiferromagnetic case one should add a dissipation term if $\boldsymbol{T}$ is high enough, and the usual theory is modified.

Ambegaokar: Mermin and I thought it might be useful to comment that in the special case of the continuous wave unsaturated (CWU) resonance Leggett's equations can be obtained rather simply and transparently, after making the adiabatic approximation introduced by Leggett. Within this approximation, one has the effective Hamiltonian

$$H_{\text{eff}} = \frac{1}{2\chi}\boldsymbol{M}^2 - \boldsymbol{M}\cdot\boldsymbol{H} + H_{\text{dip}} \tag{1}$$

where χ is the static susceptibility, $\boldsymbol{M}$ the magnetisation, $\boldsymbol{H}$ the sum of the external (static) and r.f. fields, and H_{dip} the dipole–dipole energy. Working out the second time derivative of the magnetisation on the basis of (1) and linearizing in the small (for the CWU case) quantities

$\boldsymbol{M}' \equiv \boldsymbol{M} - \boldsymbol{M}_{\text{static}}$, $\boldsymbol{H}_{\text{r.f.}}$ and H_{dip} one obtains

$$\frac{\mathrm{d}^2\boldsymbol{M}'}{\mathrm{d}t^2} = \gamma\frac{\mathrm{d}\boldsymbol{M}'}{\mathrm{d}t}\times\boldsymbol{H}_0 + \gamma\boldsymbol{M}_{\text{static}}\times\frac{\mathrm{d}\boldsymbol{H}_{\text{r.f.}}}{\mathrm{d}t} - \left[[\boldsymbol{M}, H_{\text{dip}}], \left(\frac{1}{2\chi}\right)\boldsymbol{M}^2 - \boldsymbol{M}\cdot\boldsymbol{H}\right] \tag{2}$$

Fourier transforming in time and taking expectation values one immediately has Leggett's equation:

$$(\omega^2 - \Omega^2)\boldsymbol{M}' + \mathrm{i}\omega(\boldsymbol{\omega}_L\times\boldsymbol{M}') = -\chi[\Omega^2\boldsymbol{H}_{\text{r.f.}} + \mathrm{i}\omega(\boldsymbol{\omega}_L\times\boldsymbol{H}_{\text{r.f.}})] \tag{3}$$

Here Ω^2 is the tensor quantity

$$\Omega_{ij}^2 = \frac{1}{\chi}\langle[[M_i, H_{\text{dip}}], M_j]\rangle \tag{4}$$

When written in terms of Leggett's operators $\boldsymbol{T}$, Ω^2 is seen to be exactly the quantity introduced by him.

Remarks on Anisotropic Superfluids

P. G. de Gennes

Collège de France, 11 Place Marcelin-Berthelot, 75231 Paris Cedex 05, France

Let us assume that ^{3}He in the A phase is indeed a superfluid with p state pairing. Then we are immediately faced with a number of amusing questions at the macroscopic level—in particular with the two following:

(1) What kind of permanent currents can we observe? We must have translational currents, possibly associated with an anisotropic ϱ_s. Furthermore, we expect orbital currents, as discussed long ago by Anderson & Morel [1]. Just as in a ferromagnet, where we have a spin current

$$J_{spin} = \text{curl}\, \boldsymbol{M} \tag{1}$$

We might expect to find here an orbital current

$$J_{orbit} = \text{curl}\, \boldsymbol{L} \tag{2}$$

where $\boldsymbol{L}$ is the angular momentum density. In fact, I tend to believe that eq. (2) must be replaced by something slightly more complex for reasons to be explained below.

(2) What are the textures which the anisotropic superfluid will tend to realise in practice? And, in particular, how could we make single domains?

I have tried to guess some of the answers using Landau–Ginzburg equations [2]. The results are very tentative, and leave many obscure points, but they may at least stimulate certain directions of experimental research.

The model used in [2] is the following: the pair amplitude for p state pairing is of the form

$$\langle a_{k\alpha} a_{-k\beta} \rangle = A_{pi} W_p(\alpha, \beta) k_i \tag{3}$$

where α, β are fermion spin indices and the three functions W_1, W_2, W_3 are three spin states for the spin $S = 1$ of the Cooper pairs. W_1, for instance, is an eigenfunction of the spin operator S_1 with eigenvalue 0. The A_{pi}'s form a set of nine complex parameters defining the order.

In ref. [2] I discussed only one particular form of the ordered phase: namely I assumed that A_{pi} factors into two parts

$$A_{pi} = \Delta_i V_p \tag{4}$$

where Δ_i is a (complex) vector which I tend to call the orbital state vector. I took for Δ_i the Anderson–Morel form [1] namely

$$\Delta = \Delta' + i\Delta'' \tag{5}$$

where Δ' and Δ'' are real, orthogonal, and of equal magnitude $|\Delta|$. These conditions can be summarised by imposing $\Delta^2 = 0$). If we introduce a reference frame $(x_1 x_2 x_3)$ with x_1 along Δ', x_2 along Δ'', and x_3 along the third direction, we see that the $\boldsymbol{k}$ dependence of the pairing amplitude is of the form

$$k_1 + ik_2$$

This represents a state with orbital momentum along x_3. (For this reason, it will often be convenient to introduce a unit vector $\boldsymbol{l}$ along x_3.)

The spin state vector, V_p was chosen to be real, (and we can always put $V^2 = 1$). This choice differs from the old Anderson–Morel picture [1]. In Ambegaokar's notation (p. 97), it corresponds to the Y_1 model. It also coincides with the modified model "ABM" of Anderson & Brinkman ([3] and pp 103, 116). The interpretation of the resonance shift in the A phase, either through the Leggett equations (pp 109–111) or through Anderson's assumptions on relaxation processes [4] appears to favor the ABM model against the AM. Thus (more or less by chance) my choice in [2] was probably the right one.

Let us now consider distorted states, where the order parameter A_{pi} varies (slowly) in space. Neglecting dipole–dipole forces for the moment, and assuming zero external field, the most general form for the free energy F (invariant separately by rotations in spin space and in orbital space) is

$$F = F_2 + F_4 + \tfrac{1}{2}K_L \operatorname{div} A_p \operatorname{div} A_p^* + \tfrac{1}{2}K_T \operatorname{curl} A_p \cdot \operatorname{curl} A_p^* \qquad (6)$$

Here F_2 and F_4 are the quadratic and quartic term discussed in refs [3] and [4]. A_p is the vector of components A_{pi}. Eq. (6) displays two separate invariants for the gradient terms. A third invariant $\partial_i A_{pj} \partial_i A_{pj}$ can be lumped into the first two by partial integration. I have assumed that the surface terms which this leaves out are not important for the discussion of bulk properties, but this is a delicate point and will deserve further study.

From eq. (6) we can try and derive a superfluid current J_s by the following recipe: we replace the gradients ∇ by $\nabla - i\mathcal{A}$ where $\mathcal{A}$ is a fictitious vector potential, and define J_s through

$$J_s = (\delta F/\delta \mathcal{A})_{\mathcal{A}=0} \qquad (7)$$

I shall write down the result only for cases where eq. (4) holds (with a real V_p). It is

$$J_s = 2\frac{1}{2i}\left[K_L \Delta^* \operatorname{div} \Delta + K_T \Delta^* \times \operatorname{curl} \Delta - \text{c.c.}\right] \qquad (8)$$

where the factor 2 in front (omitted in [2]) comes from the existence of pairs. To get a more physical feeling about these currents, it is convenient to go to the "London limit" where the triad (Δ', Δ'', **1**) is rotated but not locally altered. The triad can be defined then by three Euler angles, as was done in [2]. However, it is slightly more convenient to use the unit vector **1** to define two of the angles, and to define the third rotation angle φ as a rotation around **1**.

This φ is simply the "phase" as it is defined in simpler superfluids, namely the variable conjugate to the numbers of pairs. A multiplication of all the A_{pi}'s by a factor $e^{i\varphi}$ is equivalent to a simultaneous rotation of Δ' and Δ'' in their plane.

Using **1** and φ, we find that the current is naturally split into two parts

$$J_s = J_\varphi + J_1 \qquad (9)$$

The "phase current" J_φ is of the form $J_\varphi = \varrho_s \nabla\varphi$ where ϱ_s is a diagonal matrix in the $(x_1 x_2 x_3)$ frame

$$\boldsymbol{\rho}_s = 2|\Delta|^2 \begin{pmatrix} K_L + K_T & 0 & 0 \\ 0 & K_L + K_T & 0 \\ 0 & & 2K_T \end{pmatrix} \qquad (10)$$

The "orbital current" J_1 is of the form $J_1 = \boldsymbol{C} \operatorname{curl} \boldsymbol{1}$ where

$$\boldsymbol{C} = 2|\Delta|^2 \begin{pmatrix} K_L & 0 & 0 \\ 0 & K_L & 0 \\ 0 & & K_T \end{pmatrix} \qquad (11)$$

Unless K_L and K_T are equal, both ϱ_s and $\boldsymbol{C}$ are anisotropic. This anisotropy should show up in the velocities of second sound or fourth sound, measured in single domain samples [2]. Another consequence of the anisotropy is that J_1 may have a finite divergence. On the other hand, for all modes other than first sound, we may set with good accuracy $\operatorname{div} J = 0$. This leads to a coupling between J_1 and J_φ. As a result, the orbital spin waves [5] (collective modes of **1**) should have elliptical precessions and a modified spectrum, even in the absence of dipolar forces [2].

Let us now discuss briefly the problem of textures. At first sight, in an unoriented sample, we are tempted (by analogy with ferromagnets) to think of domains separated by thin Bloch walls. However, such walls occur only in systems with a discrete set of easy axes. In the present case, even including a magnetic field $\boldsymbol{H}$ and the effects of the dipole–dipole interaction, we expect that **1** has an easy plane normal to $\boldsymbol{H}$ (as explained in ref. [5]. For a continuously degenerate case like this there will not be thin walls, but rather gradual changes in orientation. This case is reminiscent of the anisotropic fluids called nematics. Here, we have a unit vector field $\boldsymbol{n}(\boldsymbol{r})$ labelling the direction of the optical axis. This field may vary smoothly, or may show singular lines (called *disclinations* by Frank [6]): these lines are frequent, and are responsible for the thread-like aspect of usual samples which originated the name of this phase (nema: thread). In the present case, we expect (*i*) singularities of φ (vortices); (*ii*) singularities in the V field; (*iii*) singularities of **1**, which I propose to call *disgyrations*. Let us start with this last type. One might expect at least two forms of disgyration lines:

(*a*) Tangential disgyrations, where the lines of force of **1** are circles surrounding the line. In this case we have permanent currents parallel to the line axis and decreasing like ϱ^{-1}, where ϱ is the distance to the line [2].

(*b*) Radial disgyrations with **1** pointing towards the line (for instance) and no permanent currents. The relative stability of the two types will depend on the ratio K_L/K_T.

There is one interesting difference between disgyrations and disclinations. The optical axis in a nematic is non polar: the states described by two unit vectors $\boldsymbol{n}$ and $-\boldsymbol{n}$ are identical. On the other hand, the state **1** and its time reversed $-\mathbf{1}$ are different. This has some practical consequences on the possible types of singularities. Consider a closed contour *C* surrounding the line, and follow the unit vector $\boldsymbol{n}$ (or **1**) attached to each point along this contour. If after one turn we recover the same $\boldsymbol{n}$, we say that the line is of integral strength. If after one turn, we find $-\boldsymbol{n}$ instead of $\boldsymbol{n}$, we say that the line is of half integral strength. Both possibilities are allowed in a nematic. On the other hand, for the orbital disgyrations, only integral strengths are allowed.

For the case of nematics, it has in fact been shown that the lines of integral strength are unstable: they disappear by an "escape in the third dimension" [7, 8]: Lines of half integral strength can never escape (for topological reasons). Returning to ^{3}He, we might ask whether the orbital disgyrations, which are of integral strength, also "escape" or not. On the basis of simple elastic calculations, I tend to believe that they do not, but this point should be analysed in much more detail.

Note incidentally that we may also have point singularities of the $\boldsymbol{1}$ field. Similar points have been found in nematics [7, 9].

The singularities of the $\boldsymbol{V}$ field, which I propose to call spin disgyrations, may also be of some physical interest. Of course, as first pointed out by Anderson, the dipolar energy H_D tends to align $\boldsymbol{V}$ parallel to $\boldsymbol{1}$ [5]. However, H_D is weak. To give a numerical meaning to this statement, it is convenient to define a characteristic length e_D (the "dipolar length"), such that, for spatial variations on the scale of e_D, the distorition energy of the $\boldsymbol{V}$ field is comparable to H_D. Returning to the Landau–Ginzburg equations, one finds

$$e_D \sim \xi_0 \left[\frac{E_F}{\mu H_L}\right]^{\frac{1}{2}}$$

where ξ_0 is the low temperature limit of the coherence length, defined as in superconductors, μ is the nuclear magnetic moment and H_L is the dipolar field due to one ^{3}He nucleus on one of its first neighbours. H_L is of order 1 Gauss, and $[E_F/\mu H_L]^{1/2} \sim 10^4$. If $\xi_0 \sim 500$ Å then we expect $e_D \sim 0.5$ mm.

The length e_D is thus rather large. In all cases where the objects considered (e.g. a spin disgyration loop) are small in size when compared to e_D, we may probably omit H_D and consider the $\boldsymbol{V}$ field as decoupled from the $\varDelta$ field: spin disgyrations may be observable.

In connection with them, it is important to observe that the two states ($\boldsymbol{V}$) and ($-\boldsymbol{V}$) are physically identical. (Remember that they are defined by the condition $\boldsymbol{S}\cdot\boldsymbol{V}=0$ where $\boldsymbol{S}$ is the spin of the pairs.) Just as in the case of nematics, this implies that, if we perform a closed contour around a spin disgyration, we may recover either the same $\boldsymbol{V}$ or its opposite: spin disgyrations may be of half integral strength. (Spin disgyrations of strength one will probably be unstable through escape in the third dimension.)

Let us end up with a few words concerning the production of single domain samples: a number of fundamental experiments will require to have single domains. In ref. [2] the possibility of aligning $\boldsymbol{1}$ by an electric field $\boldsymbol{E}$ (or by crossed $\boldsymbol{E}$ and $\boldsymbol{H}$) was mentioned. However, the atomic polarisability of helium is so low that the electric field effects are rather weak. Another possible approach would amount to use rotating magnetic fields (of typical magnitude 10–100 gauss): since $\boldsymbol{1}$ prefers to be normal to $\boldsymbol{H}$, it might then sit in a well defined direction. The rotation regimes can be analysed in principle through Leggett's dynamical equations for the state vector $\boldsymbol{V}$, the magnetisation $\boldsymbol{M}$, plus an equation for $\boldsymbol{1}$. However, the results may be seriously influenced by various relaxation processes associated with $\boldsymbol{V}$ [3] and no definite prediction can be made at the present time.

I have greatly benefited from discussions with P. W. Anderson, W. Brinkman and R. Richardson.

References

1. Anderson, P W & Morel, E, Phys rev 1961, 123, 1911.
2. De Gennes, P G, Phys lett 1973, A44, 271.
3. Anderson, P W & Brinkman, W, Phys rev lett 1973, 30, 1108.
4. Mermin, N D & Stare, C, Phys rev lett 1973, 30, 1135.
5. Anderson, P W, Phys rev lett 1973, 30, 368.
6. Frank, F C, Disc Faraday soc 1958, 25, 1.
7. Williams, C, Pieranski, P & Cladis, P E, Phys rev lett 1972, 29, 90.
8. Meyer, R B, Phil mag 1973, 27, 405.
9. Meyer, R B, Molecular crystals 1970, 16, 355.

Spin Fluctuations and Anisotropic Superfluidity in ^{3}He

W. F. Brinkman

Bell Telephone Laboratories, Holmdel, N.J. 07974, USA

Summary

In this paper the mechanism [1] by which spin fluctuations can stabilize an anisotropic superfluid state is briefly reviewed. Certain particle-particle diagrams omitted in the original calculation are then estimated and shown to give a small contribution to the susceptibility. The energetics of spin fluctuations in ^{3}He are discussed and it is shown that the variation in specific heat discontinuity with pressure is in agreement with the spin fluctuation prediction although the magnitude of the spin fluctuation predictions are somewhat larger than observed values.

The observed properties of the A phase of ^{3}He which occurs below 2.7 mK at high pressures [2] appear to indicate that this phase is an anisotropic superfluid with pairs of odd orbital angular momentum L. If one assumes, invoking the law of trying simplest cases first, that the orbital wave function is a p wave ($L=1$), one is confronted with a theorem due to Balian & Werthamer [3] which says that, for the reduced BCS Hamiltonian, the superfluid p wave state of lowest energy is an isotropic state (the BW state) in which the various possible components of the pair wave function are equally occupied. Therefore, if an anisotropic p wave state is to be stabilized one must find a strong coupling mechanism which lowers some such state relative to the BW state. Such a mechanism has recently been found [1] by considering how the introduction of a BCS type gap feeds back on the effective interactions arising from virtual exchange of spin fluctuations. It was found that the reduction of the susceptibility in the BW state due to the introduction of a gap reduces the effective interaction while if one chooses a state in which all the pairs have $S_z=0$ along some axis then the susceptibility becomes anisotropic and such that the effective interaction for the $S_z=0$ pairs actually increases. If this effect is large enough it can stabilize the anisotropic state relative to the BW state and estimates using the properties of the normal state of ^{3}He indicated that the anisotropic state used in the work of Anderson & Morel [4, 5] (the A–M state) should indeed be stable. In the paper we shall briefly review this calculation in section I. In this calculation the effect of changes in the gap in the presence of a magnetic field were not included, i.e., the particle-particle latter diagrams were not introduced into the susceptibility so there was some question as to their importance in the calculation.[1] In section II we calculate the contribution to the d.c. susceptibility due to the change in the gap in the presence of a magnetic field and show that it is of order (T_c/ε_F) compared to the terms coming from the changes in the quasiparticle spectrum. Since one expects that changes in the gap become less important at finite wavelength particle–particle scattering diagrams are not likely to have any effect on the original calculation.

In section III we discuss some aspects of the energetics of ^{3}He. In particular, using the experimental values of the effective mass and the density for pressures between 23 and 33 atm [6], it is shown that if the effective interaction stayed unchanged the transition temperature would decrease so that $T_c(23\text{ atm}) = 0.49\ T_c\ (33.4\text{ atm})$. The observed T_c decreases by $\sim 10\%$ in this region [7] so that the effective interaction must clearly be varying. The spin fluctuation theory indeed indicates that such is the case so that the slow variation of T_c is not inconsistent with the theory. One can also make a prediction of the variation of the specific heat discontinuity with pressure and it is found that the theoretical result, although somewhat larger in magnitude, varies by approximately the same amount as the recent experimental data of Webb et al. [7]. The agreement between experiment and theory app-

[1] The author would like to thank V. Ambegaokar for bringing this correction to his attention.

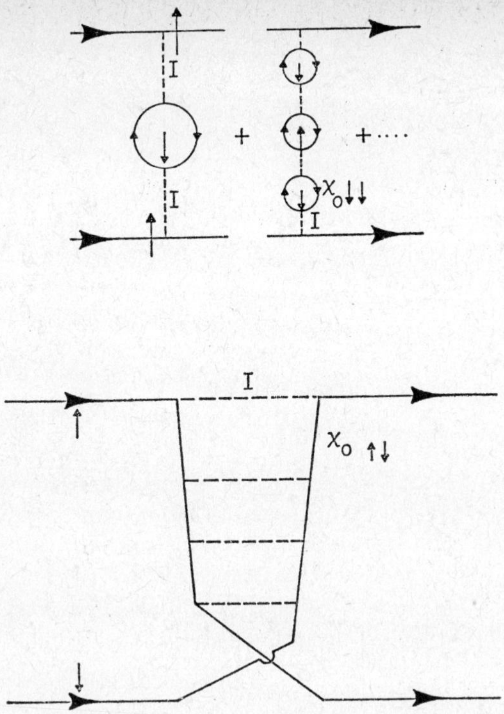

Fig. 1. The two diagrams that represent the exchange of a spin fluctuation.

ears therefore to be quite reasonable. Finally, we remark on the magnetic field dependence of the transition temperature and point out that within the spin fluctuation model one expects a relatively large quadratic field dependence of the transition temperature.

I. Summary of Spin Fluctuation Calculation

In calculating the effects of spin fluctuations on the superfluid properties we use the simple zero range model and assume that the effective interaction I and the mass can be related to the Fermi liquid parameter Z_0 and the measured effective mass m^*. This appears to be reasonably justified on the basis of the Fermi liquid results of Pethick and others [8, 9]. The effective interaction between two ^{3}He atoms is given by the two diagrams in fig. 1. If one takes the expectation value of these in the triplet state of the pairs one finds that

$$I_{\text{eff}}(q, \omega)_t = \tfrac{1}{2}\left[\frac{I}{[1 + I\chi_N^0(q, \omega)]} - \frac{I}{[1 - I\chi_N^0(q, \omega)]}\right] \tag{I-1}$$

where $\chi_N^0(q, \omega)$ is the noninteracting susceptibility. The first term is due to density fluctuations while the second is the spin fluctuation contribution. Since, for repulsive interactions ($I > 0$) the second term is much larger than the first and we simply lump the first term with all the other interactions not included in the model which we denote by V. Inserting eq. (I-1) into the expression for λ_l we obtain

$$\lambda_l = N(0)V_l - \tfrac{3}{2}\ln[1 - I\chi_N^0] \tag{I-2}$$

($\chi_N^0 = \lim_{q\to 0}\lim_{\omega\to 0} \chi_N^0(q, \omega)$) This result was originally pointed out by Layzer & Fay [10]. The spin fluctuation term is quite large. However, such calculations are renowned for their ability to over-estimate the importance of spin fluctuations so that this calculation of an effective coupling constant cannot be taken seriously. Indeed, using the experimental values of the susceptibility enhancement one finds a value of -4.5 for the contribution to λ from the spin fluctuations while the usual estimate of $\lambda_{\text{eff}} = \ln(\omega_s/T_c)^{-1} = 1/5$. $\{\omega_s = (1 - I\chi_N^0)\varepsilon_F$, the spin fluctuation frequency$\}$. This type of overestimate has plagued spin fluctuation theory since its beginning and has never been satisfactorily explained. It could be due to the problem of properly cutting off the momentum integrals [11] or it may be due to the need to introduce vertex renormalization [12]. The former problem is not as severe in the calculation we discuss next but it is not clear what effect vertex corrections might have.

We now examine the change in the effective interaction when the effect of the superfluid gap is introduced into the susceptibilities. Expressions for the susceptibilities have been written down by Balian & Werthamer [3] and we simply use their results. For the states we consider the susceptibility tensor is diagonal if q and ω are sufficiently small, i.e.,

$$\chi_{ij}^0(q, \omega) \equiv \tfrac{1}{2}\,\mathrm{F}\{\langle \mathrm{T}[M_i(q, t), M_j(-q)]\rangle_{\text{w.c.}}\}$$
$$= \delta_{ij}\,\chi_{jj}^0(q, \omega) \tag{I-3}$$

Here $M_i(q)$ is the Fourier transform of the ith component of the magnetization density and the expectation value is taken in the BCS type state. The F indicates Fourier transform with respect to time. Using eq. (I-3) the effective interactions can be written in a relatively simple form by solving the matrix equations for them as was done in the paper by Brinkman & Engelsberg [11] for a ferromagnet. The result is that

$$[I_{\text{eff}}(q,\omega)]_{\uparrow\uparrow} = [I_{\text{eff}}(q,\omega)]_{\downarrow\downarrow} = \tfrac{1}{2}\left(\frac{I}{[1+I\chi^0_{\varrho\varrho}(q,\omega)]} - \frac{I}{[1-I\chi^0_{zz}(q,\omega)]}\right) \quad \text{(I-4 a)}$$

$$[I_{\text{eff}}(q,\omega)]_{\uparrow\downarrow} = \tfrac{1}{2}\left(\frac{I}{[1+I\chi^0_{\varrho\varrho}(q,\omega)]} + \frac{I}{[1-I\chi^0_{zz}]} - \frac{2I}{[1-I\chi^0_{xx}]}\right) \quad \text{(I-4 b)}$$

We have used the additional fact that $\chi^0_{xx} = \chi^0_{yy}$ in the states to be considered and $\uparrow\downarrow$ indicates the $S_z = 0$ component of the triplet. Again the first terms are due to density fluctuations ($\varrho\varrho$ indicates density–density fluctuations) and since these will not change we drop them in further discussion. In evaluating the susceptibilities we consider two states, the BW state [3] and the AM state [4, 5]. These are defined by first writing the order parameter $\Delta_{\alpha\beta}(\boldsymbol{k}) = \boldsymbol{d}(\boldsymbol{k})\cdot(\boldsymbol{\sigma}\sigma_y)_{\alpha\beta}$ (σ_i are the Pauli matrices). Then the BW state is defined by $\boldsymbol{d}(\boldsymbol{k}) = \Delta\hat{\boldsymbol{k}}$ where $\hat{\boldsymbol{k}}$ is a unit vector along $\boldsymbol{k}$, and the AM state is given by $\boldsymbol{d}(\hat{\boldsymbol{k}}) = (\hat{k}_x + i\hat{k}_y)\Delta\hat{z}$. Then if we define

$$\delta\chi^0_{ii} = \chi^0_{ii} - \chi^0_N \quad \text{(I-5)}$$

and we find that

$$\delta\chi^0_{ii}(\text{BW}) = \delta\chi^0 = \chi^0_N \frac{C}{3}\left(\frac{\Delta}{T}\right)^2 \quad \text{(I-6)}$$

where $C = 7\zeta(3)/8$ [3]. For the AM state

$$\delta\chi_{xx} = 0;\ \delta\chi_{zz} = \chi^0_N C\left[\frac{\Delta}{T}\right]^2 \quad \text{(I-7)}$$

Inserting these results into the expressions for the effective interaction one observes that for the BW state the interactions remains independent of S_z but are all reduced while for the AM state the interaction (I-4b) is actually enhanced since χ^0_{zz} is reduced while χ^0_{xx} is unchanged. Since in this state only $S_z = 0$ pairing occurs, (I-4b) is the only interaction that matters.

In order to calculate the change in λ_1 we need to know the q and ω dependence of $\delta\chi^0$. This is obtained by noting that the integrals involved are similar to those occurring in $K(q,\omega)$, the BCS current–current response function [13, 14]. Introducing the coherence length $\xi_0 = \gamma h N_F/\pi^2 T_c$ and using the fact that only the region $q \gg \xi_0^{-1}$ is important we obtain

$$\delta\chi^0_{ii}(q,\omega) \approx \delta\chi^0_{ii}\left[\frac{3\pi}{4q\xi_0}\right] \qquad \omega \lesssim T \quad \text{(I-8)}$$

Inserting the changes in the effective interaction into the expression λ_l we obtain for the BW state [1]

$$\delta\lambda_1(\omega) = \frac{\Delta^2}{\varepsilon_F T_c[1-I\chi^0_N]^{\frac{3}{2}}} \frac{7\zeta(3)\pi^2\sqrt{3}}{64\gamma} \qquad \omega < T \quad \text{(I-9)}$$

$$\simeq \frac{0.1\Delta^2}{\omega_s T_c[1-I\chi^0_N]^{\frac{1}{2}}} \qquad \omega < T \quad \text{(I-10)}$$

For the AM state $\delta\lambda_1$ is of the opposite sign and 3 times larger. Inserting this result into the gap equation gives rise to a term proportional to Δ^4 in the expansion of the free energy [1]. If we define $(\Delta F^s)_{\text{BW}}$ to be this contribution to the free energy in the BW state and $(\Delta F^0)_{\text{AM-BW}}$ to be the difference in fourth order energies between the AM and BW states in the weak coupling theory we obtain

$$(\Delta F^s)_{\text{BW}}/[(\Delta F^0)_{\text{AM-BW}}] \equiv \delta \simeq \frac{8T_c}{\lambda_{\text{eff}}\,\omega_s[1-I\chi^0_N]^{\frac{1}{2}}} \quad \text{(I-11)}$$

Since

$$(\Delta F^s)_{\text{AM}} = -3(\Delta F^s)_{\text{BW}} \quad \text{(I-12)}$$

the two states cross at $\delta = 0.25$. Using the experimental values of the parameter in eq. (I-11) discussed further in section III we find $\delta = 0.7$ which is sufficient to stabilize the AM state.

II. Inclusion of Particle–Particle Scattering

In this section we consider the change in the susceptibilities calculated by Balian & Werthamer due to the inclusion of particle–particle scattering. If one thinks in terms of anomalous

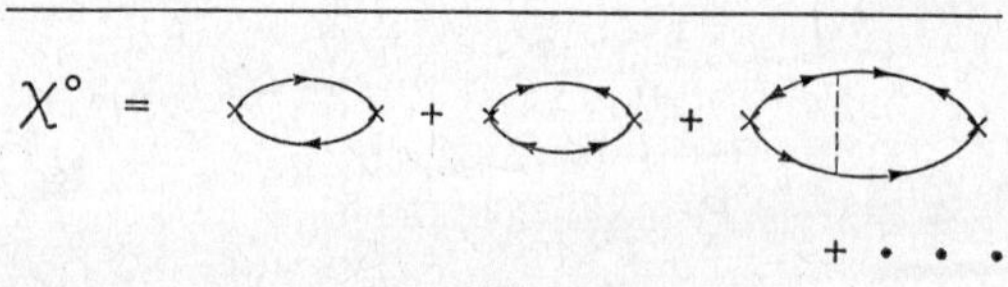

Fig. 2. The various diagrams contributing to $\chi^0(q,\omega)$. Note that the dashed line is the particle–particle interaction, not I. The first two diagrams are included in the calculation by Balian & Werthamer [3]. The calculation in the text estimates the remaining diagrams in the d.c. limit.

propagation lines in diagrams in which the arrows go in opposite directions then the Balian–Werthamer calculation only takes into account the first two diagrams in fig. 2. One knows that to properly describe the collective modes one must include the entire sum [15]. However, the collective modes are rather subtle effects and getting their dispersion correct does not mean that these diagrams have a large effect on the values of the susceptibility. To estimate the effect of these diagrams on the dc susceptibility we can simply calculate the magnetization in the presence of a field by properly solving the self-consistent mean field equations since this calculation is equivalent to the static limit of the generalized RPA [15]. In doing this we follow the notation of Balian & Werthamer [3] and consider the field to be along the axis of quantization so that the diagonal matrix ε_k is now

$$\varepsilon_k = \begin{pmatrix} \varepsilon_k + g\mu H & 0 \\ 0 & \varepsilon_k - g\mu H \end{pmatrix} = \varepsilon_k^0 + h \qquad \text{(II-1)}$$

We then must solve for the magnetization

$$M = \sum_k (\langle a_{k\uparrow}^+ a_{k\uparrow}\rangle - \langle a_{k\downarrow}^+ a_{k\downarrow}\rangle) = \tfrac{1}{2}\sum_k (\omega_{\downarrow\downarrow}^k - \omega_{\uparrow\uparrow}^k) \qquad \text{(II-2)}$$

Where as defined by BW $\langle a_{k\alpha} a_{k\beta}^+\rangle = \frac{1}{2}[\delta_{\alpha\beta} + \omega_{\alpha\beta}^k]$. The procedure here is then the standard one. One expands the selfconsistent equations for $\Delta_{\alpha\beta}(k)$ and $\omega_{\alpha\beta}^k$ to first order in H and calculates eq. (II-2) to obtain the susceptibility. The terms that are first in the field in the gap equation are

$$\Delta_1(k) = -\tfrac{1}{2}\sum_{k'} V_{kk'}\Bigg\{\frac{1}{E_{k'}}\tanh\left(\frac{\beta E_{k'}}{2}\right)$$
$$+ [\varepsilon_{k'}^0 [h\Delta_0(k') + \Delta_0(k')\, h]$$
$$+ \Delta_0^2(k')\,\Delta_1^+(k') + \Delta_0(k')\,\Delta_1(k')\,\Delta_0^+(k')]$$
$$* \frac{1}{2E_{k'}}\frac{\mathrm{d}}{\mathrm{d}E_{k'}}\left(\frac{1}{E_{k'}}\tanh\left(\frac{\beta E_{k'}}{2}\right)\right)\Bigg\} \qquad \text{(II-3)}$$

$\Delta_1(k)$ is the change in gap function in the presence of a field and it is the contribution to the susceptibility given by this change we must calculate. It is important to notice that the driving term involving h is an odd function of $\varepsilon_{k'}^0$ and therefore involves the derivative of the density of states with respect to energy. In fact, if we are only interested in the corrections to the susceptibility which are second order in Δ_0 we can solve (II-3) to obtain

$$\Delta_1(k) = \frac{\mathrm{d}\ln N(\varepsilon)}{\mathrm{d}\varepsilon}\bigg|_{\varepsilon_F} \frac{(\lambda-1)}{2(T_c - T)}[H\Delta_0(k) + \Delta_0(k)\, H]$$

In this expression $N(\varepsilon)$ is the density of states near the Fermi surface. When this result is inserted into eq. (II-2) we find a contribution to the susceptibility

$$\frac{\delta\chi}{\chi_N} = -\left(\frac{\mathrm{d}\ln N(\varepsilon)}{\mathrm{d}\varepsilon}\bigg|_{\varepsilon_F}\right)^2 \frac{(\lambda-1)^2}{2\lambda}\left(\frac{T_c}{T_c - T}\right)$$
$$\times \langle\sigma_z(\Delta_0\sigma_z\Delta_0^+ + \Delta_0\Delta_0^+\sigma_z)\rangle$$

The braces indicates that a trace over spin indices is to be taken along with an average over angles. This term is of the order $(T_c/\varepsilon_F)^2$ compared to the terms calculated by BW. It is therefore completely negligible. Note that this contribution to the susceptibility is just the result obtained if one calculated the response of the order parameter to the terms linear in field first written down by Ambegaokar & Mermin [16]. The divergence near T_c merely reflects the fact that the transition is being split into several separate transitions. Finally, it should be emphasized that these terms do not tend to decrease the anisotropy in the susceptibilities that are essential to the calculations in the previous section [1].

III. Numerical Comparison with ³He

Since the relevant Fermi liquid parameters for ³He have been determined experimentally in the range of 23 to 33 atm [6] it is of some interest to compare the variation of the superconducting properties with theoretical predictions obtained from these parameters. The first calculation one might attempt is to assume $T_c \propto \omega_s e^{-1/N(0)V_1}$ and that the only parameters that are a function of pressure are the density of states $N(0)$ and $\omega_s = (1 + \frac{1}{4}Z_0)\varepsilon_F$. If V_1 is adjusted to fit the value 2.6 mK at 33.4 atm then the resulting predicted transition temperatures are listed as T_c^* in table 1

Table 1.

P (atm)	ω_s(K)	T_c^{exp}(mK)	T_c^*	$C_</C_>$ (exp)	$C_</C_>$ (Th)
33.4	0.27	2.6	2.6	2.64	2.83
29.0	0.28	2.55	1.9	2.54	2.76
23.0	0.30	2.4	1.3	2.48	2.70

where they can be compared with the experimental values. It is, of course, seen that the theoretical values decrease much more rapidly than the experimental values. This fact is due to having not allowed V to vary. It is quite clear that V does indeed vary since Z_0 which is a product of the density of states times the antisymmetric component of the interaction varies by less than 1% between 23 and 33 atm. According to (I-2) the explicit dependence on the density of states is cancelled by the effective interaction for the contribution due to spin fluctuations in λ_1. Therefore, one expects T_c to be much less sensitive to pressure than the above calculation indicates as is indeed the case experimentally.

Using the expression for δ listed in eq. (I-11) and the recent calculation by the author and P. W. Anderson [17] we can predict the discontinuity in specific heat, i.e., the ratio of $C_<$, the specific heat just below the transition to $C_>$ the specific heat just above T_c. In terms of δ this ratio is given as

$$\frac{C_<}{C_>} = 1 + \frac{5}{3} \frac{(1.43)}{(2-\delta)} \tag{III-1}$$

if one assumes that the only strong coupling contribution to the free energy is that due to the spin fluctuations. δ can be calculated by setting $\omega_s = (1+\frac{1}{4}Z_0)\varepsilon_{F_1}$, $[1 - IN(0)] = (1+\frac{1}{4}Z_0)$, and taking $\lambda[\ln(\omega_s/T_c)]$. The theoretical values for $C_</C_>$ are listed in the last column of table 1 alongside the experimental results of Webb et al. [7]. It is seen that although the numbers are about 10% too large, their variation with pressure agrees with the experimental results. This agreement is probably somewhat fortuitous but it is encouraging.

Finally we would like to remark on the magnetic field dependence of the splitting of the transition temperature [16, 17]. If one examines the effective λ_1 due to spin fluctuations one finds that in the presence of a field the λ_1 depends on products of $N(\varepsilon_{F\uparrow})*N(\varepsilon_{F\downarrow})$ which is proportional to H^2. Therefore, one expects the linear term in the splitting of the two transitions to be considerably smaller than that predicted from the change in the density of states [16].

The author acknowledges the close collaboration and stimulating influence of P. W. Anderson. He would also like to thank J. C. Wheatley for providing unpublished values of the Fermi liquid parameters. He has also benefited from conversations with C. M. Varma, E. I. Blount, V. Ambegaokar, and N. D. Mermin.

References

1. Anderson, P W & Brinkman, W F, Phys rev lett 1973, 30, 1108.
2. Osheroff, D D, Gully, W J, Richardson, R C & Lee D M, Phys rev lett 1972, 29, 920.
3. Balian, R & Werthamer, N R, Phys rev 1963, 131, 1553.
4. Anderson, P W & Morel, P, Physica 1960, 26, 671.
5. — Phys rev 1961, 123, 1911.
6. Wheatley, J C. Unpublished.
7. Webb, R E, Greytak, T J, Johnson, R T & Wheatley, J C, Phys rev lett 1973, 30, 210.
8. Pethick, C F & Carneiro, G M. To be published.
9. Amit, D J, Kane, J W & Wagner, H, Phys rev 1968, 175, 313.
10a. Layzer, A & Fay, D, Int j mag 1971, 1, 135.
10b. Bookbinder, H, Phys rev 1971, A3, 372.
11. Brinkman, W F & Engelsberg, S, Phys rev 1968, 169, 417.
12. Hertz, J A & Edwards, D M, Phys rev lett 1972, 28, 1334.
13. Bardeen, J, Cooper, L N & Schrieffer, J R, Phys rev 1957, 108, 1175.
14. Anderson, P W & Suhl, H, Phys rev 1959, 116, 898.
15. Anderson, P W, Phys rev 1958, 112, 1900.
16. Ambegaokar, V & Mermin, N D, Phys rev lett 1973, 30, 81.
17. Brinkman, W F & Anderson, P W. To be published.

Discussion

P. W. Anderson: One slight indication that the B phase in Balian–Werthamer is given by the slope of the AB transition line, which suggests a much more favorable 6th-order (in Δ) term for the B phase than the A. If the B–W phase exists, it can only be by some mechanism which is more effective as the gap gets larger (as e.g. in "gapless superconductivity"). This would be reflected in relatively favorable high-order terms in Δ as is the case.

In general, the existence of B–W in the face of spin fluctuations is the most outstanding discrepancy, but we have some confidence that the use of a frequency-dependent $\Delta(\omega)$, (as in the case of the Abrikosov–Gorkov theory of magnetic impurities) or perhaps some more complicated way of correlating the gap with the spin fluctuations, will overcome this.

Effects of Limited Dimensionality in Solids

One-dimensional Chains in the A-15 Structure and the Properties of Superconductors of this Group

L. P. Gor'kov

L.D. Landau Institute of Theoretical Physics, Academy of Sciences, Moscow, USSR

The high temperature superconductors of the A-15 structure compounds have a number of properties which unit them in a single group. One of the most characteristic features is the "martensitic" transition which takes place in a good sample before the transition into the superconducting phase (see [1] for references). The corresponding temperatures, T_m and T_c in all known cases diffeı only by the coefficient which is not larger than two. If the "martensitic" transition is absent, there is a trend to it resulting in the softening of the elastic moduli. This softening occupies a large temperature interval up to 300–400 K. For many of these compounds an analogous temperature behaviour of the magnetic susceptibility, and of the Knight shift was observed.

It was first pointed out by Weger [2] that in these compounds three ortogonal systems of the linear atom chains of the transition elements could be considered as one-dimensional threads. This suggestion on the applicability of the tight bonding approximation spectrum of the d-electrons in the threads was then exploited in all theoretical papers [3–5]. In these models the role of d-electrons was postulated to form a large peak in the density of states just in the vicinity of the Fermi level. The existence of such a large and narrow peak could be the cause of the temperature dependence of the experimental quantities.

A most detailed theory in this direction was suggested by Labbe & Friedel [5]. In [5] the Fermi level has been placed very close to the bottom of the empty d-zone (the point Γ in the inverse lattice). The peak in the density of states $\nu(\varepsilon)$ was due to the singular behaviour of this quantity in the one-dimensional case $\nu(\varepsilon) \propto \varepsilon^{-1/2}$. By choosing the constants of the model [5] in an appropriate way it was possible to give an account for the "martensitic" transition as well. In connection with the model [5] the main objection is that there are no general arguments why the chemical potential should be so close to the bottom of the d-band. The identical order of the magnitude of the temperatures T_m and T_c is not the principal feature of the model.

Suggesting also a one-dimensional character of the electronic spectrum in the thread we consider the picture which follows in case the Fermi level is at the point X of the inverse lattice. Let us suppose that d-electrons are localized on the threads and do not mix up with s-electrons. The Luttinger theorem [6] about the filling of the energy bands makes such chemical potential position an extreme point of the system energy (see below).

The atoms of the transition elements in the structure $\beta-W$ are packed in chains with an equal distance between atoms in a single chain. The properties of a one-dimensional metal have been studied by Bychkov et al. [7]. It was shown that the electronic spectrum of this system is unstable simultaneously under the Cooper pairing and the dielectric pairing. The latter corresponds to the doubling of the period. (It was shown by Dzyaloshinsky & Larkin [8] that under appropriate conditions there is a possibility of an antiferromagnetic ordering.) In a weak interaction model it was possible to show that the corresponding transition temperatures, T_m and T_c have the following order of the magnitude:

$$T_c \sim T_m \sim \tilde{\omega} \exp\left(-\frac{1}{|g_1|}\right) \quad (1)$$

Here $\tilde{\omega}$ is a characteristic cut-off energy, and g_1 is a dimensionless constant of the effective attraction between electrons.

In the model [7] corrections of such a form have been found to the phonon frequencies at the wavelength equal to the doubled period. One of the reasons for which the picture is obtained in [7] seems to be directly connected with the properties of the A-15 compund group is the logarithmical temperature dependence of the elastic moduli [1].

$$C_{ij} = A + B \ln T$$

It is worth mentioning that a displacement of the niobium chain "sublattices" was observed in Nb_3Sn [9]. However, it is important that in the A-15 structure there are two transition element atoms for each period of the chain. Therefore, the singularities in the phonon spectrum discovered in [7] should correspond to the phonon wave vector $\boldsymbol{q} = 0$.

This guess makes it necessary to understand in more detail the character of the d-electron spectrum at the point X taking into account the symmetry of the crystal field. The group of the wave vector at the point X in these compounds has been considered by Weger [10] who mentioned that the four single valued representations $\{X_1, X_2, X_3, X_4\}$ (notation of [11]) are two-fold degenerated owing to the presence of a screw axis in the group. To get the picture similar to that of [7, 8] we are to prove the linear quasi-momentum dependence on the electron energies in the vicinity of the point X.

The splitting of the electronic levels near the point X was calculated by usual methods of the group theory [11]. The answer is that there are two types of electronic terms. Two levels of $\varepsilon_{1,2} = \pm v p_z (p_z$ is counted from $p_X = \{0, 0, \pi/a\})$ stand for the terms X_2, X_4 and the velocity v has a usual order of magnitude for d-electrons.

When the interactions between the chains are neglected no splitting for the single-valued representation of X_1, X_3 takes place. It means that this band may either be filled completely or it is empty [1]. The splitting of this term appears if one takes into account the spin-orbital interaction. Therefore in this case near $p_z = 0$ there are two small electron or hole "pockets". The velocity $\tilde{v}$ in the dispersion law $\varepsilon_{1,2} = \pm \tilde{v} p_z$ is now small so far as the spinorbital interaction is small.

The anomaly of the elastic properties in the A-15 structure compounds is supposed to be due to d-electrons localized on linear chains. Denote the electron Hamiltonian distortion caused by deformation $\lambda(\boldsymbol{A}\boldsymbol{\epsilon})$, $\boldsymbol{\epsilon}$ are some components of the strain tensor. In the second quantization form this perturbation is equal to

$$H_\varepsilon = \lambda \psi^+ (\boldsymbol{A}\boldsymbol{\epsilon}) \psi$$

Now to get the correction to the thermodynamical potential $\partial\Omega/\partial\lambda = (1/\lambda)\langle H_\varepsilon \rangle$ which is quadratic in deformations one should calculate the general expression of the form

$$\tfrac{1}{2}(\boldsymbol{A}\boldsymbol{\epsilon})^{ik}_{\alpha\beta}(\boldsymbol{A}\boldsymbol{\epsilon})^{ki}_{\beta\alpha} \int G^{(i)}(p)\, G^{(k)}(p)\, dp$$
$$+ \tfrac{1}{2}(\boldsymbol{A}\boldsymbol{\epsilon})^{ik}_{\alpha\gamma} \iint G^{(i)}(p)\, G^{(k)}(p)\, \mathcal{J}^{ilkm}_{\alpha\beta\gamma\delta}\, G^{(l)}(p')$$
$$\times G^{(m)}(p')(\boldsymbol{A}\boldsymbol{\epsilon})^{lm}_{\beta\delta}\, dp\, dp' \qquad (2)$$

Here $G^{(i)}$ are the Green's functions of the ith-zone, $\mathcal{J}$ is the vertex part for the scattering between electrons. To find out the magnetic susceptibility one has to insert an operator $(\boldsymbol{B}\cdot\boldsymbol{H})$ proportional to the components of the magnetic field into this expression. The whole structure of eq. (2) is given in fig. 1*a*. One may easily see that the integration in the loop of fig. 1*b* is logarithmic, provided $G^{(1)}$ and $G^{(2)}$ are the d-electron Green's functions corresponding to the spectrum branches with the energies $\varepsilon_{1,2} = \pm v p_z$. Let us omit the second part of eq. (2) which includes interactions between electrons. The first term of eq. (2) defines the temperature dependent corrections to the elastic moduli proportional to $\ln(\tilde{\omega}/T)$. To get them correctly one should take into account the symmetry consideration and find out the non-diagonal matrix elements of $(\boldsymbol{A}\boldsymbol{\epsilon})$. In both cases $\{X_1, X_3\}$ and $\{X_2, X_4\}$ the only essential component of the strain tensor entering is $\varepsilon_{xx} - \varepsilon_{yy}$ for a single chain in the the z-direction. The full contribution to the free energy density due to the z-chain system is equal to:

$$\delta F_z = -\frac{D_1^2}{2a^2 \pi v}(\varepsilon_{xx} - \varepsilon_{yy})^2 \ln\frac{\tilde{\omega}}{T} \qquad (3)$$

Summing over all three systems of chains one gets

$$\delta C_{11} = -\frac{2D_1^2}{a^2 \pi v} \ln\frac{\tilde{\omega}}{T}; \quad \delta C_{12} = \frac{D_1^2}{a^2 \pi v} \ln\frac{\tilde{\omega}}{T} \qquad (4)$$

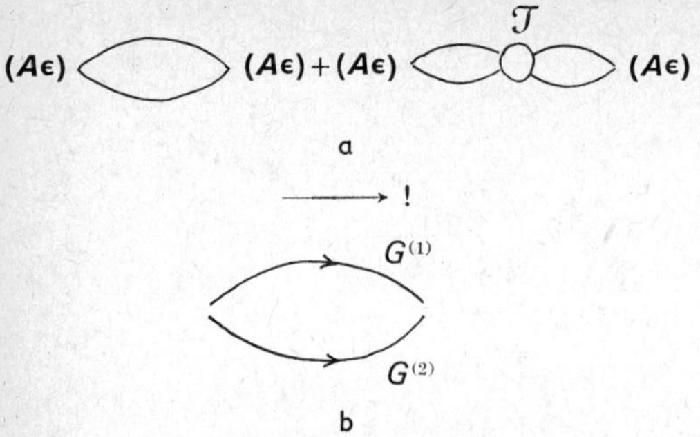

Fig. 1. (*a*) The expression for the addition to the whole elastic energy; (*b*) the first logarithmic term in (2).

The experimental data [1] show some temperature variation of C_{44}. This variation is noticeably less. For V_3Si the change of C_{11} from 300 K to 100 K is larger by a factor of ten approximately than the corresponding change C_{44}. For Nb_3Sn this factor is four or five. The absence of temperature corrections to C_{44} in (3) and (4) is the consequence of the d-electron wave function symmetry in the self consistent field approximation for a single chain. Extending the symmetry analysis in order to include interchain interactions it is possible to obtain a contribution into C_{44}. We shall not go into detail because these terms are suggested to be small.

In fact, if one describes the interchain interaction by some potential V_{12}, then deviations of the d-electron spectrum from the plane one should be of the order of magnitude (in the energy scale) V_{12}^2/E_F. On decreasing the temperature below this quantity one gets a constant cut-off in the logarithmic integrals. Considering once again the experimental results [1, 12] one sees that in any case $T_m \gtrsim V_{12}^2/E_F$. The rough estimation of V_{12} is now ($E_F \approx 1-3$ eV):

$$V_{12} \lesssim 0.1\ E_F \tag{5}$$

This order of magnitude for the interchain interaction seems to be correct [10].

Fig. 2. (*a*) A contribution to the elastic moduli due to the interaction with the other zones; (*b*) the part of the self-consisting field due to the other electrons.

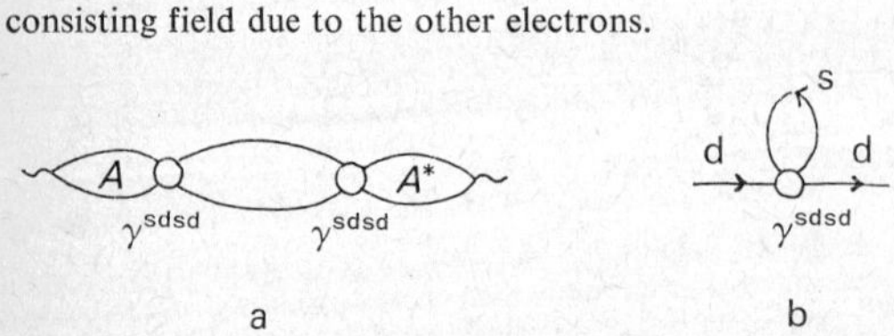

Turning to the question about the origin of the temperature corrections to C_{44} one should search therefore for some other mechanism. There are many possibilities following from the electron interactions. We shall not enumerate all of them. The right order of magnitude for $\delta C_{44} = -A^2 \ln(\tilde{\omega}/T)$ one gets, in principle, from the contribution shown diagrammatically in fig. 2*a*. The wave lines correspond to a nondiagonal component of the deformation tensor, say, ε_{xy}. The right and the left parts of the diagram (A and A*) are due to s-electrons, and, therefore, contain no logarithmic integrals. Two circles marked by γ^{sdsd} correspond to the scattering of s-electrons by the d-electrons localized on the chain. There is no reason why this γ^{sdsd} interaction should be small. For example, the diagram in fig. 2*b* takes part in forming the selfconsistent field for the chain electrons. We have mentioned above that in practice δC_{44} are smaller than δC_{11}, δC_{12}. It could be due to isotropy of s-electrons. In conclusion it is worth mentioning that δC_{44} has the behaviour similar to δC_{11}, eq. (4), and differs by the value of a coefficient.

It follows from (4) that $\frac{1}{2}\delta C_{11} = -\delta C_{12}$. The experimental data are in favour of this relation. However, it is unlikely to be a quantitative coincidence. One could point out that most of the experimental results for the elastic moduli have been obtained from ultrasonic velocity measurements [1]. The agreement with the results of static measurements is far from being the best one.

Here we would like to draw attention to the fact that the picture described above presents dispersion of elastic moduli as consequence. In other words, the dispersion law of the acoustical phonons differs from the linear one. To prove this fact it is sufficient to notice that the diagram in fig. 1*a* is equivalent to the phonon polarization operator. The contribution from a single chain, to say, in the [001]-direction contains the dependence on the *z*-component of the phonon vector. One easily gets the expression for the elastic part of the oscillation energy in the form:

$$\delta F(q) = -\frac{D_1^2}{2a^2\pi v}\left\{\varepsilon_{xx}^2\left(2\ln\frac{\tilde{\omega}}{T} + U(q_y) + U(q_z)\right)\right.$$

$$\left. + \varepsilon_{yy}^2\left(2\ln\frac{\tilde{\omega}}{T} + U(q_x) + U(q_z)\right) + \right.$$

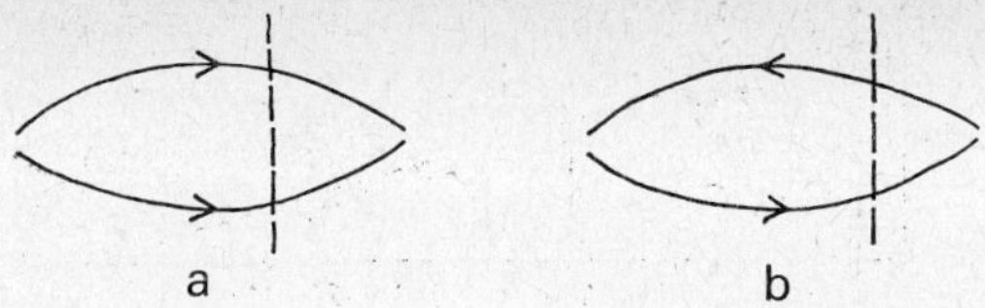

Fig. 3. Two logarithmic channels in the one-dimensional case: (*a*) the Cooper-type; (*b*) the "zero-sound" type channel.

$$+\varepsilon_{zz}^2\left(2\ln\frac{\tilde{\omega}}{T}+U(q_x)+U(q_y)\right)-$$

$$-2\varepsilon_{xx}\varepsilon_{yy}\left(\ln\frac{\tilde{\omega}}{T}+U(q_z)\right)-$$

$$-2\varepsilon_{yy}\varepsilon_{zz}\left(\ln\frac{\tilde{\omega}}{T}+U(q_x)\right)-$$

$$-2\varepsilon_{xx}\varepsilon_{zz}\left(\ln\frac{\tilde{\omega}}{T}+U(q_y)\right)\Bigg\} \tag{6}$$

For a shear mode $\boldsymbol{q}\parallel(110)$ the corresponding elastic modulus $C_s=\frac{1}{2}(C_{11}-C_{12})$ is of a form:

$$\delta C_s(\boldsymbol{q})=-\frac{D_1^2}{2a^2\pi v}\left\{3\ln\frac{\tilde{\omega}}{T}+\tfrac{1}{2}U\left(\frac{qv}{\sqrt{2}4\pi T}\right)\right\} \tag{7}$$

The function

$$U(x)=\psi(\tfrac{1}{2})-\tfrac{1}{2}[\psi(\tfrac{1}{2}+ix)+\psi(\tfrac{1}{2}-ix)] \tag{8}$$

(ψ is the digamma-function) has an asymptotic behaviour:

$$U(x)=\begin{cases}-\frac{7}{2}\zeta(3)\,x^2; & x\ll1\\ -\ln x; & x\gg1\end{cases} \tag{9}$$

Therefore in the interval $\pi/a\gg q\gg q^*=v/4\pi T$ one has from (7) and (9) the following:

$$\tfrac{1}{2}\,\delta(C_{11}-C_{12})=-\frac{D_1^2}{2a^2\pi v}\left\{3\ln\frac{\tilde{\omega}}{T}-\tfrac{1}{2}\ln\left(\frac{qv}{\sqrt{2}4\pi T}\right)\right\}$$

The neutron data on the acoustical phonon spectrum [1] confirm the fact of the phonon dispersion law distortion.

Eq. (4) shows that the shear modulus $\frac{1}{2}(C_{11}-C_{12})$ tends to zero with the decrease of temperature. The optical phonons modes are softened if they interact with d-electrons.

For example, the *z*-displacement of each second atom in the chain, u_z, has the same symmetry properties as the component $\varepsilon_{xx}-\varepsilon_{yy}$.

It is seen from fig. 1*a* that the second term including the interactions between electrons gives an additional contribution roughly of the order $g_1\ln(\tilde{\omega}/T)$. In what follows we use the system of equations for the vertex parts which has been derived in [7, 8]. The estimation of g_1 in V_3Si and Nb_3Sn using the relation (1) $|g_1|\approx1/2-1/3$ with $\tilde{\omega}$ of the order of the characteristic phonon frequences. Therefore it seems the interaction is not weak. In any case the results described below are of a qualitative character.

It was pointed out that as a whole the picture of the singularities in the electron interactions in the one-dimensional case is different from that of the BCS-theory. There are now two channels for the electron–electron scattering giving the logarithmic behaviour of amplitudes at small energies. They are shown in fig. 3*a*, *b*. The diagram (*a*) is of the "Cooper-type", the new one is the "zero-sound-type" diagram (*b*). Input of the first type diagrams into the second ones generates a set of a "parquet" equations, described in [7].

It is possible to prove that all interactions between the chain electrons at the point X can be reduced to three independent constants. There is now one to one correspondence between the equations in our problem and the model studied in [8] which differs from [7] accounting for the "Umklapp"-processes. The difference between the model [7, 8] and the chain in A-15 structure mainly consists in the physical interpretation. All the structure instabilities of [7, 8] corresponding to the doubling of the structure period are related to the phonon vector $\boldsymbol{q}=0$.

It is possible now to formulate a statement. In case an effective attraction between electrons occurs ($g_1<0$), the vertex part $\mathcal{J}$ increases:

$$\mathcal{J}\propto\frac{g_1}{1+g_1\ln(\tilde{\omega}/T)} \tag{10}$$

The pole in this expression was interpreted [7, 8] as an unstability of the linear electron spectrum near the Fermi level with respect to both the superconducting and structural transitions. In the "parquet" approximation one cannot see any difference between the temperatures T_m and T_c, but if $\mathcal{J}$ really tends to infinity at T_m^* and T_c^* both temperatures should be of the same order of magnitude.

Using the results of [8] one may easily see that the allowance for the vertex part in eq. (2) means multiplication of the expressions (4) for the elastic moduli by a factor of $\varphi(T) > 0$. Near the pole in (10) φ is proportional to $(1 + g_1 \ln(\tilde{\omega}/T))^{-\frac{1}{2}}$. Therefore, in an ideal sample the shear modulus achieves its zero value before the vertex part turns to infinity. In other words the structural transformation precedes the transition to the superconductive phase. This is just the picture one has in Nb_3Sn and V_3Si. The optical modes in this case are finite at the "martensitic" transition. However, we have already mentioned that the frequencies of optical phonons also soften. This is only the question of constants for real substancies for which softening of optical or acoustic frequencies is larger. However, the fact that transition into the superconducting phase could have occurred only after some structural transformation does not depend on it.

Impurities and stresses will strongly affect this picture. The experimental situation is that the stresses applied reduce the "martensitic" transition temperature and stiffen the elastic moduli [1]. In the framework of the theory suggested above the role of strains is assumed to split off the twofold degenerated levels at point X. The dielectric gap which is proportional to the deformation energy now cuts off all logarithmic integrations from below for the both channels of fig. 3. Let us now consider a uniaxial deformation corresponding to the structural transition into the tetragonal phase in V_3Si or Nb_3Sn. For two systems of chains this deformation acts as was described, diminishing the value of all logarithmic integral. But for the third one the deformation has no offdiagonal matrix elements and only displaces the Fermi level position near point X. This effect cuts off the contributions only from the "zero sound" channels leaving the superconducting phase transition temperature for this chain unchanged. Unfortunately there are no data on the anisotropy of conductivity near the "martensitic" transformation in Nb_3Sn. The estimation of the nonlinear dependence of the elastic moduli on the deformation gives the correct order of magnitude.

The measured values of $\partial C_{ij}/\partial P$ [1], are of the order of unity as in usual solids. The hydrostatic pressure can only change the position of the Fermi level. The absence of large anharmonic effects up to 10 kbars in the experiments [13] means in our opinion that the s- and d-electrons

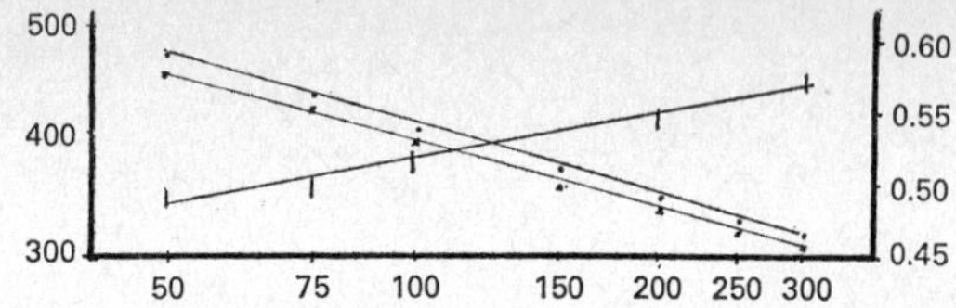

Fig. 4. *Abscissa:* T(K); *ordinate:* (*left*) $\chi(10^{-6}\ cm^3/mol\ wt)$; (*right*) Kv %. Susceptibility and the Knight-shift in V_3Si as a function of the temperature.

do not mix. Therefore, the role of impurities differs essentially from the hydrostatic pressure. A large effect of small concentrations of Al atoms in Nb_3Sn [14] on the structural transition temperature could be understood as a dopping of d-electrons in threads. The displacement of the Fermi level from the point of the electronic terms degeneration is the mechanism here which makes the lattice stable. In the "parquet" approximation the superconducting transition temperature is not changed.

Let us turn to the magnetic susceptibility. In fig. 4 it is plotted as a function of $\ln T$ for V_3Si. The data are taken from [12]. In the same graph the temperature-dependent Knight-shift is given [1]. For different-compounds the behaviour of these characteristics changes in various ways. A detailed analysis shows that, in general, there are some logarithmic contributions to the magnetic susceptibility which could account for this difference in the behaviour. As an example we may indicate the diagram in fig. 2*a*. If the scattering amplitude γ^{sdsd} has an exchange part, this diagram provides a paramagnetic contribution $\delta\chi \propto \ln(\tilde{\omega}/T)$. The contribution of the diamagnetic type was obtained in [8] as a result of corrections due to interactions:

$$\delta\chi \simeq -\chi_0 \frac{g_1^2}{2} \ln(\tilde{\omega}/T)$$

The most striking features of χ in fig. 4 are the ideal linear dependence on $\ln T$, and the large value of $\delta\chi/\chi_0 \sim 30\%$. This fact seems to be an argument in favour of choosing the term X_1 or X_3. To explain this one should remember that the linear spectrum $\varepsilon_{1,2} = \pm\tilde{v}p_z$ in this case results from taking account of the spin. By the basic functions of single valued representations ψ_1, ψ_2 the eigenvalues of the electron Hamiltonian near X correspond to the twofold systems of functions $\{\psi_{1\uparrow}, \psi_{2\downarrow}\}$ and $\{\psi_{1\downarrow}, \psi_{2\uparrow}\}$. Here the arrows denote two directions of the spin along the chain. Therefore, the usual interaction $\mu_B(\sigma \cdot H)$ for a field perpendicular to the chain gives a contribution. Summing at the given direction of the field over two systems of threads one obtains

$$\delta F = -\frac{H^2 e^2}{8\pi m^2 c^2 a^2 \tilde{v}} \ln(\tilde{\omega}/T) \qquad (12)$$

Note the large value of the susceptibility due to the velocity $\tilde{v}$ in the denominator. As in the model [5] for this last choice of terms there exist the same objections against the position of the Fermi level just near the point X. We discussed this possibility only in connection with the experimental results [12]. There are no arguments like the Luttinger's theorem.

The account of the electron interactions could be done in the way described for the elastic moduli. Using the results of [8] one sees that χ turns to infinity only if an antiferromagnetic ordering is possible. In an opposite case the magnetic susceptibility reaches maximum and then decreases after structure transformation due to developing deformations.

We believe that the deviation from the linear dependence on $\ln T$ for various quantities are mainly due to the interaction g_1 which is responsible for the phase transition. For the elastic moduli (4) and the susceptibility (12) the corresponding correction factors at high temperatures $(1 - g_1 \ln(\tilde{\omega}/T))$ and $(1 - (g_1/12)\ln(\tilde{\omega}/T))$. Probably these deviations cannot be found from the experimental data [1, 12].

The conception suggested above considers the structural instability and superconductivity as the two sides of the electron spectrum unstability. This is the main difference between three- and one-dimensional problems, the diminishing of the whole energy in the second case due to the dielectric pairing completes with the gain due to the Cooper pairing. In a three-dimensional system a small deformation does not change essentially the electronic Fermi surface.

The fact that this picture fits well in with the experimental situation for A-15 structure compounds could be considered as evidence in favour of the theoretical ideas.

From this point of view any quasi one-dimensional system characterized by a structural instability or anomalous temperature dependence of susceptibility is to be suspected of the possibility of antiferromagnetism or superconductivity.

The author is very grateful to A. Larkin, I. Dzyaloshinsky, E. Rashba for the discussions of the paper, and to N. Alekseevsky, E. Krasnoperov, I. Schegolev and A. Rabinkin for many talks about the experimental situation.

References

1. Testardi, L R, Physical acoustics, vol. 10. Academic Press, New York and London, 1973.
2. Weger, M, Rev mod phys 1964, 36, 175.
3. Closton, A M & Jacarino, V, Phys rev 1961, 121, 1357.
4. Cohen, R W, Cody, G D & Haloran, J J, Phys rev lett 1967, 19, 840.
5. Labbe, J & Friedel, J, J phys radium 1966, 27, 153, 303, 708.
6. Luttinger, J M, Phys rev 1960, 119, 1153.
7. Bychkov, Ju A, Gor'kov, L P & Dzyaloshinsky, I E, JETP 1966, 50, 738.
8. Dzyaloshinsky, I E & Larkin, A I, JETP 1971, 61, 791.
9. Shirane, G & Axe, J D, Phys rev 1971, 34, 2957.
10. Weger, M, Phys chem solid 1970, 31, 1621.
11. Bier, G & Picus, G E, Symmetry and effects of deformations in semiconductors. Nauka, Moscow, 1972.
12. Maita, J P & Bucher, E, Phys rev lett 1972, 29, 931.
13. Smith, T F, Phys rev lett 1970, 25, 1483.
14. Vieland, L J & Wicklund, A W, Phys lett 1971, 34A, 43.

Discussion

Hopfield: In a real system, the overlap between the s- and d-bands would make the Fermi energy not fall at the energy of the d-band X-point, even *without* hybridization. Hybridization will change the form of the density-of-states structure at X by introducing an energy dependence on $k_\perp$. This latter is particularly important in 4d and 5d materials. Thus I would wonder whether the theoretical logarithmic dependence of properties on T would not, except in rare accidents, be destroyed in real systems. I would especially ask whether the case of the logarithmic conductivity of Nb_3Sn can reasonably be associated with the theoretical description.

Gorkov: The calculation is based on the picture suggesting that the s and d electrons are not mixed. Then to get the right number of electrons per lattice cell you must put the chemical potential at X. If they were mixed, then according to the tight-binding approximation one would expect this mixing not to be a strong one. Experimental data show no effect of mixing. (Extracted from tape).

Doniach: In the parquet calculation there are two instabilities—lattice (or charge density) and superconducting. Can you say how the theory is changed below T_m? Presumably the lattice distortion lifts the screw degeneracy at the point X so that the charge instability no longer competes with the Cooper pairing?

Gorkov: I am especially talking about the high-temperature interval, because in the parqeut equation you cannot go far. I don't know if it is possible to get an answer like in the BCS theory for the expression for T_c, which is correct to a numerical coefficient. The only thing that one can say with this parquet equation is the order of magnitude with the accuracy of some sort of numerical coefficient. So, you cannot divide T_c by T_m, but you can divide T_c and T_m and the temperature of the antiferromagnetic ordering, because T_c and T_m are due to the constant g_1 and the temperature for the anti-ferromagnetic ordering is connected with (as shown by Dzyaloshinsky and Larkin) g_3, which is independent of g_1. (Extracted from tape).

(TTF) (TCNQ): A One-Dimensional Organic Metal[1,2]

A. F. Garito and A. J. Heeger

Department of Physics and Laboratory for Research on the Structure of Matter,
University of Pennsylvania, Philadelphia, Pa 19174, USA

Summary

The physics of the organic charge transfer salts is discussed from the point of view of achieving and stabilizing the one-dimensional metallic state. The particular case of (TTF) (TCNQ) is discussed in detail. Conductivity, infrared properties, magnetic susceptibility, nuclear relaxation rates and thermoelectric power data are summarized, and analyzed in terms of the properties of this one-dimensional metal. The giant conductivity peak is discussed and interpreted in terms of superconducting fluctuation phenomena. The sources of electron-phonon interaction in such narrow band ionic metals are clarified, including the effect of electronic screening.

The class of solids known as organic charge transfer salts has been of considerable interest in recent years. This interest arises from the novelty of the systems themselves as well as from the fact that the flat planar molecules involved lead to anisotropic structures and therefore to pseudo one-dimensional electronic properties. More generally, the existence of a class of materials made up of organic molecules as fundamental units allows the possibility of utilizing the flexibility of organic chemistry to design and synthesize molecules with particular characteristics. The over-all goal would be to create solid state materials with a desired bulk property by the preliminary design of the constituent molecules.

The discovery of the extraordinary electrical conductivity maximum in (TTF) (TCNQ) and the interpretation of these data as arising from superconducting fluctuation phenomena in these pseudo one-dimensional organic solids places these phenomena at the focus of future work in this area [1].

[1] Supported by the National Science Foundation through the Laboratory for Research on the Structure of Matter and GP-29583, and the Advanced Research Projects Agency through DAHC 15-72C-0174.

[2] Read by A. J. Heeger.

The organic charge transfer salts are composed of a donor molecule, D, and an acceptor molecule, A [2]. In contrast to conventional molecular crystals (that is, crystals made up of neutral organic molecules bonded by van der Waals forces), these salts have unpaired electrons on the acceptor or donor or both as a result of the basic charge transfer reaction:

$$DA \rightarrow D^+A^- \tag{1}$$

In the simplest cases, the energy required for charge transfer is given by:

$$\Delta E_{ct} = (I_D - A_A) - (E_M + E_{ex} + E_{pol}) \tag{2}$$

where the neutral DA system is taken as the zero of energy. I_D is the donor ionization potential and A_A is the electron affinity of the acceptor. The final three terms are of electrostatic origin: E_M is the total Madelung energy arising from the sum over the long range Coulomb forces in the ionic D^+A^- solid, E_{ex} is the exchange contribution, and E_{pol} represents the interaction of the ionic charges with the induced dipoles arising from the molecular polarizability. E_M and E_{ex} are well known; E_{pol} may be important in systems containing molecules with large π-electron polarizabilities.

The charge transfer to the ionic configuration is itself an important and difficult problem. The quantity $(I_D - A_A)$ is in all known cases greater than zero, so that stability rests on the magnitude of the electrostatic terms. However, even in the limit $\Delta E_{ct} > 0$ some virtual charge transfer stability is added into the basically neutral ground state via admixture of ionic configurations through perturbation theory. As, for example, $(I_D - A_A)$ is reduced, this virtual charge transfer increases and is aided by the internal

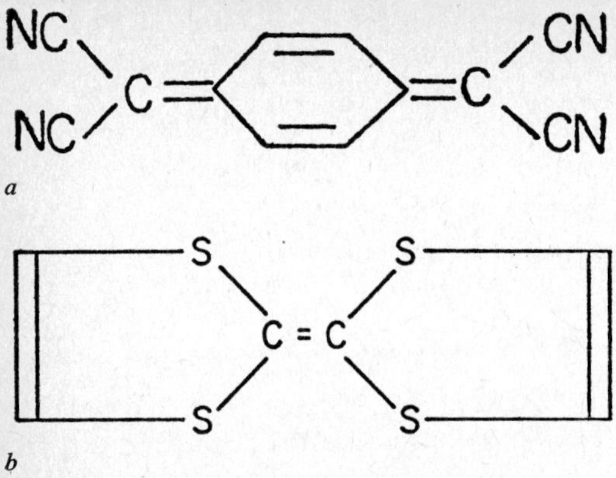

Fig. 1. Electronic structure of the (*a*) TCNQ; (*b*) TTF molecule.

fields arising from the Madelung and polarization energies. The charge transfer process is thus a bona fide many-body problem. McConnell and co-workers [3] considered this in detail in their early work. Using a Hartree mean field factorization of the electron–electron terms, it was concluded that a genuine phase transition occurred. Although this may be the case, since long range Coulomb forces are involved, the question requires further study using more sophisticated modern techniques, especially from the point of view of clarifying the role of fluctuations in smearing the transition. A detailed theoretical understanding of the charge transfer process for systems near the neutral-ionic boundary is of importance, for many of the most interesting compounds under study appear to be in this category.

When $\Delta E_{ct} < 0$, the resulting system will be nearly fully ionic with unpaired electrons outside a filled core. Generally in such molecules, the affinity level corresponds to the lowest unfilled π-level of the neutral acceptor molecule so that one is dealing primarily with the $2p_z$ wave functions of C and N. The directionality of the p_z wave functions, together with the chain-like crystal structures, leads to the pseudo one-dimensional electronic properties of these compounds.

Much of the work done in this area has utilized the excellent acceptor tetracyanoquinodimethan (TCNQ) [1], the structure of which is shown in fig. 1*a*. The molecular physics of this system has been discussed in detail elsewhere [4]. We note here briefly only the most important features.

The TCNQ molecule is a closed-shell planar quinoid system having four highly electron-withdrawing cyanide groups located at the terminal methylene carbon atoms. These characteristics account for its very large electron affinity and associated behavior of taking one electron when placed in contact with virtually any electron donor to form the open-shell $TCNQ^-$ monanion radical. The single unpaired electron on $TCNQ^-$ occupies the lowest energy empty π-level and is expected to reside mainly localized on the terminal dicyano methylene carbon groups in order to take maximum advantage of the strong electron affinity of the cyanides. This picture of the electronic structure is confirmed by the theoretical results of Lowitz [5], by the small hyperfine coupling to the ring protons in solution studies [6], and by analysis of the x-ray bond lengths associated with $TCNQ^-$ in solid salts [7].

The most direct experimental information on the charge density in the $TCNQ^-$ anion comes from ^{14}N nuclear quadrupolar resonance (NQR) studies carried out in our laboratory [8]. NQR measurements are directly sensitive to the local charge density on the nitrogen. These experiments indicate that the added charge density is 0.14 per nitrogen, so that the excess electron resides almost completely on the four highly electron attracting C≡N groups.

In the event of an ionic fluctuation with two excess electrons on a single TCNQ molecule (as would be the case in a metallic state or near the metal-insulator transition) [4], there is a clear tendency for the two electrons to localize at opposite ends of the molecule and correlate to stay apart in order to reduce their mutual Coulomb repulsion. This Heitler–London correlation, together with the polarizability of the cation, makes the resultant effective Coulomb repulsion sufficiently small that the metallic state can be achieved in some of the TCNQ compounds. The crucial role of the cation polarizability has been demonstrated in our earlier work [4].

Despite their apparent complexity, these systems represent a relatively simple class of ionic solids. Only a single narrow electron energy band is involved (the other molecular levels being several volts higher in energy) and tight-binding theory is an excellent approximation.

Some insight into the relative simplicity of these solids can be obtained from studies of the low temperature heat capacity.

The systematic dependence of the Debye temperature, θ_D, for the (alkali)$^+$ (TCNQ)$^-$ salts on $(MV_c)^{-\frac{1}{2}}$, where M and V_c are the molecular mass and volume of the chemical unit cell, suggests that the expected ionic Madelung contribution is dominant [9]. The somewhat larger relative values of θ_D for (NMP) (TCNQ) (Quinolinium) $(TCNQ)_2$, and (TTF) (TCNQ) ($\theta_D = 89$ K) imply the existence of significant polarization contributions since the relatively large cation polarizability in these salts has been shown to be an important factor in achieving the metallic state. This conclusion is strengthened for (NMP) (TCNQ) by the direct calculation of the Madelung constant by Metzger [10], who shows that the resulting Madelung energy is too small to account for the ionic ground state. The obvious suggestion is that the polarization contribution arising from the large cation polarizability is dominant, and stabilizes the structure as well as the ionic configuration [11].

I. The Metallic State

The achievement of a metallic state in an organic solid is all the more remarkable since we are used to thinking of molecular solids as weakly interacting collections of molecules. There are three basic requirements: (1) The existence of unpaired electrons; (2) a uniform crystal structure such that, in absence of electron–electron interactions, the electronic band structure of the system would be metallic; (3) relatively weak electron–electron repulsive interactions.

The first requirement needs little comment. Without unpaired electrons, the system would remain insulating until the band widths exceeded the intramolecular level splittings, $\Delta E_{\pi-\pi^*}$. Since this would require intermolecular lattice spacings comparable with those between atoms within the molecule, such a state could only be achieved at very high pressures or if a true polymerization occurred. Were this the case, the resulting metallic state would be analogous to that in the alkaline earth metals. The existence of unpaired electrons in the charge transfer salt greatly facilitates the achievement of the metallic state, for in the absence of interactions any weak intermolecular transfer integral, t, leads to formation of energy bands and delocalization of the electronic wave functions. The observation that $t/\Delta E_{\pi-\pi^*} < 1$ simply means that the band structure may be calculated in the tight-binding approximation with high accuracy. However, it is clear that such a band structure need not be metallic. For example, in a one-dimensional system, any weak dimerization of the structure into pairs leads to an energy gap at the Fermi surface (the Peierls instability) and a resulting semiconducting state; hence requirement (2), the need for a uniform structure.

The role of the Coulomb interaction in the metal-insulator transition has been discussed in detail elsewhere [4]. Basically, the Coulomb interaction tends to cause correlation of the electrons so as to stay apart. For the case of a half-filled band system, in the limit of large Coulomb repulsion, the electrons localize one per site to form a magnetic Mott–Hubbard insulator. This tendency toward localization competes with the transfer integral which tends to lower the energy of the system by delocalizing the electronic wave functions and forming energy bands. Hubbard has attempted to describe the relevant physics in terms of the model Hamiltonian [12],

$$H = -t \sum_{\langle i,j \rangle} c_{i\sigma}^{+} c_{j\sigma} + U \sum_{i} n_{i\uparrow} n_{i\downarrow}, \tag{3}$$

where t is the near neighbor transfer integral and U represents the Coulomb repulsion between two electrons on the same site. The detailed solutions of the Hubbard model as applied to the TCNQ salts have been discussed elsewhere. Qualitatively, $U/t \gg 1$ corresponds to the insulating limit and $U/t \ll 1$ to the metallic limit. The metal-insulator boundary as given qualitatively by the Hartree–Fock approximation occurs at $U/4t = 1$. In these semi-qualitative terms requirement (3), the need for weak Coulomb repulsion is clear.

Requirements (1) and (3) are relatively easy to achieve and control. A wide class of charge transfer salts exist, and flexibility in choice and design of both the cation and the anion are possible. Because of the structure of the TCNQ molecule (and other related systems), the bare Coulomb repulsion is considerably reduced from the atomic value by the Heitler–London correlation described above. The final step in achieving the reduced Coulomb interaction is via the cation polarizability. Since this polarizability arises from virtual excitation of the $\pi-\pi^*$ singlet excitons in the cation or anion chain, the problem can be reduced to that of interacting excitonic polarons [11, 13]. Detailed calculations indicate the existence of an indirect attractive interaction via these virtual excitons, resulting in a total effective interaction which may be written as [11, 13],

$$U_{\text{eff}} = U - 2E_{\text{B}}, \tag{4}$$

where $E_{\text{B}} = \Gamma^2/\hbar\omega_0$ is the excitonic polaron binding energy arising from the coupling Γ of the electronic system to the cation polarizability expressed in terms of cation excitons at energy $\hbar\omega_0$. This indirect attractive interaction may indeed be large, and under favorable circumstances can reduce the net interaction to the point where the metallic state is accessible.

Requirement (2) is more difficult. Since the overall energy $|\Delta E_{\text{ct}}|$ is relatively small, and since isotropic Coulomb forces are involved, multiple crystal phases are likely and several examples have been observed experimentally. Generally, however, the crystal structures consist of stacks of molecules with separations between molecules within a chain of 3.2–3.5 Å and with the individual chains widely separated. As far as the electronic properties are concerned, the obvious anisotropy of such a structure is enhanced by the directionality of the π-electron wave functions with the result that interchain coupling can be negligible. Two basic kinds of structures have been observed. In the first, the anions stack in linear chains separated by parallel cation chains. The anion stacking may be uniform as in (NMP) (TCNQ) or dimerized as in $Rb^+(TCNQ)^-$. The uniform anion chains are potentially metallic; the dimerized systems are necessarily semiconducting. The second kind of structure can be described as chains of the form $D^+A^-D^+A^-$ ... where each cation has nearest neighbor anions along the chain, and vice versa as is the case in (TMPD) (TCNQ). The latter type of structure is unlikely to be metallic, as an accidental near degeneracy of the D^+ and A^- levels would be required. Obviously, the structure is not easy to control, although experimentally without exception the uniform structures are found with flat planar organic cations.

One final comment on structure is in order. It has been noted that in several of the highly conducting salts, there is disorder in the cation chains. Although it has been shown that this disorder does not dominate the electronic properties [14], it may act to stabilize the desirable uniform stacking in anion chains by removing the logarithmic singularity in the dielectric constant through residual disorder. Although speculative, this possibility deserves further study in the context of gaining some control on the lattice structure.

II. The (TTF)(TCNQ) System

As part of a continuing effort to stabilize the metallic phase, a variety of donor cations have been considered. There are two basic requirements: (1) large cation polarizability; (2) relatively small size.

The former was discussed above; the latter follows from the fact that the interaction between the conducting electron and the induced polarization dipole varies as r^{-4}. Based on these arguments, it would appear that molecular systems like tetrathiofulvalene (TTF) (see fig. 1*b*) would be prime candidates for stabilizing the metallic state, being both relatively small and possessing a large molecular polarizability as a result of the low lying excited states of the sulfur heteroatoms. That the resultant TTF cations are also open-shell systems and thus potentially metallic provides an additional attractive feature.

Above 60 K, (TTF) (TCNQ) is metallic with electron–electron interactions apparently playing a negligible role. The crystal structure consists of chains of $(TCNQ)^-$ anions in parallel with chains of $(TTF)^+$ cations as shown in fig. 2 [15]. The parallel chain structure and relatively close packing of the molecules are consistent with expectations based on the importance of the polaron polarization bond to the stability of the system [11].

The x-ray structure analysis shows uniform chains [15] with each molecule on average equidistant from its nearest neighbors along the chain. However, the possibility of a dynamic distortion (with little interchain correlation) on the one-dimensional chains cannot be excluded. Such a dynamic Peierls distortion has in fact

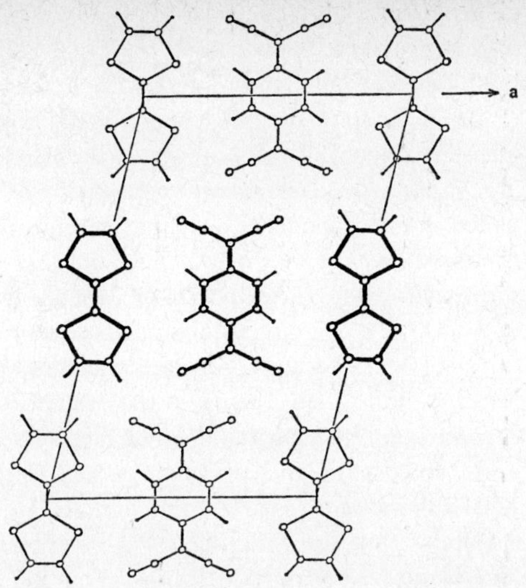

Fig. 2. X-ray crystal structure of (TTF) (TCNQ) from [15]. The lattice constant along the chain axis is 3.819 Å.

been observed in the mixed valence Pt salts, $K_2Pt(CN)_4Br_{0.3}\cdot 3H_2O$, by diffuse x-ray scattering and inelastic neutron scattering experiments [16].

The conductivity at room temperature for (TTF) (TCNQ) is the highest of any organic salt [$\sigma > 10^3$ (Ω-cm)$^{-1}$] [11] and shows the negative temperature derivative expected for metallic behavior. The thermoelectric power is negative and linear with temperature (above 100 K), again indicating metallic behavior [17].

Fig. 3. *Abscissa:* λ (μm); *ordinate:* $R_{||}$ (%).
Reflectance of (TTF) (TCNQ) for light polarized parallel to the conducting axis. The solid curve is derived from polycrystalline data, the dashed curve from single crystal data [18].

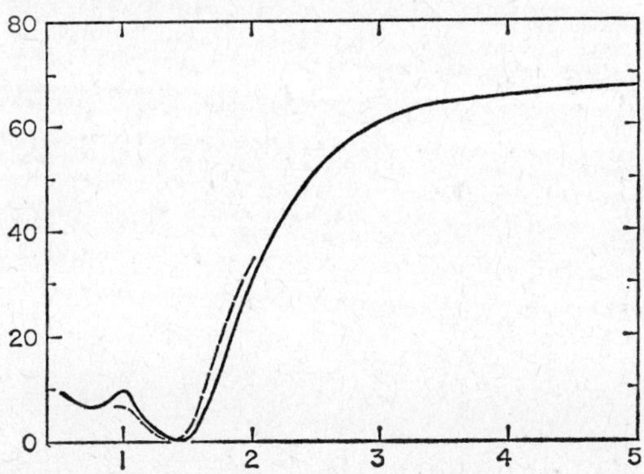

Reflectivity studies [18] on single crystals and polycrystallic compactions of (TTF) (TCNQ) have recently been completed. The single crystal reflectivity in the infrared is anisotropic, showing metallic reflection when the electric vector is polarized along the conducting (crystallographic *b*) axis. The reflectance curves are shown in fig. 3. The plasma edge is clearly evident at 1.4 μ followed by the characteristic rise in reflectivity at longer wavelengths. The peak at 1.0 μm (which again shows up only for polarization parallel to the axis) has been assigned to the onset of an interband transition which dominates the optical properties in the visible.

The dielectric function $\varepsilon(\omega)$ for a metal (ignoring interband transitions) is given by the expression,

$$\varepsilon(\omega) = \varepsilon_{\text{core}} - \frac{\omega_p^2}{\omega^2 - i\omega/\tau} \equiv \varepsilon_1 + i\varepsilon_2, \tag{5}$$

where

$$\omega_p^2 = \frac{4\pi N e^2}{m^*}, \tag{6}$$

$\varepsilon_{\text{core}}$ is the residual dielectric constant at high frequency arising from core polarizability, τ is the electronic relaxation time, ω_p is the plasma frequency, N is the electron density, and m^* is the anisotropic optical effective mass. In terms of the real and imaginary parts of ε, the reflectance is given by,

$$R = \frac{1 + |\varepsilon| - [2(|\varepsilon| + \varepsilon_1)]^{\frac{1}{2}}}{1 + |\varepsilon| + [2(|\varepsilon| + \varepsilon_1)]^{\frac{1}{2}}}, \tag{7}$$

where $|\varepsilon| = (\varepsilon_1^2 + \varepsilon_2^2)^{\frac{1}{2}}$. The parameters $\varepsilon_{\text{core}}$, ω_p, and τ were determined by fitting eqs (5) and (7) to the magnitude and frequency dependence of the experimental data. The effect of a contribution to ε_2 from the interband transition discussed above is unimportant for wavelengths greater than 1 μm. The best fit at room temperature was obtained with $\varepsilon_{\text{core}} = 3$, $\omega_p = 1.8 \times 10^{15}$ s^{-1} ($\hbar\omega_p = 1.2$ eV), and $\tau = 3 \times 10^{-15}$ s. At 85 K, $\varepsilon_{\text{core}}$ and ω_p remained the same, but the relaxation time increased to $\tau \simeq 5 \times 10^{-15}$ s. The value of $\varepsilon_{\text{core}}$ is consistent with expected core polarizability contributions for molecular crystals.

Several interesting properties of (TTF) (TCNQ) may be calculated from these values. The conventional expression relating the plasma frequency and the d.c. conductivity for a metal is,

$$\sigma = \omega_p^2 \tau / 4\pi \tag{8}$$

Using the above values for ω_p and τ, one obtains an estimate of $10^5(\Omega m)^{-1}$ for the conductivity along the b-axis, in good agreement with measured d.c. values at room temperature.

The effective mass, m^*, can be found from eq. (6). Using the value of $N = 4.7 \times 10^{27}$ m^{-3} calculated from the x-ray data, m^* is found to be $5m_e$, where m_e is the free electron mass. The optical effective mass may, in turn, be used to estimate the bandwidth. The electrical conductivity for a one-dimensional tight-binding band is given by

$$\sigma = \frac{4Ne^2\tau ta^2}{\pi \quad \hbar^2}, \tag{9}$$

where t is the tight-binding transfer integral, and a is the lattice constant. Comparing eqs (8) and (9), the effective mass is seen to be,

$$m^* = \pi\hbar^2/(4ta^2) \tag{10}$$

This expression agrees with the conventional definition, $(1/m^*) = \hbar^{-2}(\partial^2\varepsilon_k/\partial k^2)$, averaged over the occupied states of the tight-binding band. Eq. (10) is strictly valid only in the single band tight-binding limit. Deviations from this limit are expected to lead to differences between the optical and band effective masses. Using the experimental value, $m^* = 5m_e$ as obtained from ω_p, leads to a value for t of 0.13 eV. This is in agreement with estimates obtained from the spin susceptibility and the thermoelectric power data.

In the metallic region (above 60 K) the magnetic susceptibility of (TTF) (TCNQ) shows a relatively weak temperature dependence [19, 20]. Subtracting the core diamagnetism, one obtains a Pauli spin susceptibility of about 6×10^{-4} emu/mole at room temperatures, decreasing smoothly to 3×10^{-4} emu/mole at 60 K. That this is in fact a spin susceptibility has been confirmed by quantitative spin resonance experiments [21]. There is no evidence of the Curie or Curie–Weiss behavior which is the signature of localized moment formation. From the magnitude of the room temperature susceptibility, one can estimate the Fermi energy as $E_F \simeq 0.1$ eV, assuming identical tight-binding bands on the cation and anion chains. The temperature dependence of χ may be a simple band structure effect, although the factor of two change appears to be too large to be understood on the basis of a single band one-dimensional system with E_F of the appropriate value. An alternative explanation of the temperature dependence of χ has recently been proposed by Lee, Rice & Anderson [22] who show that such a temperature dependence is expected for a one-dimensional metal at temperatures below the mean-field Peierls transition temperature. In their theory, the temperature dependence is dominated by the large fluctuations present in the one-dimensional system. This suggestion is particularly interesting, for it implies that the mean-field Peierls transition temperature is above room temperature and that a dynamic distortion at $q = 2k_F$ is present at (and below) room temperature. The possibility of different behavior on the cation and anion chains is currently being studied by NMR techniques.

Studies of the proton nuclear spin-lattice relaxation time indicate metallic behavior with negligible electron–electron interaction [23]. In fig. 4, we plot the product $(\chi^2 T_1 T)$ for (TTF) (TCNQ). T_1 was measured directly using pulsed NMR techniques. The result, $\chi^2 T_1 T$ = constant, is consistent with the Korringa relation,

Fig. 4. *Abscissa:* T(K); *ordinate:* X^2T_1T.
The product $\chi^2 T_1 T$ vs T. The temperature independence and magnitude indicate that the electron–electron Coulomb repulsion is not important in the metallic state.

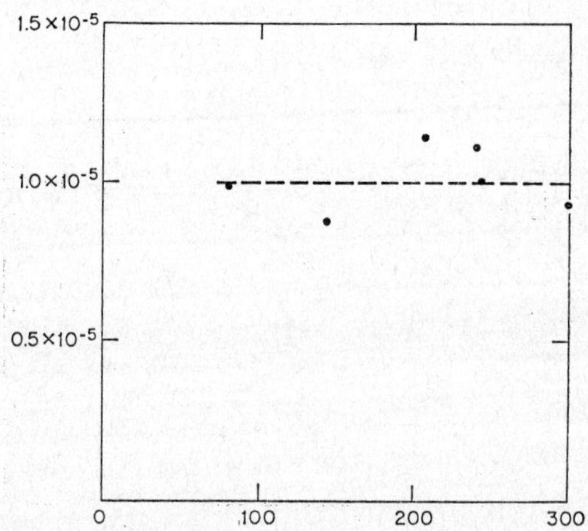

$$\frac{1}{T_1} = \frac{\pi}{\hbar\mu_B^2}\left(\frac{A}{g\mu_B}\right)^2 \chi^2 k_B T, \tag{11}$$

in the limit where electron–electron interactions are unimportant [24]. We note that even the magnitude of $\chi^2 T_1 T$ is consistent with an estimate based on using hyperfine constants [6] obtained from solution ESR studies. Earlier work on (NMP) (TCNQ) [24] has demonstrated that because of the incipient logarithmic singularity in the Lindhard function for one-dimensional systems, nuclear relaxation studies are a particularly sensitive probe into the importance of electron–electron interactions. For (NMP) (TCNQ), the rate was enhanced by nearly two orders of magnitude and a characteristic temperature dependence (related to the logarithmic divergence at $q = 2k_F$) was evident. By contrast, the simple behavior shown in fig. 4 provides strong evidence that the increased cation polarizability of (TTF) (TCNQ) has successfully reduced the strength of the effective electron–electron interaction to the point where this system may be considered a one-dimensional tight-binding metal.

At approx. 60 K, there is a phase transition to an insulating ground state, and below this temperature (TTF) (TCNQ) has the electronic properties of a non-magnetic semiconductor. The d.c. conductivity and spin susceptibility become exponentially small, characteristic of single particle excitations across a semiconductor band gap [25]. The low temperature specific heat shows only the T^3 Debye lattice contribution with no term linear in T [26]. Finally, nuclear relaxation studies at 4.2 K yield proton relaxation times of order several minutes [27]. The insulating ground state presumably arises from the Peierls instability [28] in one-dimension wherein a lattice distortion with wavenumber $q = 2k_F$ lowers the overall energy of the electron-lattice system by placing an energy gap at the Fermi energy. The sharpness of the transition suggests that a three-dimensional ordering of the distortions in different chains is involved.

The extraordinary conductivity maximum recently observed in samples of (TTF) (TCNQ) (fig. 5) has been interpreted as arising from superconducting fluctuations in this compound [1]. The magnitude of the peak conductivity simply cannot be understood on the basis of conventional metallic conduction. The unit cell volume of (TTF) (TCNQ) is approx. 840 Å^3 [15]. Allowing for four carriers per unit cell (an electron on $TCNQ^-$ and a mobile hole on TTF^+), the resulting carrier density is 4.7×10^{21} or roughly a factor of 20 times smaller than that of copper metal. For comparison, the room temperature conductivity of Cu is 6×10^5 $(\Omega - \text{cm})^{-1}$, and the mean free path has been determined as $\lambda_{Cu} = 420$ Å. Thus, if the bandwidth were the same as in Cu, a maximum mean free path greater than 1 μ would be implied. The actual bandwidth is almost certainly much less than that of a nearly free electron metal, as indicated above, and the structure is highly anisotropic. For the tight-binding one-dimensional limit, the mean free path is given by [29],

$$\lambda = v_F \tau = \frac{\sigma\pi\hbar}{2Ne^2a}, \tag{12}$$

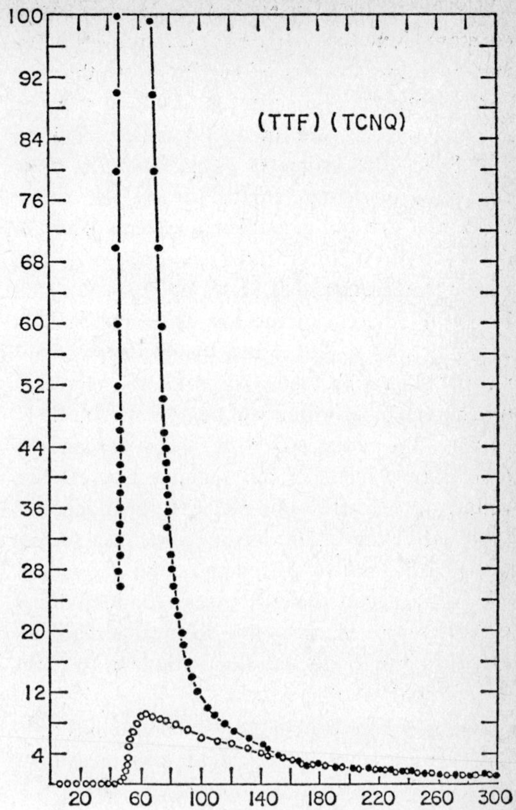

Fig. 5. Abscissa: T(K); *ordinate:* $\sigma/\sigma_{R.T.}$.
Temperature dependence of the conductivity of (TTF) (TCNQ) single crystal (●—●) and of (TTF) (TCNQ) typical crystals (○—○) [1].

where N is the carrier density, a the lattice constant along the chain, e the electronic charge, and σ the measured conductivity. Note that the band parameter, i.e., the tight-binding transfer integral, cancels out of the equation. Taking $a \simeq 3.8$ Å, one estimates $\lambda_{max} > 4\,000$ Å. Such enormous scattering lengths (nearly 10^3 lattice constants) are not possible in systems such as these at the temperatures in question. The Debye temperature of (TTF) (TCNQ) as obtained from the T^3 term in the low temperature heat capacity is 89 K [26]. Thus, the strong electron-phonon scattering that limits the room temperature conductivity would still be operative at 60 K [witness the variation of $\sigma(T)$ for Cu or any other simple metal in the vicinity of its Debye temperature where a simple T^{-1} dependence is observed]. These quantitative arguments, together with the qualitative sharpness of the σ versus T anomaly, suggest that the excess conductivity is to be interpreted as arising from superconductivity or superconducting fluctuations in the one-dimensional metallic system.

The enormous conductivity maximum has been observed in only a few of the samples measured. Typical crystals show a conductivity which follows the apparently divergent curve at high temperatures, but breaks away as the temperature is lowered, exhibiting a weaker maximum near 60 K. In fact, there is no "typical" behavior, for the values of σ_{max}/σ_{RT} vary from 5 to 30. It appears that extreme microscopic crystal perfection is required to obtain the highest conductivity. There is, in addition, some indication of strain sensitivity, with characteristically higher values found with smaller diameter gold wire contacts and smaller silver paint contract areas. However, taken at face value, the data indicate that the divergent behavior is characteristic of only rare microscopically perfect crystals of (TTF) (TCNQ).

Even the so-called typical data provide indirect evidence in favor of the paraconductivity interpretations. First of all, the values of σ_{max}/σ_{RT} vary from sample to sample in a range from 5 to 30. Such variability is not observed, for example, in (NMP) (TCNQ), where the normalized conductivity reproduces to within a few percent. Furthermore, even values of $\sigma_{max}/\sigma_{RT} \sim 20$ are difficult to understand using conventional arguments which would predict a T^{-1} dependence. It is interesting in this context that the conductivity perpendicular to the chain axis does vary as T^{-1} [30], whereas the parallel conductivity consistently shows a stronger dependence.

The varying results for σ_{max}/σ_{RT} from sample to sample suggest that the conductivity is being limited by defects (in effect, series "resistors") in the one-dimensional chains, but that the phenomenon in question is taking place in all samples. Studies of the thermoelectric power provide evidence that this is, in fact, the case [17].

Since the one-dimensional chains are extremely sensitive to imperfections, defects, and impurities, a conductivity measurement on a highly conducting chain would be limited by the resistance of any interruptions. It is therefore plausible that the measurement of an extremely high conductivity would only be observed in near-perfect crystals. The thermoelectric power, however, is a zero current transport measurement and would not be limited by breaks in the chains, assuming that most of the temperature drop occurred across the unbroken regions (which of necessity for high conductivity are much longer than the insulating regions). We would therefore expect and indeed observe only one type of behavior for the samples measured.

At high temperature 140 K $< T <$ 300 K, the Seebeck coefficient is linear with temperature [17], extrapolating through zero to within 0.5 μVK^{-1} (see fig. 6). The linear temperature dependence is characteristic of a degenerate electron gas and metallic conduction. The most sensitive

Fig. 6. *Abscissa:* T(K); *ordinate:* μV/K.
Thermoelectric power of several crystals of (TTF) (TCNQ) as a function of temperature [17].

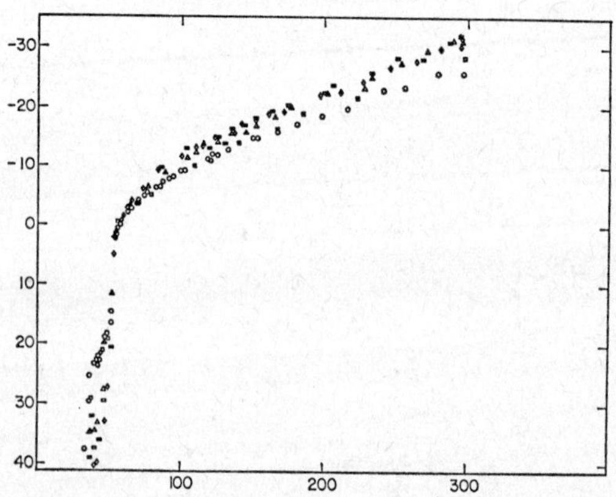

electron transport property of a metal is its thermopower. A simple calculation assuming free electrons and energy independent scattering yields the result

$$S = \frac{\pi^2 k^2 T}{3\,eE_F} \tag{13}$$

The slope of the linear portion of our data can then be used to obtain the Fermi energy. Our result of $E_F = 0.27$ eV is in agreement with the estimate obtained from the high temperature (Pauli) susceptibility and the effective mass as obtained from the plasma frequency.

The most interesting temperature region is from 140 to 57 K, where the giant conductivity maximum is observed. It is clear from the TEP data that there is a phase transition at about 56 K. It is not yet possible to determine whether the thermopower or its temperature derivative is discontinuous as this point. Below 140 K there is a deviation from the linear T dependence, with S decreasing to zero at the transition temperature. At this point, the absolute thermopower of the (TTF) (TCNQ) is zero to a precision of better than 0.2 μVK^{-1}. (During the experiment the reading near this point was about 1 μVK^{-1}, corresponding to the Au leads.)

The gradual appearance of a semiconducting gap as temperature is lowered is not a reasonable explanation of the data. In this region (140 K > T > 57 K) the conductivity in all samples is still increasing. In addition, a variety of other measurements also indicate that the samples are metallic above the transition. The thermopower itself indicates that the transition is sharp rather than gradual and hence that we are not merely following a smooth transition from electron to hole conductivity. The abrupt formation of a semiconducting gap would not explain why the thermopower tends to zero above the transition unless one invokes specific forms of electron correlations at higher temperatures which self-consistently lead to the formation of the gap.

Since the thermopower is a measure of the entropy per carrier, it is most reasonable to ascribe the tendency toward zero as the transition is approached from above to the formation of strong electron–electron correlations. It is well known that when measured statically (low frequency), the thermopower of a superconductor is zero. However, there has been no theoretical or experimental work on the effect of superconducting fluctuations on the thermopower, nor has the thermopower been analyzed for a one-dimensional superconducting system including fluctuations below the mean field transition temperature.

Below the transition, the thermopower is impurity sensitive [17] as expected for a semiconductor. However, in the region above 60 K, the thermopowers in pure and impure samples are identical.

The thermopower can be written in terms of the Boltzman transport coefficients,

$$S = K_1/(eTK_0) \tag{14}$$

where $K_n = -\frac{1}{3}\int v_k^2 \tau_k (\varepsilon_k - \mu)^n (\partial f_{0k}/\partial \varepsilon_k)\,dk$ and $\sigma = e^2 K_0$. Thus since S is the ratio of K_1 over K_0, both of which are temperature dependent, and since only one temperature dependence is observed, it is reasonable to conclude that all samples have the same intrinsic $\sigma(T)$. One might then naturally argue that the measured conductivities differ widely due to imperfections along the chains.

The transport properties in (TTF) (TCNQ) clearly require further study. If the intrinsic conductivity is indeed as large as that found in the few samples, a many-body effect is clearly involved. On the other hand, in crystals with anisotropic conductivity, inhomogeneous current distribution can lead to a small voltage drop at the measuring voltmeter which is not related to the intrinsic conductivity. This possibility, although implausible, can only be absolutely ruled out by acquiring a mass of data and by going to contactless (e.g., microwave) techniques.

III. The Electron–Phonon Interaction and the Mechanism for Paraconductivity

Obviously, a detailed understanding of the mechanism for paraconductivity in these compounds is required before appropriate molecular features can be designed into the molecular structure guided toward stabilization of the superconducting state. Two possible mechanisms have been proposed. The first is conventional BCS pairing in which the high T_c might be obtained via a treatment of the coupled electron–lattice system in the small polaron limit [1]. It has been shown that when the indirect attractive electron–electron interaction arises from a site diagonal electron–phonon coupling (i.e., phonon modulation of the ionization potential), the usual strong-coupling renormalization may not appear in the expression for T_c [1]. As a result, high transition temperatures might be expected, especially in the presence of soft lattice modes. In this point of view, the particular importance of the one-dimensional electronic structure lies in the soft lattice mode expected near the Peierls transition (T_p) [28]. As $T \to T_p$, the phonon spectrum goes soft because of a giant Kohn anomaly located at $2k_F$ where k_F is the Fermi wave number of the one-dimensional band. Although direct evidence of a soft phonon at $2k_F$ is not available for (TTF) (TCNQ), the observation of the giant Kohn anomaly in the phonon spectrum of the one-dimensional mixed-valence Pt salt, $K_2Pt(CN)_4Br_{0,3} \cdot 3H_2O$, has recently been reported [16]. A straightforward analysis of the one-dimensional system yields,

$$\omega_q = 2[(K - \gamma^2 F(qT))]/m]^{\frac{1}{2}} \sin \tfrac{1}{2}qa \qquad (15)$$

where K is the spring constant, m the molecular mass, $F(q, T)$ the one-dimensional tight-binding Lindhard function, and γ the electron–phonon coupling constant. For a detailed numerical analysis of $F(q, T)$, see [19]. Thus, as $T \to T_p$ and $\omega(2k_F) \to 0$, the lattice response diverges and an anomalously large indirect attractive electron–electron interaction might be expected. This is particularly significant for a one-dimensional system because $2k_F$ spans the Fermi surface, and would thus lead to pair–pair scattering near E_F as required for superconductivity.

It was originally proposed that the Peierls transition temperature for (TTF) (TCNQ) was 60 K. More recently, the suggestion has been made [31] that T_p is well above room temperature, and that the 60 K transition is simply a three-dimensional ordering of the distortion via interchain coupling. In this view, the one-dimensional superconductivity is limited by fluctuations into the normal state. As noted above, the temperature dependence of χ follows naturally on inclusion of the fluctuations below the mean field transition temperature [22].

An alternative suggestion was proposed by Bardeen [32], who suggested that the origin of the paraconductivity is to be associated with Fröhlich's 1954 one-dimensional model of superconductivity in which the electrons and lattice vibrations are self-consistently coupled into a travelling wave. The Fröhlich mechanism does not involve pairing. The essence of the Fröhlich idea lies in the fact that a lattice distortion at $2k_F$ places an energy gap at the Fermi energy in a one-dimensional system. If the distortion moves with the electrons when the distribution function is displaced in k-space (so as to give a current), the energy gaps then move with the Fermi surface and superconductivity might be expected. However, as pointed out by Bardeen [32], the Fröhlich superconducting state is probably unstable, relaxing to a Peierls semiconducting state below T_p. Although Fröhlich [33] has argued that in a jellium model the self-consistently coupled electron-lattice wave can have superconducting properties, consideration of the discrete tight-binding lattice suggests that the collective modes would not propagate freely. This follows since the movement of the lattice wave in a discrete lattice requires that the molecules pass through their equilibrium positions in the undistorted lattice, so that activation over (or tunneling through) an energy barrier with height of order kT_p per molecule would be involved. Only relatively near T_p would the order parameter be small enough to give significant paraconductivity. Thus, with the mechanism, the observed effect would at best be paraconductivity arising directly from interaction of the electrons with the soft lattice mode just above T_p. The Fröhlich mechanism is currently being studied theoretically, and its conductivity properties are not yet understood in detail.

The one-dimensional metal problem has been studied theoretically by Bychkov, Dzyaloshinsky & Gorkov [34]. In this paper the authors show that the electronic spectrum is generally unstable simultaneously to BCS pairing and the Peierls instability.

Whether the paraconductivity is of the conventional BCS pairing type, or relates to the 20-year-old mechanism of Fröhlich, is not yet clear; only extensive experimental work can provide the answer. However, the experiments already indicate that the electron–phonon interaction is of clear importance in leading to the giant conductivity. Other mechanisms (e.g., the indirect attraction via excitons as proposed by Little [35]) may be playing a rôle, especially since the large π-electron molecular polarizability is built into the metallic chains. However, we focus attention on the electron–phonon interaction in the following paragraphs.

The two important sources of electron–phonon coupling in the organic charge transfer salt metals are as follows: Phonon modulation (*a*) of the intermolecular electron transfer integral; (*b*) of the solid state ionization potential.

Phonon modulation of the intermolecular transfer integral is a necessary consequence of a tight-binding (LCAO) metal. The intermolecular transfer integral scales with the overlap integral, and since the latter is expected to vary exponentially with distance between molecules, a strong electron–phonon contribution is implied. Denoting $t(x)$ as the transfer integral as a function of intermolecular spacing, one can write,

$$t(x) = t_0 + (\partial t/\partial x)\,\Delta x + \ldots \simeq t_0 + \gamma_a\,\Delta x \qquad (16)$$

where γ_a gives the strength of the electron–phonon coupling. Note that as the bandwidth narrows, both t_0 and γ_a decrease, so that in the limit of very narrow bands this contribution is less important.

The phonon modulation of the solid state ionization potential also is to be expected in such ionic metals in which the polaron polarization bond plays an important role in the cohesive energy. The polaron energy, E_B, is again a strong function of distance, r^{-4}, indicating a special sensitivity to lattice vibrations. Thus again, and for different reasons, the cation and/or anion polarizability takes on special significance, and large polarizability is desirable. It is to be noted that the phonon modulation of the polaron polarization bond [11] energy, E_B, which leads to a relatively strong contribution to the electron–phonon interaction and thereby potentially to superconductivity, is not an excitonic mechanism, for the molecular polarizability enters in a quasi-static manner. An excitonic mechanism is possible only if the bandwidths become comparable with the intramolecular $\pi - \pi^*$ splittings of the molecules. Denoting $I(x)$ as the ionization potential as a function of intermolecular spacing, one can write

$$I(x) = I_0 + (\partial I/\partial x)\,\Delta x + \ldots = I_0 + \gamma_b\,\Delta x \qquad (17)$$

These two contributions to the electron–phonon interaction have the mathematical form,

$$H_{ep} = H_{ep}^{(a)} + H_{ep}^{(b)}$$

$$H_{ep} = \gamma_a \sum_l (c_{l+1}^+ c_l + c_l^+ c_{l+1})(U_{l+1} - U_l) + \gamma_b \sum_l c_l^+ c_l (U_{l+1} - U_l) \qquad (18)$$

where l is the site index and all the quantities have standard definitions. The important point to note is that $H_{ep}^{(a)}$ necessarily scales on the bandwidth whereas $H_{ep}^{(b)}$ can be large even in the limit of very narrow bands.

The electron–phonon contributions described above are viewed in the "rigid ion" approximation; i.e, the modulation of the transfer integral and ionization potential are considered while holding the molecular charges fixed. In fact, charge can redistribute simultaneously and thereby reduce the overall interaction by screening. The self-consistent treatment of the electron–phonon coupling in the tight-binding approximation has been considered by Barisic [36]. In contrast to the nearly free electron case where the rigid-ion result is reduced by the electronic dielectric constant, $\varepsilon(q, \omega)$ (simple screening), Barisic finds the rigid-ion result is reduced only by a small additive term which is itself reduced by the electronic dielectric constant. For the nearly half-filled band case, Barisic shows that the added term is unimportant, so that the rigid-ion treatment gives correct results.

Barisic's treatment of the self-consistent electron–phonon problem is particularly important in the context of the conductivity results described above for (TTF) (TCNQ). The bond lengths obtained from x-ray analysis indicate nearly complete charge transfer, so that the bands in question are approximately half-filled. As a result, one can hope to view these systems in the relatively simple rigid-ion approximation.

Independent studies of the electron–phonon interaction itself, as well as the electron–phonon scattering, are clearly of value. To this end we have analyzed the infrared optical properties described above following the suggestion of Hopfield. The dimensionless coupling constant, $\lambda_{ph} = N(0)\,V_{ph}$ (where V_{ph} is the indirect attractive electron–electron interaction via virtual phonons and $N(0)$ is the electronic density of states at the Fermi energy), may be obtained directly from the infrared data ($T > \theta_D$) through the relation [37],

$$\lambda_{ph} = \frac{\hbar}{2\pi\tau_{ph}} \frac{1}{kT}. \quad (19)$$

Using the room temperature data, one finds the strong coupling value of $\lambda_{ph}^{RT} \approx 1.3$. That this should be the correct value may be inferred from the good agreement of eq. (8) with the room temperature d.c. conductivity. This strong coupling value is consistent with the fact that an estimate of the mean free path at room temperature using eq. (12) above yields a magnitude of order one lattice constant. Thus, strong electron–phonon scattering is implied. Note that given λ_{ph} and E_F, the Peierls mean field transition temperature can be estimated directly. For a tight-binding one-dimensional metal [38],

$$k\,T_p = (4\gamma E_F/\pi e^{3/2})\,e^{-1/\lambda}, \quad (20)$$

where $\gamma = 1.7811$ is an Euler constant. Using $E_F \approx 2 \times 10^3$ K, one obtains $T_p \sim 450$ K. This intriguing result, if taken literally, once again suggests the possibility of a dynamic distortion on either cation or anion chains similar to that found recently in the mixed valence Pt salts, and is consistent with the proposal of Anderson et al. [31] that the mean field transition temperature is well above room temperature.

Conclusion

It is already quite clear that the organic charge transfer salts are an interesting and potentially important class of materials. Work on metallic systems in this class has only just begun, and we can look forward to a variety of new compounds in the future.

The materials problems are severe and deserve careful attention. Ultra-high purity must be required of any experimenter, for it has already been demonstrated [24] that impurities can give a misleading picture of the physics involved.

The mechanism for the giant conductivity can only be clarified through a wide range of experimental studies. To this end, work in progress in our laboratory includes studies of magnetic susceptibility, infrared reflectivity and absorption, spin resonance, transport properties, and nuclear magnetic relaxation rates. The goal is clearly to provide a detailed understanding of the mechanisms in order to guide the synthesis of new materials.

References

1. Coleman, L B, Cohen, M J, Sandman, D J, Yamagishi, F G, Garito, A F & Heeger, A J, Solid state comm 1973, 12, 1125.
2. For a review of the early work on the highly conducting charge transfer salts, see LeBlanc Jr, O H, in Physics and chemistry of the organic solid state (ed M M Labes) p. 123. Interscience, New York, 1967.
3. McConnell, H M, Hoffman, B M & Metzger, R M, Proc natl acad sci US 1965, 53, 46.
4. Epstein, A J, Etemad, S, Garito, A F & Heeger, A J, Phys rev 1972, B5, 952.
5. Lowitz, D A, J chem phys 1967, 46, 4698.
6. Reiger, P H & Fraenkel, G K, J chem phys 1962, 37, 2795.
7. Shibaeva, R P & Atovmyan, L O, Zh strukt khim 1972, 13, 546.
8. Murgich, J & Pissanetzky, S, Chem phys lett 1973, 18, 420.
9. Etemad, S, Garito, A F & Heeger, A J, Phys lett 1972, 40A, 45.
10. Metzger, R M, J chem phys 1972, 57, 1870; 1972, 57, 1876; 1972, 57, 2218.
11. Chaikin, P M, Garito, A F & Heeger, A J, J chem phys 1973, 58, 2336.
12. Hubbard, J, Proc roy soc London 1963, A276, 238; 1963, A277, 237; 1964, A281, 401.
13. Chaikin, P M, Garito, A F & Heeger, A J, Phys rev 1972, B5, 4966.
14. Ehrenfreund, E, Etemad, S, Coleman, L B, Rybaczewski, E F, Garito, A F & Heeger, A J, Phys rev lett 1972, 29, 269.
15. Phillips, T E, Kistenmacher, T J, Ferraris, J P & Cowan, D O. To be published; Daly, J J & Sanz, F. To be published.
16. Renker, B, Rietschel, H, Pintschovius, L, Gläser, W, Brüesch, P, Kuse, D & Rice, M J, Phys rev lett 1973, 30, 1144.
17. Chaikin, P M, Kwak, J F, Jones, T E, Garito, A F & Heeger, A J, Phys rev lett 1973, 31, 609.
18. Bright, A A, Garito, A F & Heeger, A J, Solid state comm 1973, 13, 943.
19. Ferraris, J, Cowan, D O, Walatka Jr, V & Perlstein, J H, J Am chem soc 1973, 95, 948.
20. Scott, J C, Garito, A F & Heeger, A J. To be published.
21. Etemad, S, Garito, A F & Heeger, A J. To be published.
22. Lee, P A, Rice, T M & Anderson, P W. To be published.
23. Rybaczewski, E F, Garito, A F & Heeger, A J, J Am phys soc 1973, 18, 450.
24. Ehrenfreund, E, Rybaczewski, E F, Garito, A F & Heeger, A J, Phys rev lett 1972, 28, 873.

25. Coleman, L B, Cohen, M J, Scott, J C, Garito, A F & Heeger, A J. To be published.
26. Wei, T, Etemad, S, Garito, A F & Heeger, A J. Phys lett A 1973. In press.
27. Rybaczewski, E F, Garito, A F & Heeger, A J. To be published.
28. Peierls, R E, Quantum theory of solids, p. 108. Oxford University Press, London, 1955.
29. Coleman, L B, Cohen, J A, Garito, A F & Heeger, A J, Phys rev 1973, B7, 2122.
30. Perlstein, J H, Ferraris, J P, Walatka Jr, V V, Cowan, D O & Candela, G A, Proc 18th int conf on mag and magnetic materials, Denver, 1972, p. 1494. In press.
31. Anderson, P W, Lee, P Z & Saitoh, M, Solid state comm 1973, 13, 595.
32. Bardeen, J, Solid state comm 1973, 13, 357.
33. Fröhlich, H, Proc roy soc London 1954, A223, 296.
34. Bychkov, Yu A, Gor'kov, L P & Dzyaloshinskii, I E, JETP 1966, 23, 489.
35. Little, W A, Phys rev 1964, A134, 1416.
36. Barisic, S, Phys rev 1972, B5, 932; 1972, B5, 941.
37. Hopfield, J J, Comments on solid state physics 1970, 3, 48.
38. Sham, L J. Private communication.

Key Words

organic metal
one-dimensional
superconducting fluctuations

Discussion

Hopfield: The usual mechanism for the conductivity relaxation time in metals is single electron scattering from phonons. The Drude tail of the optical absorption, for $\omega \gg$ phonon frequencies, is usually from single particle transitions in which the photon energy is equal to the difference between the electron initial and final state energies, and the phonon absorbs the necessary momentum. Such optical absorption cannot exist for $\hbar\omega >$ the band width.

The experiments shown have indicated that $\tau_{\text{optical}} \approx \tau_{\text{d.c.}}$ for photons of energy equal to about ~ 0.7 eV, while the bandwidth estimate is rather less. It seems to me that either the band width must be ~ 1 eV, or the agreement between the room temperature and the optical τ must have a rather more complicated meaning.

Schrieffer: (comment on Hopfield's question). It appears that the optical edge will be broadened even if there are no free particle-hole pair states in this range, simply by the plasmon inelastically scattering off the lattice, i.e. phonon emission. One should make a calculation to see if this effect, which basically involves the same interaction that produces the low frequency resistivity, can account for the observed width.

Hopfield: Normally in the optical conductivity the existence of plasmons is essentially irrelevant After all you get the same relaxation time a little bit below the plasma frequency as above, and below ω_p you could not possibly have a plasmon in the final state.

Fröhlich's Model of One-Dimensional Superconductivity

J. R. Schrieffer

Department of Physics, University of Pennsylvania, Philadelphia, Pa 19104, USA

The recent experimental results of Heeger, Garito et al. ([1] and p. 129) on organic charge transfer salts show a dramatic rise of the electrical conductivity σ in the material ATTF-TCNQ in the vicinity of 60 K. These results have reopened the question of the relation between superconductivity and lattice distortions in quasi one-dimensional systems. In this case, the structure corresponds to the parallel one-dimensional chains of the ATTF and of the TCNQ molecules in the crystal. The above authors have suggested that the sharp rise in σ results from paraconductivity fluctuations into a pairing type of superconducting state, in which the pairing interaction is enhanced by a soft phonon mode at the Brillouin zone boundary $Q=\pm\pi/a$, where a is the intermolecular spacing along the chains. It was proposed that this soft mode corresponds to the Peierls instability [2] which leads to a doubling of the unit cell along the chain axis. Below the Peierls temperature $T_p \sim 58$ K, the system is observed to become a narrow gap semiconductor, with the lattice distortion presumably being locked in place so that all even numbered molecules in a given chain are displaced in one direction and the odd numbered molecules in the other direction.

Some twenty years ago, Fröhlich [3] proposed on the basis of a one-dimensional coupled electron–phonon model that a superconducting state could occur in which a periodic lattice distortion of wave number $Q=2k_F$ caused by the electron–phonon interaction, moves with constant speed carrying with it an electron current. As in the Peierls state, an energy gap occurs at the Fermi surface, however, the gap persists in the current carrying state, when viewed in the translating frame. Fröhlich argued that the resence of the gap in the moving frame would lead to persistent current carrying states. Bardeen [4] has suggested that the experiments of Heeger, Garito et al. could be interpreted on the basis of the Fröhlich model, in that fluctuations to the moving Fröhlich states above the critical temperature could give rise to a large enhancement of σ, even though below this temperature lattice imperfections would likely lock the phase of the wave and render the system a semiconductor, as is observed.

We would like to point out that the Fröhlich Hamiltonian implicitly describes a structureless "one-dimensional jellium" and that when proper account is taken of periodicity effects in actual systems of interest, there is (1) a strong tendency for the period of the distortion to be commensurate with the period of the lattice and (2) a corresponding tendency to lock the phase of the distortion relative to that of the lattice. For nearly one valence electron per molecule, presumably the case in the above experiments, this phase locking suppresses the conducting Fröhlich states and only the Peierls state is physically relevant. Similar effects occur for other electron concentrations.

1. Electron-Phonon System

The Hamiltonian for a one-dimensional electron–phonon system in a periodic lattice is

$$H=\sum_{k,\sigma}\varepsilon_k n_{k\sigma}+\sum_q \omega_q N_q + \sum_{q,k,K,\sigma} g_{qkK}\, c^+_{k+q+K,\sigma}\, c_{k\sigma}\, \varphi_q$$

where $K=0, \pm 2\pi/a$ are the reciprocal lattice numbers and q and k are restricted to the first Brillouin zone ($-\pi/a$ to π/a). g_{qkK} is the electron–phonon matrix element and φ_q is the phonon coordinate,

$$\varphi_q = (a_q + a^+_{-q}) \tag{2}$$

with a and a^+ being the usual phonon operators. The first two terms in (1) describe the uncoupled electrons and phonons respectively. Fröhlich took g to be independent of k and did not include Umklapp processes, i.e. only $K=0$ was considered. This results in the Born–Oppenheimer energy for the model, i.e. the energy of system for clamped phonon coordinate $\langle\varphi_Q\rangle$, being independent of the phase of the wave corresponding to this lattice distortion relative to rest frame of the jellium. Thus, the wave can travel freely without potential barrier effects as the distortion moves. Physically, it is clear that this cannot be the correct situation for a chain of discrete molecules, particularly for one valence electron per site. In this case, the distortion occurs at $Q=2k_F=\pi/a$ and corresponds either to all even numbered molecules moving in the +direction and odd molecules moving in the −direction (denoted as +distortion), giving a low energy due to the Bragg scatterings from this displacement. Equally well, the reverse displacements can occur. Symmetry guarantees that the energy of two inverted distortions are equal. For $Q=\pi/a$, it is clear that in going from the + to − distortion, the phonon coordinate $\langle\varphi_Q\rangle$ goes from + to −, passing through zero. When $\langle\varphi_Q\rangle$ is small, the Bragg scattering is not effective in lowering the energy and a potential barrier results, the height of which is the total stabilization energy of the distorted state. Thus, rather than free translation of the distorted state, a tunneling process must occur in which the phase of the wave shifts by π between successive minima, passing over a barrier at odd multiples of $\pi/2$. In other words, for the distorted state at $Q=\pi/a$ to propagate, all the molecule pair bonds must be simultaneously broken and then new bonds formed, which requires an impossibly high energy.

Alternatively, one might consider distortions incommensurate with the lattice. While these states in principle could translate without a barrier, they are not energetically stable and will spontaneously lock their phase to be commensurate with that of the lattice. In particular, if the Fermi wave number is somewhat less than (or greater than) $\pi/2a$, Umklapp will lock the distortion to $Q=\pi/a$. For sufficiently large deviations from one electron per molecule, the lattice distortion wave number Q will jump away from π/a to a value closer to $2k_F$, subject to being a rational fraction of π/a. The stronger is the Peierls instability, the larger deviations of k_F from $\pi/2a$ will be tolerated and still maintain $Q=\pi/a$ as simple tight binding models show. Phase locking of the distortion occurs for $Q\neq\pi/a$ as for $Q=\pi/a$, in that a barrier will exist for moving the distortion pattern, as a consequence of the commensurate structure stabilization.

It is possible that a wavepacket state of the Fröhlich type moving distortion could be constructed which moves with small activation energy, say as a fluctuation above T_p. However, unless the packet is extended in space, the Peierls–Fröhlich gap will be smeared as a consequence of the imperfect Bragg scattering off of the local distortion pattern. Thus, the wavepackets would be expected to have either a rapid decay rate and not lead to large increase of conductivity or if extended in space, would have a high activation energy for motion and would be readily phase locked to the lattice.

In view of the above situation, it would appear that the pairing state is more likely the proper explanation of the observed conductivity rise in the experiments of Heeger, Garito et al. A more complete theoretical and experimental investigation of quasi one-dimensional systems is required before firm conclusions can be drawn.

References

1. Coleman, L B, Garito, A F & Heeger, A J, Solid state comm 1973, 12, 1125.
2. Peierls, R F, Quantum theory of solids, p. 108. Oxford, 1955.
3. Fröhlich, H, Proc roy soc (London) 1954, A 223, 296.
4. Bardeen, J, Phys rev lett 1973. In press.

Discussion

Dzyaloshinsky: I would like to present a rather simple argument that it is only commensurate states which are possible. Consider for instance any kind of Ginzburg–Landau theory. If a state Ψ is incommensurate, that is if $\Psi \sim e^{iqx}$ with $q \neq 2\pi n/(ma)$, n and m being integers and a the period of the crystal, then under translation by a the function Ψ is multiplied by e^{iqa}, qa being an irrational number. It is clear then that only terms of the type $(\Psi^*\Psi)^N$, where N is integer, can enter the Ginzburg–Landau expression. On the other hand, when q is commensurate, $q = 2\pi n/(ma)$, the state Ψ goes to $e^{i2\pi/m}\Psi$ under translation, and a new kind of invariants, $\Psi^m \pm (\Psi^*)^m$, can be added to the Landau–Ginzburg energy.

It means that the Landau–Ginzburg energy as a function of q is discontinuous at every point, commensurate states (q "rational") having lower value of energy than any incommensurate state (q "irrational") which is arbitrary near to this commensurate state. Therefore, a system will stick at every rational value of q avoiding irrational values.

Kohn: Quite apart from the problem of the lattice locking in a stationary Peierls-distorted state, Patton & Sham conclude that the Fröhlich 1954 "superconductor" invoked by Bardeen is not really a superconductor. A straightforward calculation of its transport properties leads to coherence vectors like those of the (non-superconducting) excitonic insulator and hence to a finite conductivity.

Continuous variation of q with temperature is found experimentally in the case of Cr.

Schrieffer: The locking of the period of the distortion so that it is commensurate with that of the lattice is analogous to the situation for chromium doped with manganese. For pure chromium, the period of the spin density wave is several percent off from being twice the lattice period. When manganese is added, with a concentration increasing to several percent the period snaps discontinuously to just twice the lattice period. Thus, for an electron concentration slightly off from one electron per atom in the one-dimensional chain, the Peierls distortion will remain at $q = \pi/a$, while for larger deviations, the distortion will presumably shift away from $q = \pi/a$, with a consequent decrease of the gap Δ, but a compensating increase of the energy lowering since the gap will again be at (or near to) the Fermi surface.

Peierls superconductors and all that

P. W. Anderson

Bell Telephone Laboratories, Murray Hill, N.J. 07974, USA, and Cavendish Laboratory, Cambridge, UK

This lecture was based on two papers which at the time of the Symposium were to appear in Phys rev lett and Solid state commun. For the benefit of these Proceedings, the abstracts of these two papers are presented below.

1. *Fluctuation effects at a Peierls transition*

P. A. Lee, T. M. Rice and P. W. Anderson

The effects of fluctuations on the Peierls transition in one-dimension are calculated by taking a functional average over variations in the order parameter. It is found that the transition is suppressed to a temperature of approximately one quarter of the mean-field transition but remains fairly sharp. The coherence length and density of states are calculated as a function of temperature and brief comparison is made to experimental systems.

This paper has been published in Physical review letters 1973, 31, 462.

2. *Remarks on giant conductivity in TTF-TCNQ*

P. W. Anderson, P. A. Lee and M. Saitoh

We comment on some features of the recently observed giant conductivity in TTF-TCNQ that are independent of any particular mechanism. We point out that a mean field T_c of several hundred degrees is required to explain the magnitude and suggest a formula based on an activation process that fits these data as well as a $(T-T_c)^{-3/2}$ law.

This paper has been published in Solid state communications 1973, 13, 595.

Work at the Cavendish Laboratory supported in part by the Air Force Office of Scientific Research Office of Aerospace Research, US Air Force, under grant No. 73-2449.

Remarks on the Theory of the Superfluid Phase Transition in 2-Dimensional He Films

S. Doniach

Department of Applied Physics, W. W. Hansen Laboratories of Physics, Stanford University, Stanford, Calif. 94305, USA

The existence of a thermodynamic phase transition for 2-dimensional systems with degenerate order parameter (superfluids, Heisenberg magnet) has long been debated since it is known that no long range order can exist in such systems [1–4].

In this summary I will describe some work [5] on the existence of a phase transition in a Ginzburg–Landau model field theory with n-component order parameter which suggests that $T_c(n) = O(1/n^2)$ for large n. I will also describe an extrapolation to $n=2$ and suggest an interpretation of recent measurements on ^{4}He films by Bretz in terms of a genuine 2-dimensional superfluid phase transition.

To discuss the phase transition, we consider the large n limit of Ferrel & Scalapino [6] and S–K Ma [7] for the model considered by Wilson in $d>2$ dimensions. The pair correlation function in the high temperature phase may be written

$$g^{-1}(k) = k^2 + \alpha(T) + \Sigma(k) \tag{1}$$

where $\alpha = \xi_0^{-2}(T - T_c^0)/T_c^0$ and $\Sigma(k)$ is the contribution from the sum of all self energy diagrams. In the large n limit Σ may be written for $d>2$ as a sum of a Hartree term plus the shielding approximation corrections of Ferrell & Scalapino (FS)

$$\Sigma(k) \approx \Sigma_H + \Sigma_{FS}(k) \tag{2}$$

where

$$\Sigma_H = T\sum\nolimits_{k<k_c} g(k) \tag{3}$$

where k_c is a cut-off parameter.

The condition for T_c becomes

$$\alpha + \Sigma(k=0) = 0 \tag{4}$$

However, as $d\to 2$ one cannot estimate (3) using the bare $g(k) = 1/k^2$ since the infrared divergence makes the sum blow up. We therefore use the renormalized g at the critical point

$$g(k) = k_c^{-2}(k/k_c)^{2-\eta(n)} \tag{5}$$

where $\eta(n)$ is the critical index of the pair correlation function.

One then finds

$$\Sigma_H(T=T_c) = T_c k_c^{-2}/\eta(n) \tag{6}$$

and, with the same assumption about g,

$$\Sigma_{FS}(k=0) = O(\eta(n))/n) \tag{7}$$

The dominant term in (4) thus becomes the Σ_H term, and one finds that T_c depends on the critical index (unlike 3-dimensional systems) through

$$\frac{T_c}{T_c^0}(1 + q^{-1}/\eta(n)) = 1 \tag{8}$$

where q is a parameter measuring the range of the interaction in the Wilson model (in terms of k_c).

In order to estimate $\eta(n)$ for $d=2$, large n, we expand in powers of $1/n$ and use the fact that Ma has shown the coefficient of the $(1/n)$ term vanishes as $d\to 2^+$ to guess that

$$\eta(n) = b/n^2 + \text{higher terms}$$

for large n. So in the large n limit (8) becomes

$$T_c/T_c^0 = qb/n^2 \tag{9}$$

which agrees with the spherical model limit of $T_c\to 0$ as $n\to\infty$. This argument suggests the existence of a thermodynamic phase transition despite the fact that the system has no long range order. Presumably below T_c there exists some kind of "quasi off diagonal long range order" (or

QODLRO) as suggested by Berezinskii [8]. We note from (7) that the Ferrell Scalapino (FS) corrections only come in to order $1/n^5$ in T_c/T_c^0.

As n decreases η will start to increase (we report elsewhere [5] an interpolation for η based on Wilson's ε-expansion) and one would like to compare the above formula for T_c with the results for the Ising model ($n=1$) with long range forces.

These may be expressed in the form

$$\frac{T_c}{T_c^0}(1+q^{-1}\ln q)=1 \tag{10}$$

following Thouless [9], where q now measures the number of neighbors exchange coupled to a given spin. Thus as $\eta(n)$ increases we may expect the form (8) to bend over to the form (10). The weakness of the FS corrections then suggests that the T_c for $n=2$ (superfluids) will be rather close to those for $n=1$ (Ising).

It would be interesting to compare the above suggestion with the results of the simple physical picture of Kosterlitz & Thouless [10, 11] for the nature of the quasi ordered superfluid state in $d=2$. In their picture one has pairs of vortex lines forming "molecules" at low temperatures. Then at high temperatures the free energy of the vortex line resulting from its entropy becomes sufficient to dissociate the "molecules" at some critical temperature. In order to compare with the above picture one would like to generalize the vortex concept to $n>2$. Possibly some of the ideas discussed by de Gennes at this meeting for "disclinations" in liquid crystals could apply to this. So far as I know this has not been done yet.

Finally, the results of Bretz [12] on the heat capacity of He films of 3–11 layers on a graphite substrate may be interpreted in terms of the above rough estimates of T_c. Bretz finds that the λ-point of bulk ^{4}He becomes "cut-off", or truncated, in these films. In the vicinity of the bulk λ-point the specific heat curve bends over and has a rather flat region to a lower temperature, $T_1(N)$, where $N=$ number of layers. Experimentally Bretz & Dash find that the onset of superflow and large thermal conductivity are associated with T_1. Below T_1 the specific heat falls off rather sharply.

T_1 varies quite rapidly with N in a manner which agrees qualitively with the N-dependence predicted for the large n limit discussed above [5]. We argue that T_1 is a true 2-dimensional λ-transition, and that the observed specific heat maximum in the vicinity of T_λ (bulk) results from short range fluctuations whose wavelength is smaller than the layer thickness. Presumably, if further improved substrates could be found, the specific heat kink at T_λ would sharpen up into a singularity.

Recent work by Elgin & Goodstein [13] suggests that a superfluid transition might also occur in a partly filled second He monolayer on graphite (the first is solidified by attraction to the substrate) at temperatures below 0.7 K.

Research supported by US Army Research Office, Durham, N.C., and by NSF through center for Materials Research at Stanford University.

References

1. Stanley, H & Kaplan, T, Phys rev lett 1966, 17, 913.
2. Mermin, N & Wagner, H, Phys rev lett 1966, 17, 1133.
3. Hohenberg, P, Phys rev 1967, 158, 383.
4. Reatto, L, Phys lett 1968, 26A, 400.
5. Doniach, S. Submitted for publication.
6. Ferrell, R & Scalapino, D, Phys rev lett 1972, 29, 413.
7. Ma, S-K. Preprint 1972.
8. Berezinskii, V, Sovjet phys JETP 1971, 32, 493.
9. Thouless, D, Phys rev 1969, 181, 954.
10. Kosterlitz, J & Thouless, D, J phys 1972, C5, L124.
11. — Ibid 1973, 6, 1181.
12. Bretz, M, Proc low temp conf, p. 13. Boulder, 1972 (and preprint 1973).
13. Elgin, R & Goodstein, D, Symposium on He films, Stevens Institute, 1973.

Discussion

Fisher: I would like to mention the series extrapolation calculations by Ritchie and myself (Phys rev 1973, B7, 480) on two-dimensional Heisenberg ($n=3$) lattice layers of thickness $l=1, 2, 3...$ and a series of experiments on layered magnetic materials by De Jongh & Miedema (discussed in their forthcoming review). From these studies I believe one can draw the conclusion thus—if there really is a non zero transition temperature T_c in the ideal $d=2$ case, then it is numerically a good deal lower (compared with the mean field T_c) than, say, in the Ising $n=1$ case. Furthermore most of the spin entropy is taken up above T_c, the specific heat thus having a rounded maximum some way above T_c (and at most a small anomaly at T_c). However, I do feel that it is still hard to think of reasons why an $n \geqslant 3$ system should actually remain stable in some sort of "ordered" state at finite temperature.

Martin: (1) There are many arguments for believing that, for $2 \leqslant n \leqslant \infty$ there is a phase transition. For $n=2$ the $1/n$ series seem no more convincing than the ε-series which give $\eta \approx 0.22$ for $n=1$ and $n=2$ and a value of γ that seems to agree with series expansions. The series expansions have been stressed by Stanley and theoretical arguments supporting this view have been presented by Dyson, Berezinskii, Sarma, and Thouless. I find it hard to believe that T_c is determined by η since this is false for $n=1$. I believe the only connection is that for $d=\varepsilon=2$ and $n=\infty$ there is no transition in which case there is also no meaning to η but you can take it equal to 0. For n finite, $T_c \neq T_c(n)$, since there are terms like these for $n=1$.

(2) It has been proven that in infinite two-dimensional systems the order parameter vanishes for $T<T_c$ but the susceptibility is ∞. It is probable that ϱ_s (which is not the order parameter) is finite. Whether on those films one can think of the order parameter as also finite in the sense that coherence extends to greater distances than the film size is less clear.

(3) In an infinite film, even though ϱ_s might be finite below T_c the critical velocity is probably zero. The onset of superflow may therefore be particularly sample dependent.

(4) It can be proven quite generally that $\varrho_s/\varrho > |\psi|^2/n$, that is, the superfluid fraction is larger than the condensate fraction. Hence de Gennes' proposed explanation of the Rednick experiment seems incorrect.

Layered Superconductors[1,2]

A. Luther, M. R. Beasley, and R. A. Klemm

Lyman Laboratory of Physics and Division of Engineering and Applied Physics,
Harvard University, Cambridge, Mass. 02138, USA

Along with the excitement surrounding the discovery that certain layered superconductors could be intercalated with organic molecules [1] came hard questions about the possibility of any unique or potentially useful properties which they might have. It was not clear how these materials differ from anisotropic type-II superconductors, beyond the obvious dimensional-crossover behavior found, for example, in layered magnets. We wish to examine briefly some of the properties of this class of materials which are, in fact, unique and outside the framework of ordinary anisotropic type-II superconductivity.

The superconductors of this class are indeed highly anisotropic. This follows from their basic structure, which can be visualized as consisting of thin (~6 Å) metallic layers separated by insulating barriers of organic molecules. For the intercalates of TaS_2, which is the most thoroughly studied case, the insulating region can be made from 3 to 50 Å wide depending on the organic molecule used. Also, the metallic layers are presumably only very weakly coupled to each other by Josephson tunneling. With this picture in mind, it is reasonable to expect that the two-dimensional character of the metallic layers can be important.

In oder to understand the consequences of this two-dimensional character, it is necessary to have a more detailed description of the superconductivity [2]. The phenomenological model introduced by Lawrence & Doniach [3] has proven to be a convenient starting point in these purposes. Using this model, the superconducting fluctuation contributions to the diamagnetic susceptibility [4], conductivity, and NMR relaxation [5] above T_c have been computed, along with studies of the establishment of the three-dimensional ordering at T_c [4]. Characteristically, the results indicate that the fluctuation contributions can be two-dimensional well above T_c, and become three-dimensional at a crossover temperature $T_0 > T_c$, which depends on the weak inter-layer coupling.

This characteristic behavior is precisely that expected by analogy to layered magnets. However, there are several important differences. An external magnetic field couples to the fluctuations in a superconductor in a different fashion than in a magnet—consequently the magnetization of a superconductor has no analog. It turns out, in fact, that the magnetization as a function of the external field has important quantum corrections. Comparing the contribution of these corrections with experiment provides one rather good test of the model used to describe these fluctuations. We return to this point below.

Under appropriate conditions, it also appears possible to enter a "two-dimensional" region well below T_c, at least for certain properties of the superconductor. Although the ordering remains three-dimensional, the response to an external probe can be essentially two-dimensional. An example is the upper critical field in the direction parallel to the layers, $H_{c2,||}$, described below. If the inter-layer coupling is sufficiently weak, the equation found for $H_{c2,||}$ reduces to that of independent two-dimensional layers, giving an infinite $H_{c2,||}$. Of course the Pauli paramagnetism of the Cooper pairs will eventually give an upper limit, but it appears from recent measurements [6] that the usual simple estimates of this limit are far exceeded.

---

1 Supported, in part, by the National Science Foundation.
2 Read by A. Luther.

Fluctuation-Induced Diamagnetism

The fluctuation contribution to the magnetization is a good test of any model for layered superconductors, since it can be measured with precision and samples a large part of the fluctuation spectrum. A detailed calculation of this quantity, based on the Lawrence–Doniach model, has been given elsewhere [7]. We wish here to present only simple physical arguments which lead to the same conclusions. Our starting point is the linearized equation describing the order parameter,

$$\frac{\partial \psi}{\partial t} = D\left\{p_{\|}^2 + \frac{2m}{Ms^2}(1-\cos p_\perp s)\right\}\psi + \frac{8}{\pi}(T-T_c)\psi \tag{1}$$

where ψ is the order parameter, $p_{\|}$ and $p_\perp$ are canonical momentum operators parallel and perpendicular to the planes, $p_i = (-i\nabla_i + 2eA_i)$, s is the distance between the layer centers, and D is the diffusion constant, $\frac{1}{3}v_F^2\tau$ with v_F the Fermi velocity and τ the scattering rate. If we consider the limit as $s\to 0$, corresponding to a continuous medium, eq. (1) reduces to the Ginzburg-Landau equations for a dirty anisotropic superconductor, with pair "masses" of m and M. We have chosen the dirty limit because experiment indicates that to be appropriate. Eq. (1) can be derived microscopically from the BCS model in the limit $\tau^2 \ll Ms^2/mv_F^2$, that is, when the scattering time for electrons in a layer is short compared to the time required for an electron to tunnel to the adjacent layer.

For the external magnetic field perpendicular to the layers, eq. (1) leads to the fluctuation spectrum, given by Landau levels with free propagation perpendicular to the layers,

$$\varepsilon_{n,k_\perp} = 4DeH(n+\tfrac{1}{2}) + \frac{2mD}{Ms^2}(1-\cos k_\perp s) + \frac{8}{\pi}(T-T_c) \tag{2}$$

The lowest eigenvalue, $\varepsilon_{0,0}=0$, defines the critical field near T_c, $H_{c2,\perp} = 4(T_c - T)/\pi eD$. Fitting to experiment [6] indicates $(E_F\tau)\sim 20$ for TaS_2 (Pyridine)$_{\frac{1}{2}}$, where $E_F = \frac{1}{2}mv_F^2$.

In order to compute the fluctuation magnetization, we use the expression of Maki & Takayama [8] for the free-energy due to fluctuations, and differentiate twice with respect to H. From eq. (2), however, we can guess the field at which the layered nature of the structure will become important. For the case $T = T_c$, it is given by the condition $eH_0/m \sim (Ms^2)^{-1}$, that is, when the difference between the $(1-\cos k_\perp s)$ and its small momentum expansion is significant. As the field is lowered, eventually only the long wavelength fluctuations are important, $(1-\cos k_\perp s)$ can be replaced by $\frac{1}{2}k_\perp^2 s^2$, and the magnetization is characteristic of a three-dimensional anisotropic superconductor. This determines a "crossover field", H_0 which is a new scale field in the layered problem.

The situation is slightly different for the magnetization in a strong field, however. In that case, yet another scale enters the problem. This scale can be understood most readily at $T = T_c$. There is then a field strength for which the characteristic fluctuation spectrum lies higher in energy than T_c. Examination of eq. (2) indicates that the field is roughly $H_{c2,\perp}(0) \sim 4T_c/\pi eD$, and for these large fields, it is no longer possible to use the classical approximation in the evaluation of the fluctuation contributions.

We thus find two scales important for the fluctuation-induced magnetization, considered as a function of H at T_c. It is worth inquiring what the implications are for experiments and their interpretation. In this respect it is useful to define a parameter $r \equiv 2mD/Ms^2T_c \approx (8/\pi)H_0/H_{c2,\perp}(0)$. Large r implies that the "crossover" field is larger than the "dynamic" scale field. This means that, as H is increased from zero, the "dynamic" cutoff field is reached before the magnetization enters the two-dimensional region. Conversely, small r means that the two-dimensional region is reached before the "dynamic" cutoff. The experimentally interesting region is for fields not too far in excess of the "dynamic" cutoff, since the magnetization decreases beyond that.

The available data on the much-studied TaS_2 (pyridine)$_{\frac{1}{2}}$ indicates that $r\sim 2$, and therefore that three-dimensional fluctuation behavior should be expected over essentially the entire range of fields (and also temperatures) over which the fluctuations are readily measurable. This is in agreement with the latest data on the fluctuation-induced diamagnetism in TaS_2 (pyridine)$_{\frac{1}{2}}$ which shows that the temperature dependence of the diamagnetism is virtually identical to that observed earlier in ordinary three-dimensional alloys [9]. Currently, measurements on other intercalates of TaS_2 with larger layer separations are underway in an attempt to observe the expected crossover behavior.

Upper Critical Fields and Pauli Paramagnetism

It is interesting to speculate about the properties of the layered compounds below T_c based on eq. (1). In particular, we consider the upper critical field for the case where the external field is parallel to the layers. For this purpose we are interested in the zero eigenvalue, and choose a gauge such that

$$\left\{\frac{-\nabla_x^2}{2m}+\frac{1}{Ms^2}(1-\cos 2eHxs) + \frac{4}{\pi mD}(T-T_c)\right\}\psi=0 \tag{3}$$

which is recognized as the equation for a particle in a periodic potential. For small H, eq. (3) reduces to the harmonic oscillator equation and yields the usual result $H_{c2}\propto(T_c-T)$. However, solving in the limit of large H gives,

$$H_{c2,\parallel}^2 \simeq \frac{m}{4Me^2s^4}\left[\frac{8}{\pi r}\left(\frac{T}{T_c}-1\right)+1\right]^{-1} \tag{4}$$

Thus, provided $r<8/\pi$, there exists a temperature $T_\infty = T_c\ [1-\pi r/8]$ at which $H_{c2,\parallel}\to\infty$. For all $T<T_\infty$, $H_{c2,\parallel}$ is infinite, reflecting the fact that the lowest eigenvalue of eq. (3) becomes field independent.

The physical interpretation of this result is that apparently for $T<T_\infty$ it is energetically favorable, in this model, for the normal core of a vortex to be confined to the region between the superconducting layers.

The present discussion has omitted the effects of the external field on the spins of the Cooper pairs. Eventually, when the Zeeman energy gained by flipping one spin in the BCS singlet, equals the condensation energy per pair, the system will become normal. This Pauli paramagnetism limit is given for simple systems by the condition $H_p = 18.4\ T_c$ (kOe). The limiting field for $TaS_2(\text{pyridine})_{\frac{1}{2}}$, with $T_c = 3.5$ K, is approx. 64 kOe. Measurements of $H_{c2,\parallel}(T)$ in large fields [6] yield $dH_{c2,\parallel}/dT$ of 93 kOe/K, and extrapolate to $H_{c2,\parallel}(T=0)\sim 230$ kOe, far in excess of the Pauli limit. For $TaS_2(\text{S Coll})_{\frac{1}{6}}$ [6], the numbers are $H_p = 58$ kOe and $H_{c2,\parallel}(T=0)\sim 310$ kOe.

If spin-orbit scattering is now introduced into the picture, it is possible to "explain" this effect at the expense of introducing a very short spin-orbit scattering time, τ_{so}. Estimates of $E_F\tau_{so}$ indicate that it must be comparable to or less than $E_F\tau$ deduced above from $H_{c2,\perp}$ data. The fact that $\tau\sim\tau_{so}$ presumably indicates strong orbital effects are present and further raises questions about the ability of existing theory to correctly describe them.

It is thus by no means certain that experiments can be interpreted by this simple model. Even if it is inadequate, however, the experimental facts still indicate peculiar behavior associated with the layered structures which remains to be understood.

We wish to acknowledge many helpful discussions with P. C. Martin. Also, one of us (M.R.B.) would like to acknowledge the generous hospitality of the Department of Applied Physics at Stanford University and the Department of Physics at the Massachusetts Institute of Technology during the time in which part of this work was carried out.

References

1. Gamble, F R, DiSalvo, F J, Klemm, R A & Geballe, T H, Science 1970, 168, 568.
2. Kats, E I, Zh eksp teor fiz 1969, 56, 1675. Translation: Soviet phys JETP 1969, 29, 89.
3. Lawrence, J & Doniach, S, Proc conf low temp phys (LT-12) p. 361. Kyoto, 1970.
4. Tsuzuki, T, J low temp phys 1972, 9, 525.
5. Cyrot, M, Solid state comm 1972, 11, 747.
6. Foner, S et al., Bull Am phys soc 1972, 17, 289. (Also, Foner, S. Private communication.)
7. Klemm, R A, Beasley, M R & Luther, A, Phys rev B. To be published.
8. Maki, K & Takayama, H, J low temp phys 1971, 5, 313.
9. Beasley, M R, Prober, D & Schwall, R. To be published.

Surface Physics

Theory of Surfaces of Simple Metals

W. Kohn

University of California, San Diego, La Jolla, Calif. 92037, USA

In this paper I shall try to convey the current view of the electronic structure of simple metals from the standpoint of the density functional formalism [1, 2] of the inhomogeneous electron gas. This formalism, using quite simple approximations for many-body effects, has been applied with, perhaps surprisingly, good results both to bare metal surfaces [3, 4, 5] and to surfaces with adsorbed metal atoms [6–10]. It appears to give a useful description of such systems which is free of adjustable parameters, conceptually elementary, and of semi-quantitative (typically 10–25%) reliability.

The subject matter of this paper is, of course, very large and therefore comprehensiveness is out of the question. A much more detailed discussion can be found in a forthcoming review by N. Lang [11]. In the following three sections I shall only touch on some of the key points: (1) a brief characterization of the density functional formalism, including its present limitations; (2) a short status report of applications to bare metal surfaces; (3) a similar report on applications to surfaces with adsorbed metal atoms.

1. The Density Functional Formalism and its Limitations

This formalism is derived from a basic theorem of Hohenberg & Kohn [1] applying to the ground state of any non-magnetic system of electrons in an external potential $v(\boldsymbol{r})$. Let $n(\boldsymbol{r})$ be the electronic density. Then there exists a universal functional of $n(\boldsymbol{r})$, $G[n(\boldsymbol{r})]$, (not explicitly depending on $v(\boldsymbol{r})$) in terms of which one may define the energy functional

$$E_v[n(\boldsymbol{r})] \equiv \int v(\boldsymbol{r})\, n(\boldsymbol{r})\, d\boldsymbol{r} + \tfrac{1}{2}\int \frac{n(\boldsymbol{r})\, n(\boldsymbol{r}')}{|\boldsymbol{r}-\boldsymbol{r}'|}\, d\boldsymbol{r}\, d\boldsymbol{r}' + G[n(\boldsymbol{r})]. \quad (1.1)$$

This functional, $E_v[n]$, has the following property: Its minimum value is the correct ground state energy of the system and is attained when $n(\boldsymbol{r})$ is the correct ground state density.

From this general and abstract starting point one may proceed in two directions for practical applications.

1A. *Direct minimization of $E_v[n]$*

One may adopt simple approximations for $G[n]$ and directly minimize $E_v[n]$ with respect to $n(\boldsymbol{r})$. In practice a gradient expansion [1] of $G[n]$ has been found useful.

$$G[n(\boldsymbol{r})] = \int d\boldsymbol{r}[g_0(n(\boldsymbol{r})) + g_2(n(\boldsymbol{r}))\,|\nabla n(\boldsymbol{r})|^2 + \dots]. \quad (1.2)$$

Here $g_0(n)$ is the (kinetic + exchange + correlation energy) per unit volume of a uniform gas of density n. This function is now known with considerable confidence over a wide range of densities [12]. The coefficient $g_2(n)$ of the following term, in atomic units, is given by

$$g_2(n) = \frac{1}{72n} - 0.0083\frac{1}{n^{\frac{4}{3}}} + 0.0042\frac{1}{n^{\frac{4}{3}}}. \quad (1.3)$$

Here the first term comes from the kinetic energy [13], the next from the exchange energy [14, 15] and the last from the correlation energy [16]. The first two terms are valid for arbitrary density, the third only in the high density limit.

There are two difficulties with the use of this expansion. The lesser is that the correlation term in (1.3) is known only in the high density limit and it is unlikely to be meaningful at realistic electron densities in the metal interior, let alone in the tenuous tail of the surface density distribution. More serious is the use of the gradient expansion altogether in the surface region, even for the jellium model. The presumed criterion for the validity of this expansion is

$$|\nabla n/n| \ll k_F, k_{TF}, \tag{1.4}$$

where k_F is the Fermi momentum and k_{TF} the Thomas–Fermi screening constant

$$k_{TF} = (8k_F/3\pi)^{\frac{1}{2}}, \tag{1.5}$$

which, for realistic densities have similar magnitudes. However, in the surface region one finds

$$|\nabla n/n| \approx k_{TF}, \tag{1.6}$$

so that (1.4) is certainly not well satisfied.

Nevertheless use of this approach by Smith [3] has yielded surprisingly good results for the surface energy of the jellium model. The reason appears to be that the classical electrostatic interactions, which are treated exactly in this approach, and the zeroth order term in the gradient expansion (1.2), which is accurately known, are the dominant parts of the total energy,[1] so that even rather crude ways of allowing for the non-uniformity of the density yield quite acceptable results. This matter requires further analysis.

In a more realistic model of a half metal, including the ionic potentials, there are, of course, rather rapid variations of the electron density in the region of the ion cores, again violating the criteria (1.4). For simple metals describable by weak pseudopotentials this problem can be quite effectively overcome by replacing the true ions by pseudo-ions and the true density by the slowly varying pseudodensity. But for metals like the noble and transition metals this approach does not work very well. We have been exploring different ways of dealing with the surfaces of these important classes of metals but are not yet ready to report on this work.

1B. *Self-consistent Equations*

From the minimal property of $E_v[n]$, eq. (1.1), one can derive an exact set of self-consistent equations [2, 16]

$$\{-\tfrac{1}{2}\nabla^2 + v_{\text{eff}}[n(r); r]\}\psi_i(r) = \varepsilon_i\,\psi_i(r) \tag{1.7a}$$

$$n(r) = \sum_{i=1}^{N} |\psi_i(r)|^2, \tag{1.7b}$$

$$v_{\text{eff}}[n(r); r] = v(r) + \int \frac{n(r')}{|r-r'|}\,dr' + \frac{\delta E_{xc}[n(r)]}{\delta n(r)}; \tag{1.7c}$$

here $E_{xc}[n]$ is derived from $G[n]$ by subtracting the kinetic energy $T_s[n]$ of a *non-interacting* electron gas of density distribution $n(r)$.

$$E_{xc}[n] \equiv G[n] - T_s[n]. \tag{1.8}$$

The total energy is given by

$$E = \sum_{1}^{N} \varepsilon_i - \tfrac{1}{2}\int \frac{n(r)\,n(r')}{|r-r'|}\,dr\,dr' + E_{xc}[n] - \int \frac{\delta E_{xc}[n]}{\delta n(r)}\,n(r)\,dr. \tag{1.9}$$

For a given functional $E_{xc}[n]$, this procedure is exactly identical to the Hartree method except for the last terms in eq. (1.7*c*) and in eq. (1.9). The difficult problem is an adequate representation of $E_{xc}[n]$. In practice what has been done is to use the zeroth order term of a gradient expansion (cf eq. (1.21)).

$$E_{xc}[n] = \int dr\, g_{xc}(n(r)), \tag{1.10}$$

where

$$g_{xc}(n) = n\varepsilon_{xc}(n) \tag{1.11}$$

and $\varepsilon_{xc}(n)$ is the exchange and correlation energy per particle of a uniform non-interacting electron gas of density n. The advantage of this approach is that not only electrostatic energies but also the kinetic energy is treated exactly, including all inhomogeneity effects, and only the dependence of the exchange and correlation energies on density gradients is neglected. As a result this more accurate treatment which was used in refs [4, 5, 6] not only presumably gives more dependable values for surface energies, work functions etc., but also correctly describes the Friedel oscillations of the density near the surface (fig. 1) which are lost in approach A.

2. The Bare Metal Surface

2A. *Charge distribution and surface energy*

We shall here briefly characterize the results obtained by the self-consistent field method [4].

[1] This is definitely the case in the high density limit.

The charge distribution and surface energy were first calculated for the jellium model. The results for the surface energy are shown as the dashed line of fig. 2. Agreement for the low-density alkali metals is quite good. However, at high densities ($r_s \leqslant 2.4$) the calculated surface energy becomes *negative* while the observed surface energies continue to rise.

Is this calculated negative surface energy real or an artifact of our calculation? This question can be definitively answered, since for high densities the surface energy can be written in powers of r_s^{-1}, with the leading term being the Thomas Fermi result [18]

$$\sigma = -0.0763\, r_s^{-9/2} + \dots \tag{2.1}$$

The negative sign has its origin in the reduction of the kinetic energy of the electrons due to their spilling out across the surface of the charge background.

When one passes from the jellium model to the more realistic pseudo-ion problem one finds two additional contributions to the surface energy: One coming from the difference between the interaction of the electrons with the pseudopotentials and with the uniform charge background respectively; the other a classical Madelung-type energy describing the work done when the metal is cleaved and the two halves are separated (without allowing the electrons to spill out). These contributions are positive and very important for high density metals. When they were added to the jellium surface energy good agreement with experiment was obtained for all but one of the simple metals which were studied (see fig. 2). The one exception was Pb whose calculated surface energy is about twice the best reported measurements.

In recent months the suggestion has been made [19, 20] that apart from minor corrections the surface energy of simple metals can be explained as due to the change of the zero point energy of the plasmon modes when the metals are cleaved. For the jellium model a rough estimate of this change by the authors gives

$$\sigma = +Cr_s^{-\frac{5}{2}} \tag{2.2}$$

where C is a positive constant. Quantitative agreement with experimental data is good. In spite of the tempting simplicity of this proposal we do not consider it sound. Our reasons are given in some detail in a forthcoming note [21]. Here we merely remark that the proposed jellium surface energy (2.2) is in obvious contradiction with the exact high density result (2.1).

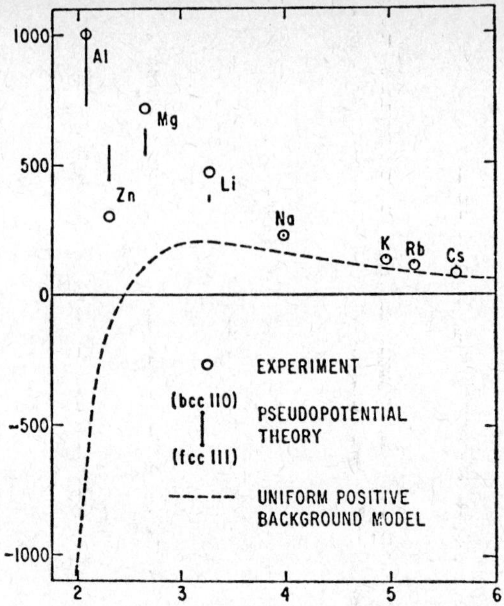

Fig. 2. *Abscissa:* r_s; *ordinate:* surface energy (ergs/cm²). Comparison between theoretical and experimental values of the surface energy, after [4].

Fig. 1. *Abscissa:* distance (Fermi wavelengths); *ordinate:* electron density.
Electron density near a metal surface for two values of r_s, after [4]. One Fermi wavelength $= 2\pi/k_F$. For $r_s = 2$ it is 3.5 Å, for $r_s = 5$ it is 8.7 Å.

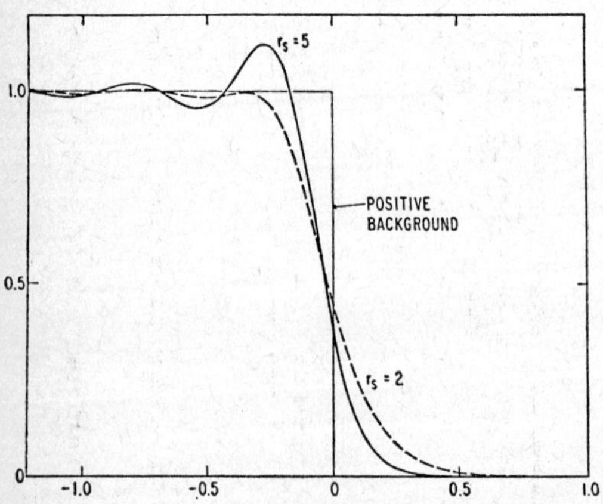

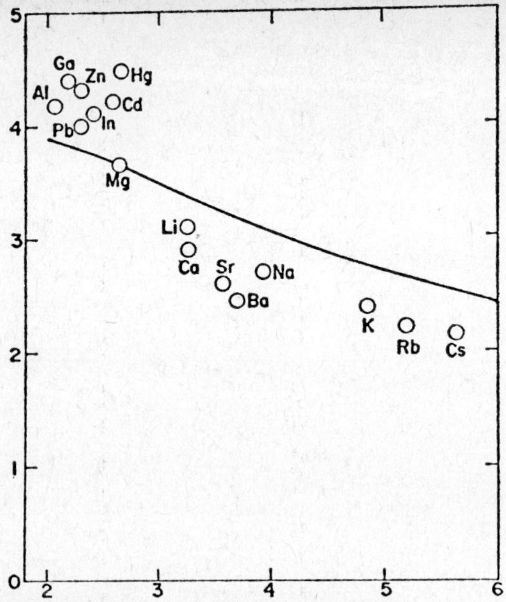

Fig. 3. *Abscissa:* r_s; *ordinate:* Φ (eV).
Comparison of theoretical work function using the jellium model, after [5] (solid line) and measured work functions for polycrystalline samples.

2B. *Work function and dipole barrier*

Work functions Φ and dipole barriers $\Delta\varphi$ were first calculated for the jellium model [3, 5]. As r_s ranges from 2 to 6, calculated values of Φ range from 3.9 to 2.4 eV while those of $\Delta\varphi$ range from 6.8 to 0.04 eV. Agreement of the calculated values of Φ with those observed in polycrystalline materials is rather good (see fig. 3).

When the effects of the pseudo-ions are included [5] the resulting (equal) changes of Φ and $\Delta\varphi$ are found to be rather small (generally less than 0.5 eV) and provide estimates of variations of the work function from one crystal face to another. The inclusion of these effects also leads to some improvement of agreement with experiment on polycrystalline samples.

The calculated values of Φ for the jellium model, when extrapolated to $r_s = 0$, strongly suggest a finite limiting value in the range

$$\lim_{r_s \to 0} \Phi(r_s) \approx 4\text{–}5 \text{ eV}. \tag{2.3}$$

Precise knowledge of this limiting value would be of great interest. Unfortunately, the Thomas Fermi theory and simple extensions thereof only yield the result

$$\lim_{r_s \to 0} r_s^2 \Phi = 0 \tag{2.4}$$

consistent with the finite limiting value in (2.3) but otherwise not informative. A recent estimation by Peuckert [22] gives the surprisingly low value of $\Phi(0) = 1.3$ eV.

Interaction with Metal Atoms

3:A. *The image force*

A preliminary problem in this area is the well-known image potential. It has been shown [8] that for an infinitesimal point charge q at a distance x, far from the surface, it is given by

$$U = -\frac{q^2}{4(x - x_0)}, \tag{3.1}$$

where x_0 is the center of mass of the surface charge induced by a uniform electric field on the metal surface. This surface charge is shown in fig. 4 for $r_s = 2$. The value of x_0 relative to the jellium edge varies between 1.5 and 1.2 a.u. as r_s varies from 2 to 6. These results are pertinent to an understanding of the relationships between the location of an adsorbed ion, its dipole moment, and its adsorption energy.

For finite values of q, such as $q = 1$, there are of course non-linear modifications of (3.1) as $x \to x_0$. These have been approximately examined

Fig. 4. *Abscissa:* x(a.u.); *ordinate:* $\delta n(x) / \int_{-\infty}^{\infty} \delta n(x)\, dn$ (a.u.).
Screening charge induced at a metal surface by a weak uniform normal electric field, after [5].

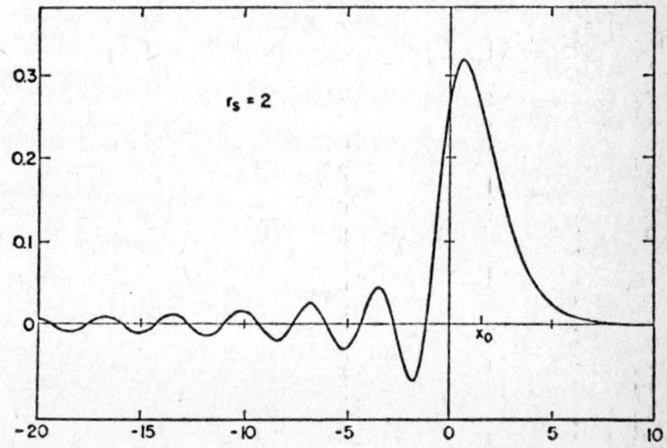

by Appelbaum & Hamann [7] and found to be quite small even when $(x - x_0)$ is as small as 1.5 a.u.

3:B. *Hydrogen adsorption*

The following model system has recently been studied by means of the electron density functional formalism [9, 10]: A metal surface with uniform charge background and an infinitesimal point charge q at an arbitrary distance x (positive or negative) from the surface. This is a complicated, three-dimensional problem. It was solved to first order in q using the direct minimization of $E_v[n]$, as described on p. 154 (sect 1:A). The results are then optimistically applied to $q = +1$ (proton) and yield remarkably sensible conclusions. For a value of $r_s = 1.5$, which gives a reasonable representation of the (110) surface of tungsten, the results for the energy of the entire system (metal surface + proton) are shown in fig. 5 as function of x. A definite minimum is found at about $x_0 = 1.0$ a.u. and the corresponding ionic adsorption energy is $E_I = 9$ eV. This should be compared with the experimental value of 11 eV. Another interesting result is the additional dipole moment as function of x. It will be seen in fig. 6 that this moment is small and changes sign near $x = x_0$. This is in qualitative agreement with the observations which give very small dipole-moments for H adsorbed on W, whose sign is still controversial.

3:C. *Submonolayers of adsorbed alkalis*

Work function changes associated with submonolayers of adsorbed alkali atoms have been studied

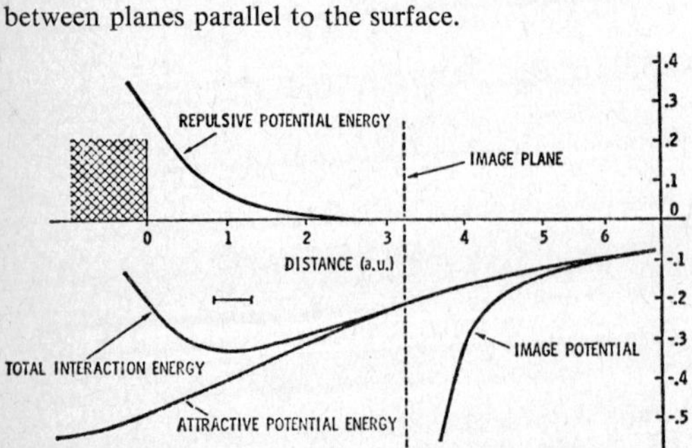

Fig. 5. *Ordinate:* energy (a.u.).
Hydrogen ion-metal interaction energy vs separation distance, after [10]. The nuclei of the surface plane of the metal are located at $-d/2$, where d is the distance between planes parallel to the surface.

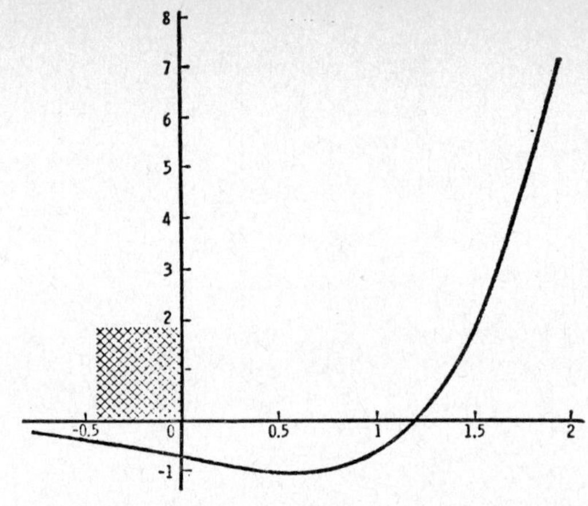

Fig. 6. *Abscissa:* distance (a.u.); *ordinate:* dipole moment $\times 10^2$ (a.u.).
Hydrogen adatom dipole moment versus the distance between the proton and the metal substrate, after [10].

as function of the fractional coverage θ. It is found that Φ drops at first linearly in θ but, at $\theta \approx 0.4$, it surprisingly begins again to rise.

This system has been modelled by Lang [6] as follows. The substrate is represented by a positive jellium and the submonolayer of adsorbed alkalis by a slab of positive charge of fixed thickness d, suitably chosen, and variable charge density $\varrho(\theta)$ such that for each value of θ, $\varrho(\theta)d$ represents the number of alkali-ions/unit area. For this model Φ was calculated and found to reproduce quite well the experimental behavior of Φ as function of θ, including the minimum, see fig. 7.

Concluding Remarks

We have reviewed the current status of density functional theories of simple metal surfaces with and without adsorbed metal atoms. In spite of the simple approximations made in these theories they appear to provide a useful semiquantitative guide for the understanding of a wide variety of experimental facts.

References

1. Hohenberg, P & Kohn, W, Phys rev 1964, 136B, 864.
2. Kohn, W & Sham, L J, Phys rev 1965, 140A, 1133.

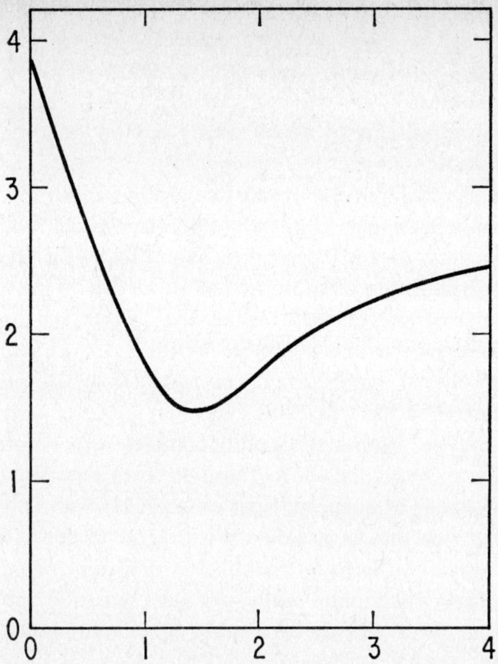

Fig. 7. *Abscissa:* r_s*; ordinate:* Φ (eV).

Calculated work function Φ of the tungsten (110) surface as function of the number of adsorbed cesium atoms.

3. Smith, J R, Phys rev 1969, 181, 522.
4. Lang, N D & Kohn, W, Phys rev 1970, B1, 4555.
5. — Ibid 1971, B3, 1215.
6. Lang, N D, Phys rev 1971, B4, 4234.
7. Appelbaum, J A & Hamann, D R, Phys rev 1972, B6, 1122.
8. Lang, N D & Kohn, W, Phys rev 1973, B7, 3541.
9. Ying, S C, Smith, J R & Kohn, W, J vac sci and tech 1972, 9, 575.
10. Smith, J R, Ying, S C & Kohn, W, Phys rev lett 1973, 30, 610.
11. Lang, N, in Solid state physics (ed H Ehrenreich, F Seitz & D Turnbull). Academic Press, New York. In press.
12. Pines, D, in Elementary excitations in solids, ch. 3. Benjamin Press, New York, 1963.
13. Kampaneets, A S & Pavlovski, E S, Zh eksp teor fiz 1956, 31, 427. (Soviet phys JETP, 1957, 4, 328.)
14. Sham, L J, in Computational methods in band theory (ed P M Marcus et al.). Plenum Press, New York, 1971.
15. Ma, S K & Ortenberger, I B. Unpublished, cited ibid, p. 469.
16. Ma, S K & Brueckner, K A, Phys rev 1968, 165, 18.
17. Kohn, W, in Tokyo summer lectures in theoretical physics, part I. Many body theory (ed R Kubo). Syokabo, Tokyo & Benjamin Press, New York, 1966.
18. Samailovich, A, Zh exptl teor fiz 1946, 16, 135.
19. Craig, R A, Phys rev. 1972 B6, 1134.
20. Schmit, J & Lucas, A A, Solid state comm 1972, 11, 415.
21. Kohn, W, Solid state comm 1973, 13, 323. (See also Feibelman, P, Ibid 1973, 13, 319.)
22. Peuckert, V. Private communication.

Discussion

Suhl: The functional formalism you propose is obviously very powerful in problems with no marked spin-density dependence. Have you considered the possibility of generalizing it to such problems?

Kohn: Yes, it has been formally generalized in papers by L. J. Sham and myself, L. Hedin and N. March. I have also heard here of recent calculations in Sweden where good results were obtained. However, I do not yet know of an effective practical way of extending this formalism to transition metals. I believe that Professor March has done some work on this problem.

B. Lundqvist: Gunnarsson and I have recently applied the spin-density-functional formalism to some systems, for which the spin dependence has shown to be very important. To test the formalism within the local-density approximation we applied it to the trivial case of atomic hydrogen. The spin of the hydrogen electron is indeed non-zero, and thus the exchange and correlation effects present in the formalism should be simulated much better by a ferromagnetic electron gas giving different potentials for electrons with different spins ("unrestricted" case) than by a paramagnetic electron gas ("restricted"). Indeed, we obtain the energy of the hydrogen atom (−13.6 eV) within 1.5% from the spin-density formalism, while a "restricted" calculation gives a result 10% from the exact value. In an identical application to sodium we find that the use of an "unrestricted" rather than a "restricted" scheme is enough to account for the discrepancy between the earlier result for the cohesive energy of the sodium metal obtained by Tong (Phys rev 1972, B6, 1189) and the experimental value.

Schrieffer: The density variable approach developed by Professor Kohn and his coworkers provides, in my view, a powerful formalism in which one can treat correlation effects in inhomogeneous electron systems. The great simplicity of dealing with a one-body effective potential allows one to attack many problems which are at present inaccessible by other methods.

I would, however, like to raise two points. Firstly, in establishing the merit of a specific theoretical approach by comparing its predictions with experiment, care must be taken to compare quantities to judge percentage error only after large constant terms have been subtracted. For example, if one calculated the binding of what is initially a positive helium ion to a metal surface, clearly most of the binding energy corresponds to the binding of the electron (or electrons for He^{2+}) in the 1s orbital of neutral He when it sits on the surface. The physically interesting quantity is the weak residual physisorption bond between the neutral He atom and the metal surface. It would be misleading to compare the total binding energy of some 23 eV of the initially positive ion with a theoretical prediction, since all but a small fraction of an eV is due to the atomic binding and a large error in the physically relevant physisorption bond could occur within a small percentage error in the total binding energy of the ion. Chemisorption is an intermediate case; however if one is dealing with hydrogen chemisorption it would seem desirable to compare the theoretical calculation with the heat of atomic chemisorption, which is of order 2–3 eV on transition metals.

Secondly, while the density variable approach is quite general, the exchange-correlation energy functional $E_{xc}(n(\boldsymbol{r}))$ is usually treated within a local density approximation, say proportional to $n^{1/3}(\boldsymbol{r})$. There are many situations in which this is a good approximation, although this is not always the case. In particular, if the Coulomb interactions on the adsorbate are quite strong, as is the case for hydrogen, there is a tendency for a localized moment to occur, which as the adsorbate moves away from the surface goes over to the unpaired electron spin of neutral hydrogen. In unrestricted Hartree–Fock (URHF) theory, the magnitude of the localized moment decreases as the adsorbate approaches the surface, and for sufficiently strong hybridization of the substrate and adsorbate orbitals the moment disappears. We know, however, from dilute alloy theory, that the actual process of spin quenching is not that described by the URHF scheme, which improperly breaks rotational invariance in spin space, but rather by a process which begins as a Kondo-like effect for large spacing and goes over to what one might call an induced covalent bond regime corresponding to a very strong Kondo situation in which charge fluctuation effects on the adsorbate remain small but spin fluctuations are strong, since a singlet state is locally formed in the vicinity of the adsorption site. Finally, if the hybridization of substrate and adsorbate orbitals is so strong that the virtual level widths on the adsorbate are comparable to the intra atomic Coulomb repulsion U on the adsorbate, a restricted Hartree–Fock approach is presumably suitable. With a sufficiently complete expression for $E_{xc}[n(\boldsymbol{r})]$, the above transition from a localized moment to quenched spin can be obtained. It is my view that in cases like H with large U, the spin correlation effects are likely important in a proper description of the chemisorbed state and that the exchange correlation functional should be constructed to take explicit account of these effects.

Kohn: Professor Schrieffer's first point is that the theory should give an accurate description of the atomic desorption energy of hydrogen which is quite small (approx. 2 eV) rather than the ionic adsorption energy which is about 11 eV. He suggests that most of the ionic energy consists of the trivial binding energy of the electron in a hydrogen atom. In response to this I would like to remark the following. First of all, by its nature, our method of calculation yields specifically the ionic binding energy. The fact that this energy correctly comes out as similar in magnitude to the hydrogen binding energy does, I believe, represent a genuine success of this method. That there is no general rule of this kind may be verified by examining the adsorption energies of the other alkali metals which behave quite differently.

As for Professor Schrieffer's second point, I find his comments very illuminating in emphasizing the important role which correlation effects can play in principle. Our point of view has been that it is difficult to ascertain a priori how important these effects would be quantitatively and that it was therefore worthwhile to carry through the very simple minded calculation on which I reported, and compare it with experiment.

The fact that both for the adsorption energy and the dipole moment we obtain results in rather good agreement with experiments, suggests that our approach may be adequate for a semi-quantitative understanding of the hydrogen problem and encourages us to pursue this direction of work further. Sooner or later we expect to run into the limitations discussed by Professor Schrieffer and we would like to find out just when this happens.

Dzyaloshinsky: If you compare your results with liquid metals (Al), should you not have any Madelung correction?

Kohn: The Madelung energy arises from a summation of multipole terms and depends only on short-range correlations between the ions. These are not very different in the liquid and solid phases. For calculational convenience, we have always computed this energy for solid surfaces but in many cases have had to compare our results with experimental data on liquid surface tension since no data for surface energies of a solid phase was available.

Halperin: I think there is a simple viewpoint which makes it clear why jellium has a negative surface energy at high densities. Jellium is only stable at zero pressure at one value of r_s. If one wishes to go to small r_s one must apply pressure, which increases the internal energy. Eventually, one will have a negative cohesive energy, and eventually it will be favorable, from the point of view of internal energy to split the jellium up into small pieces. Aluminum is stable at small r_s because of the energy gain from the Madelung energy, and in the term

$$\int (V(\mathbf{r}) - \bar{V}) n(\mathbf{r}) \, d^3r.$$

These of course are the same terms which Professor Kohn took into account in order to get reasonable results for the surface energy. Since these terms are very necessary for an understanding of the cohesive energy, especially at small r_s, it is not surprising that they must be taken into account in order to obtain reasonable surface energies.

Hedin: Regarding the question by Professor Suhl about the spin polarized case, I would like to mention that this generalization was carried out recently by von Barth and myself (J phys 1972, C5, 1629). Our theory gives the potential as a functional of the spin density matrix $\varrho_{\gamma\beta}(\mathbf{r})$ and could possibly predict say spin density waves in chromium and spiral spin structures. How accurate the result would be for say a split band calculation on a transition metal is however still an open question. The spin-polarized case has also been considered recently by Gunnarsson, Lundqvist & Lundqvist (Solid state comm 1972, 11, 149) and by Rajagopal & Callaway (Phys rev 1973, B7, 1912).

Brinkman: I believe it is misleading to say that the long wavelength part of the correlation energy is dominated by the plasmons. In particular, it is not dominated by the zero point motion of the plasmons since that term is repulsive and the total correlation energy is attractive at all wavelengths. It appears to be possible to interpret the largest contribution (the so-called self-energy of the single electron) in terms of second order perturbation theory in which the excited states contain one plasmon and a particle-hole pair. This term is clearly negative and dominates at small q.

Hedin: Professor Schrieffer raised the question about energy levels of ad-atoms on surfaces. I would like in this connection to remind about a paper by Kohn & Majumdar (Phys rev 1965, 138 A, 1617). There it is shown that for a point charge Z in an electron gas, the charge density varies completely smoothly with Z while bound energy levels appear suddenly at certain values of Z. A linear response theory used for a finite (fairly large) Z-value can thus very well give a reasonable approximation for the charge distribution while it gives no prediction at all for ad-atom energy levels. If we go beyond linear response in the Kohn–Sham theory the charge distribution should improve. The energy levels that may appear, have however no experimental significance.

Doniach: One place where I feel the neglect of spin correlation which appears to be implicit in your approach, will be important is in bulk transition metals such as Ni. Here one has some evidence, from the fact that the photoemission is little changed above the Curie temperature, that the spin fluctuation forces are important in determining the band structure.

At least for single particle excitations far from the Fermi level, one may guess that the electron sees the Coulomb potential of the opposite spin electrons in a manner rather like a random alloy. This will tend to broaden the resulting band relative to what one would calculate from a spin compensated density representation such as in your theory.

March: In the one-body potential V_{eff}, you have the functional derivative of E_{xc} with respect to the density $n(\boldsymbol{r})$. Is it possible to say completely generally the conditions under which this functional derivative exists? For the single–particle kinetic energy T_s, the functional derivative can be shown to be defined to all orders in perturbation theory, but in the manybody case, some similar perturbative assumptions (i.e. existence of an expansion in the displaced charge) appear to be involved.

Kohn: I do not believe that there is an essential difference between the single particle and the many body case. For example, in both cases one can explicitly demonstrate that in the case of a weak perturbing potential the functional derivative exists to all orders in the perturbation. This leads me to believe that this functional derivative will exist in all but some pathological conditions. However, I cannot state just what these conditions are nor can I give a mathematical proof which does not depend on the convergence of a perturbation expansion.

Renormalized Surface Atoms*

A. Luther[1], *R. E. Watson*[2] *and P. Fulde*[3]

[1] Harvard University, Cambridge, Mass. 02138, [2] Brookhaven National Laboratory, Upton, N.Y. 11973, USA and [3] Institut Max von Laue–Paul Langevin, D-8046 Garching, Germany

The complexities involved in understanding the electronic properties of d-electrons in transition metals might discourage attempts to study their properties at a surface. There are, however, some simple arguments exploiting the atomic character of the d-functions which lead to some insight into this more complicated problem.

It has been observed [1] that the renormalized atom procedure for calculating the position and width of d-levels in bulk agrees well with full band structure calculations. This procedure takes atomic d-functions, renormalized to the Wigner-Seitz sphere, for constructing the potential. As a consequence of the renormalization, charge is transferred from outside to inside the Wigner-Seitz sphere, to preserve neutrality. This results in an increase of the d-levels in bulk compared with the free atom. For Ni, this increase is approx. 0.4 Ry.

A simple physical picture emerges from these calculations for bulk, which associates an upward shift of the d-levels with the charge renormalization. It is plausible that a surface site will undergo less of an upward shift than a bulk site, since that portion of the s-function [2] as well as the d-function which extends into the vacuum is free to be atomic-like, without the necessity of renormalization. For the f.c.c. lattice, a surface site has on average only two thirds the nearest neighbors of a bulk site. One thus would guess the surface d-levels to be shifted upwards roughly two thirds the shift of a bulk site, or 0.27 Ry for Ni.

Similarly, the loss of nearest neighbors would be expected to increase the amount of time a d-electron spends at a particular site, corresponding to a narrowing of the surface "bandwidth" by approx. two thirds of the bulk value. For Ni, the reduction is from 0.3 Ry to 0.2 Ry. There results, for the neutral surface site, a lowering and narrowing of the d-states compared with bulk, and in general, a transfer of charge from s to d. Finally, charge will flow onto the surface site from bulk to fill these lowered states, and a new equilibrium will be established as the levels are pushed again upwards by the Coulomb repulsion so that the chemical potential of the surface atom equals that of the bulk metal.

This line of reasoning has been applied to Ni [3], with a resulting value for the charge transfer of approx. 0.1 electrons per spin, decreasing the hole count from the bulk value of 0.3 electrons per spin by roughly one-third. It was found quite essential to use Coulomb integrals which include relaxation, rather than the bare integrals appearing in the work of Levin et al. [2].

This approach assumes, of course, that the surface site may be treated as a bulk site, with bulk symmetry, while the d-bandwidth and position are chosen to be appropriate for the surface. In fact there will be lifting of bulk degeneracies at the surface, which can have important consequences. For example the t_{2g} states in Ni will split into a singlet and doublet at a (100) surface, and we estimate that the doublet will be completely filled, while approx. $\sim$0.1 holes per spin reside in the singlet. The total hole count is unchanged from the previous estimate, but the redistribution results in a modification of the Hartree-Fock equations used, for example, to discuss surface magnetic properties. These points are discussed in more detail in ref. [3].

It is a pleasure to acknowledge helpful discussions with H. Ehrenreich, H. Fukyama, and L. I. Hirst.

A.L. has been supported, in part, by the National Science Foundation.

Brookhaven National Laboratory is operated under the auspices of the US Atomic Energy Commission.

References

1. Hodges, L, Watson, R E & Ehrenreich, H, Phys rev B 1972, 5, 3953.
2. Levin, K, Liebsch, A & Bennemann, K H, Phys rev B 1973, 7, 3066.
3. Fulde, P, Luther, A & Watson, R E, Phys rev B 1973, 8, 440.

---

* Read by A. Luther.

Dynamical Image Charge Theory

G. D. Mahan

Physics Department, Indiana University, Bloomington, Ind. 47401, USA

At the interface between the vacuum and the plane surface of a dielectric, the electrostatic interaction of a fixed charge with the surface may be represented by image charges. The solution to this problem is presented in every elementary textbook in electricity and magnetism. The present discussion is concerned with a variation on this problem—the interaction between the charge and the vacuum-dielectric interface when the charge is moving. This is called the dynamical image charge theory [1–4]. For the sake of simplicity we will restrict the discussion to the case of charges coming normal to the surface. The treatment of non-normal incident particles is very much more complicated [5, 6]. Because of the time limitation, only metal surfaces will be treated in detail. The results for insulators will be introduced only to help guide our intuition in discussing metals.

The classical results for the image potential for a charge e both outside and inside the interface are

Outside $(z<0)$

$$V=+e^2/4z \qquad \text{(metal)} \tag{1}$$

$$V=+\frac{(\varepsilon-1)\,e^2}{(\varepsilon+1)\,4z} \qquad \text{(insulator)} \tag{2}$$

Inside $(z>0)$

$$V=0 \qquad \text{(metal)} \tag{3}$$

$$V=\frac{(\varepsilon-1)\,e^2}{(\varepsilon+1)\,4\varepsilon z} \qquad \text{(insulator)} \tag{4}$$

An explanation is needed for (3). In a metal, screening causes the potential of a point charge to have the form of a Yukawa potential except for the Friedel oscillations at large distances. The potential is very short ranged. We will always assume distances large compared to a screening length, and in this limit the image potential inside the metal is zero.

Surface plasmons and bulk plasmons are the normal modes of the metal which interact strongly with our test charge. We shall consider these the only important normal modes in the metal. When the test charge moves, it induces polarization changes in these modes. These modes, when polarized, make electric fields which act back upon the electron. This is the source of the image charge potential. The first step in the theory is to recall how these modes interact with electrons.

Bulk modes are given by the expression that $\varepsilon=0$, while surface modes are determined by $\varepsilon=-1$. These expressions are valid only for when the radiation, or retardation, corrections may be omitted. This is a satisfactory approximation in image charge theory. For a simple metal, the dielectric function $\varepsilon=1-\omega^2/\omega_p^2$ results in the frequencies ω_p for bulk plasmons and $\omega_s=\omega_p/\sqrt{2}$ for surface plasmons.

The interaction of charged particles with surface modes is described by the Hamiltonian [7–9]

$$H=H_{\text{electron}}+H_{s0}+H_i \tag{5}$$

$$H_{s0}=\sum a_k^+\,a_k\,\hbar\omega_s \tag{6}$$

$$H_i=\sum_k \Gamma_k\,e^{-kz}(a_k+a_{-k}^+)\,e^{ik\cdot\rho} \tag{7}$$

$$\Gamma_k=\left(\frac{\pi e^2\hbar\omega_s}{kA}\right)^{\frac{1}{2}}$$

where $\boldsymbol{k}$ is a two dimensional wave vector in these summations.

About the easiest problem to solve using (5) is the image potential self energy of an electron fixed a distance z from the surface. Since a fixed particle has no kinetic energy, the effective Hamiltonian for the problem is just $z<0$)

$$H=\sum_k\{\hbar\omega_s a_k^+ a_k+\Gamma_k e^{ik\cdot\rho}e^{kz}(a_k+a_{-k}^+)\}$$

This may be solved by just completing the square.

$$H=\sum_k\left\{\hbar\omega_s\tilde{a}_k^+\tilde{a}_k-\frac{\Gamma_k^2}{\hbar\omega_s}e^{2kz}\right\}$$

The last term gives the self energy

$$-\sum_k\frac{\Gamma_k^2}{\hbar\omega_s}e^{2kz}=-\pi e^2\int\frac{d^2k}{(2\pi)^2}\frac{e^{2kz}}{k}=-\frac{e^2}{4z}$$

which is just the classical image charge result. The fixed electron has an electric field which acts upon the surface modes, and polarizes them. These modes, when polarized, create electric fields which act back upon the electron. This is the source of the image potential. The bulk modes are not involved, as long as the electron is outside [10].

The above calculation, for a charge fixed in space, is rather easy. The case which is more useful experimentally and more difficult theoretically, is when the charged particle is moving. This is called the dynamical image charge problem, which was discussed by Ray & Mahan [1]. Their solution followed an earlier treatment of Lucas et al. of a relaxed problem [8, 9]. Of course, the problem cannot be solved exactly. An approximation which, once made, permits an exact solution, is to solve the case where the particle is traveling at the surface with a constant velocity.

The wave function of the surface modes, in the interaction representation, has a time development of

$$I\hbar\frac{\partial}{\partial t}\psi(t)=H_I(t)\psi(t)=\sum_k\Gamma_k e^{kz(t)}(a_k e^{i\omega_s t}+a_{-k}^+e^{i\omega_s t})\psi(t)$$

where $z(t)=vt$. If the particle started at $t=-\infty$ $z=-\infty$, the solution at zero temperature is

$$\psi(t)=\prod_k\exp[-\tfrac{1}{2}|I_k(t)|^2-iI_k(t)^*a_k^+]|0\rangle$$

$$I_k(t)=\frac{\Gamma_k e^{t(vk+i\omega_s)}}{vk+i\omega_s}$$

Use this wave function to evaluate the expectation value of the Hamiltonian

$$V(z,v)=\langle\psi|H|\psi\rangle$$

$$\langle\psi|\sum_k\hbar\omega_s a_k^+ a_k|\psi\rangle=\sum_k\hbar\omega_s|I_k(t)|^2 \tag{8}$$

$$\langle\psi|H_i|\psi\rangle=-2\sum_k\hbar\omega_s|I_k(t)|^2$$

The results may be collected into the final form $(t=z/v)$

$$V(z,v)=-\pi e^2\int\frac{d^2k}{(2\pi)^2}\frac{e^{2kz}}{k}\frac{\omega_s^2}{\omega_s^2+k^2v^2} \tag{9}$$

This integral can not be done exactly, but it may be expressed in the standard form

$$V(z,v)=-\frac{e^2\omega_s}{2v}f(2|z|\omega_s/v)$$

$$f(x)=\int_0^\infty d\alpha\exp(-\alpha x)/(1+\alpha^2)$$

Since $f(0)=\pi/2$, the potential is finite everywhere. It does not diverge at $z=0$, as in the static case. At large values of z the image potential approaches its static value $e^2/4z$. Fig. 1*a* compares the constant velocity potential with the static potential, for an electron with 1.0 eV kinetic energy, approaching a surface with a plasmon energy $\hbar\omega_p=5.0$ eV. The dynamical image potential has also been derived by Sunjic et al. [2], by Heinrichs [3] and by Richmond & Chan [4].

The image potential comes from the interaction between the electron and the surface modes. It is velocity dependent because the surface modes take time to adjust to the motion of the electron. The image potential depends not only upon z and v, but also upon the history of the motion of the particle. For example, a different result is obtained if the particle starts at $z=0$ and goes outward from the surface.

The average number of surface modes excited may be obtained from (8)

$$N(z)=\langle\psi|\sum_k a_k^+ a_k|\psi\rangle=-V(z,v)/\hbar\omega_s$$

So $N(0)$ is the number of modes excited when the electron comes from infinity to the surface. Define a quantity Q

$$Q=2N(0)=-2V(0,v)/\hbar\omega_s \tag{10}$$

Powell [11] and Ibach [12] have measured the properties of surface modes by inelastic scattering electrons from the surface. The probability of the electron exciting n excitations at zero temperature is

$$P_n = e^{-Q} Q^n / n!$$

where Q is defined as the average number of modes which are excited. This is the same Q as above, and the factor of two comes in because the electron goes from infinity to the surface and back to infinity. This gives the result for Q which was first obtained by Stern [13]

$$Q = \frac{\pi}{2} \frac{e^2}{\hbar v} \tag{11}$$

One advantage of (10) is that any improvement in estimating the dynamical image potential provides an identical improvement in the estimate of Q. For example, it is possible to estimate the maximum value of Q. According to (11), Q increases without limit as v is reduced. This is physically unreasonable. If a very slow particle approaches the surface, the strong image potential causes it to accelerate. Because $V(z, v)$ is smaller for larger v, the acceleration reduces the image potential. This problem may be solved self consistently, and provides an estimate of Q_{max}. A particle with initial velocity v_0 has initial kinetic energy $\frac{1}{2} mv_0^2$.

$$\tfrac{1}{2}mv_0^2 = \tfrac{1}{2}mv^2 - \frac{e^2 \omega_s}{2v} f(2|z|\omega_s/v) \tag{12}$$

For a constant value of v_0, the equation may be solved to obtain $v(z)$ and hence the self-consistent potential. The dotted line in fig. 1a shows this potential for our previous example.

The maximum potential depth is obtained at $z = 0$ for a particle with $v_0 = 0$. Solution of (12) in this case yields

$$v_{max} = \left[\frac{\pi e^2 \omega_s}{2m}\right]^{1/3}$$

This gives for the maximum well depth

$$V_{max} = -\left[\frac{\pi^2 e^4 \omega_s^2 m}{32}\right]^{\frac{1}{3}} = -\left(\frac{3\pi^2}{8}\right)^{\frac{1}{3}} \frac{1}{r_s} \quad \text{Rydberg}$$

and for the maximum value of Q

$$Q_{max} = \pi^{2/3} 2^{-1/2} 3^{-1/6} r_s^{-1/2} \approx 1.25\, r_s^{-1/2}.$$

Next we wish to derive the dynamical image potential for the constant velocity particle inside of the solid surface. Here we have to consider the dynamical response of both the surface and the bulk modes. Let us begin by first deriving the result obtained from the classical theory of Heinrichs [3] and Richmond & Chan [4]. One starts with Poisson's equation for a charge moving with constant velocity

$$\nabla \varepsilon(z) \nabla \phi(\boldsymbol{r}) = 4\pi e \partial(x)\, \partial(y)\, \partial(z - vt)$$

After Fourier transforming in time, and spatially in the x and y directions,

$$\left(\frac{\partial}{\partial z} \varepsilon \frac{\partial}{\partial z} - \varepsilon k^2\right) \phi = \frac{4\pi e}{v} e^{+i\omega z/v}$$

This equation may be solved in a straightforward way by matching fields at the surface

$(z < 0)\ \phi(z, \boldsymbol{k}, \omega)$

$$= \frac{4\pi e v}{\omega^2 + k^2 v^2} \left[\varepsilon^{+i\omega z/v} - \left(\frac{\varepsilon(\omega) - 1}{\varepsilon(\omega) + 1}\right) \varepsilon^{kz}\right] \tag{13a}$$

$(z > 0)\ \phi(z, \boldsymbol{k}, \omega)$

$$= \frac{4\pi e v}{\omega^2 + k^2 v^2} \frac{1}{\varepsilon(\omega)} \left[\varepsilon^{+i\omega z/v} + \left(\frac{\varepsilon(\omega) - 1}{\varepsilon(\omega) + 1}\right) \varepsilon^{-kz}\right] \tag{13b}$$

The second term gives the induced potential, and the dynamical image potential is just 1/2 of this. The factors of $\varepsilon(\omega)$ are the same as (1)–(4) in the insulator case. For $z < 0$, do the frequency integral first. Close the contour in the upper half plane, and pick up the pole at $\omega = +ikv$. The poles of $\varepsilon + 1$ are infinitesimally in the lower half plane, and are not included. This gives the result

$$V(z, v) = -\pi e^2 \int \frac{d^2 k}{(2\pi)^2} \frac{\varepsilon^{2kz}}{k} \frac{\varepsilon(ikv) - 1}{\varepsilon(ikv) + 1}$$

which has been derived before [2, 3, 4]. This is identical to the earlier result (9). So the classical analysis gives the same result as the earlier one.

Inside the surface, with $z > 0$ and $t > 0$, we

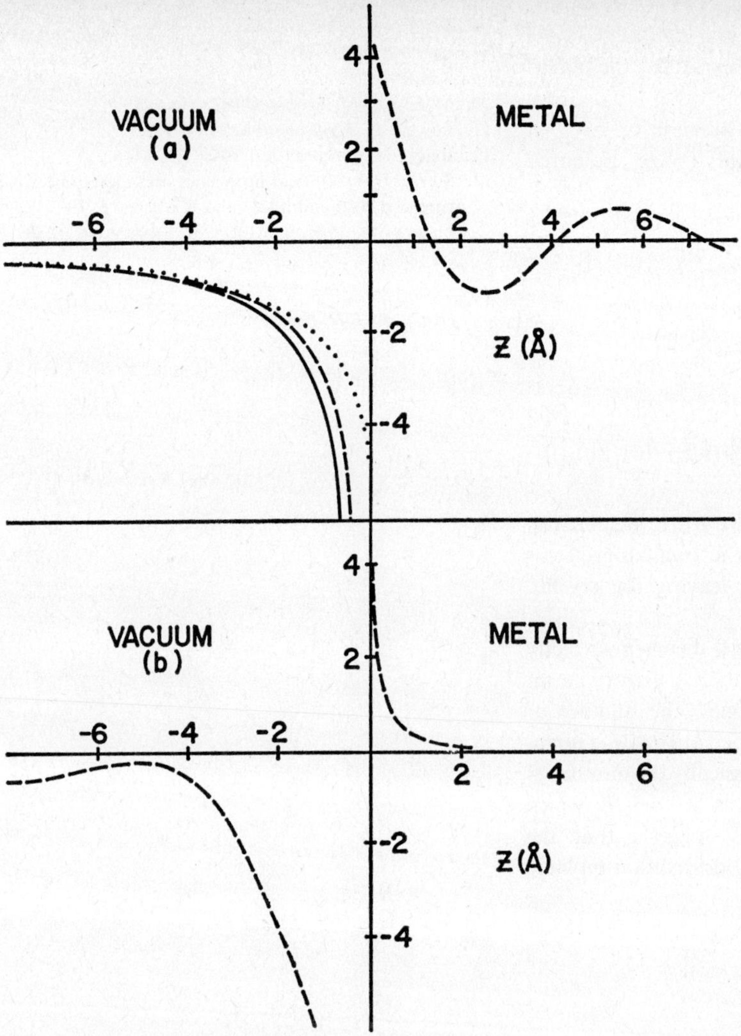

Fig. 1 *Ordinate:* image potential (eV).
The classical result for the dynamical image potential for an electron of kinetic energy 1 eV approaching a metal surface, with a volume plasmon energy of 5 eV. (*a*) The electron is going from vacuum to metal. ——, $e^2/4z$; ---, dynamical image potential; · · · ·, self consistent potential; (*b*) here the electron is going from metal to vacuum.

close the contour in the lower plane, and pick up the poles of $\varepsilon+1$, and also ε.

This gives the result

$$V(z, v) = \pi e^2 v \int \frac{d^2k}{(2\pi)^2} \left[\frac{\omega_s^2 kv}{(\omega_s^2 + k^2v^2) \cdot (\omega_p^2 + k^2v^2)} + \frac{2\omega_p \sin(\omega_p z/v)}{\omega_p^2 + k^2v^2} - \frac{2\omega_s \sin(\omega_s z/v)}{\omega_s^2 + k^2v^2} \right]$$

$$V(z, v) = \frac{e^2}{2} [(f(2z/z_B) + 2g(z/z_B) \sin(z/z_B))/z_B - (f(2z/z_s) + 2g(z/z_s) \sin(z/z_s))/z_s] \quad (14)$$

where

$$g(x) = \int_0^\infty \frac{\alpha \, d\alpha}{1+\alpha^2} e^{-x\alpha}$$

$$z_{s,B} = v/\omega_{s,B}$$

This has also been obtained by Heinrichs [3]. It is plotted in fig. 1*a* for $z > 0$. The dynamical image potential oscillates for $z > 0$. This occurs because the electron, in approaching the surface, creates real surface modes. The electron continues to interact with these modes as it leaves the surface.

This is best illustrated by calculating the potential for an electron going the other way—from inside the metal out to the vacuum, or $z = -vt$. Reevaluating the dynamical image potential gives

$(z>0)$

$$V(z, v) = \frac{e^2}{2}[f(2z/z_B)/z_B - f(2z/z_s)/z_s]$$

$(z<0)$

$$V(z, v) = -\frac{e^2}{2z_s}[f(2|z|/z_s) + 2g(z/z_s)\sin(|z|/z_B)]$$

Now the oscillatory terms are when the electron is outside of the surface. The oscillatory terms occur when the electron is leaving the crystal. This is plotted in fig. 1*b*.

We believe that the classical theory incorrectly represents the strength of the oscillatory term. In a quantum mechanical view the number of bulk and surface modes which are made is quantized to integer values. Classically the number of surface and bulk modes equals $Q = \pi e^2/2\hbar v$ which need not be integer. This suggests that the correct quantum mechanical description replaces (14) by

$$V(z, v) = \frac{e^2}{2}[f(2z/z_B)/z_B - f(2z/z_s)/z_s] + \frac{N_B \hbar\omega_B}{\pi} g(z/z_B)\sin(z/z_B) - \frac{N_s \hbar\omega_s}{\pi} g(z/z_s)\sin(z/z_s)$$

where N_B and N_s are the actual (integer) number of modes which are excited.

References

1. Ray, R & Mahan, G D, Phys lett 1972, 42A, 301.
2. Sunjic, A, Toulouse, G & Lucas, A A, Sol state comm 1972, 11, 1629.
3. Heinrichs, J, Phys rev 1973, B8, 1346.
4. Richmond, P & Chan, D. In press,
5. Sak, J, Phys rev 1972, B6, 3981.
6. Evans, E & Mills, D L, Sol state comm 1972, 11, 1093.
7. Wang, S Q & Mahan, G D, Phys rev 1972, B6, 4517.
8. Lucas, A A, Kartheuser, E & Badro, R G, Phys rev 1970, B2, 2488.
9. Sunjic, M & Lucas, A A, Phys rev 1971, B3, 719.
10. Mahan, G D, Phys rev 1972, B5, 739.
11. Powell, C, Phys rev 1968, 175, 972.
12. Ibach, H, Phys rev lett 1970, 24, 1416.
13. Stern, E A, Optical properties and electronic structure of metals and alloys (ed F Abeles) p. 396. Wiley, New York, 1966.

Collective Cohesion of Macroscopic Surfaces

A. A. Lucas

Institute of Physics, University of Liège, Belgium

Summary

Some comments are made regarding the contribution of collective excitations to the surface tension of solids and to the dispersion forces between voids.

By macroscopic surface we mean a surface whose principal radii of curvature are everywhere much larger than the interatomic spacing. The surface may be the external boundary of a plain sample or the walls of an internal cavity or void, but the concept does not strictly apply to such material inhomogeneities as small atomic vacancy clusters, the edge of a crystal, the tip region of a crack, etc.

The surface free energy or surface tension $\gamma(T)$ of a macroscopic surface may be defined as the excess free energy per unit area that a finite sample possesses over the same amount of matter in an infinite medium. For pure substances, it is composed of two terms

$$\gamma(T) = \sigma(T) - TS_s \tag{1}$$

where $\sigma(T)$ is the surface (internal) energy and TS_s the latent heat of formation accompanying the creation of a unit area of the surface. S_s is the specific surface entropy, configurational, vibrational and otherwise.

Accurate values of the surface tension of the elements have been obtained for their liquid phase over some temperature range. These data, extrapolated to lower temperatures, have also provided estimates of γ for elemental solids. Fig. 1 gives the experimental result as a function of atomic number, as reviewed by Allen [1]. One observes an overall periodicity of this surface property similar to that of the bulk cohesion. Fig. 2 indeed confirms the narrow linear correlaion that exists between γ and the bulk cohesive energy E_c, both expressed in eV/atom. However, an important point to notice is the systematic departure of the actual proportionality coefficient from the value that one would obtain from a simple model of broken bonds. If nearest neighbor bonds only are included, the relationship should be

$$\gamma = \frac{Z_b - Z_s}{Z_b} E_c \quad \text{(eV/atom)} \tag{2}$$

where Z_b and Z_s are the bulk and surface coordination numbers, respectively. For fcc crystals $(Z_b - Z_s)/Z_b$ takes on values $\frac{1}{4}, \frac{1}{3}, \frac{5}{12}$ for the principal faces (111), (100) and (110) respectively. The average coefficient $\frac{1}{3}$ is nearly twice as large as the observed correlation coefficient 0.18, indicating a systematic overestimation of the surface tension in this model.

This observation suggests that there is no a priori necessity for the various contributions to the surface cohesion to have the same relative values as their corresponding terms in the bulk cohesion problem. A particularly striking illustration of this is provided by the case of ionic crystals. Their bulk binding energy is essentially contributed by the Madelung electrostatic energy, the repulsion terms amounting to some 10% and the van der Waals dispersion energy giving but 2 or 3% of the total cohesion. In contrast, the van der Waals term contributes no less than 40% to the surface energy of the cleavage planes (100) and (110) [2]. These large differences are easily traced down [2] to the fact that the definition of the surface binding energy involves cleaving the crystal into neutral pieces whereas the bulk cohesion refers to splitting the solid into positive and negative ions. Manifestly in this case, bulk and surface cohesions are two qualitatively different problems.

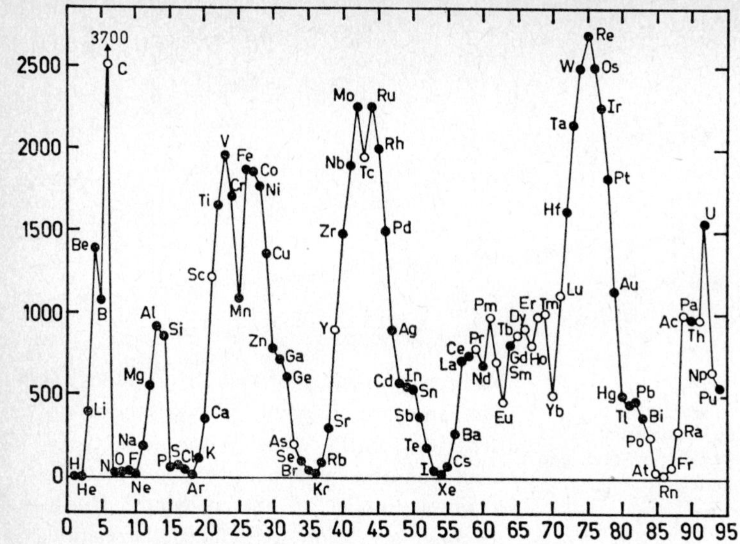

Fig. 1. *Abscissa:* atomic number; *ordinate:* γ (erg/cm²). ○, estimated values.

The surface tension of elements at their melting temperature.

A more relevant case is provided by rare-gas solids. For solid Argon, say, the ratio between the total many-atoms dispersion energy and the pair interaction energy is 95% [3–5], leaving some 5% contribution from the more-than-two-body interactions. At the surface on the other hand, the latter contribution rises to as much as 15% of the total surface tension [6]. This can be understood by the following qualitative argument. Let E_2, E_3, ... designate binding energies for pair, triplet, ... of neighboring atoms. If Z is the coordination number, the number of multiplets of n atoms that can be formed with an atom and its neighbors is of order Z^{n-1}. This means that in forming the surface, a number of order $Z_b^{n-1} - Z_s^{n-1}$

Fig. 2. *Abscissa:* cohesion energy (eV/atom); *ordinate:* σ (eV/atom).

Correlation between surface tension and cohesive energy of elements. The straight lines correspond to the prediction of a broken bond model for different crystal faces.

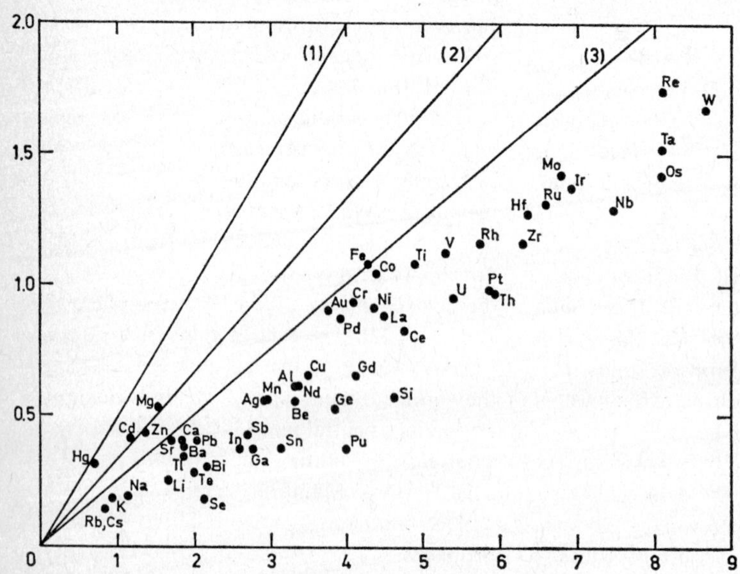

of "multiplet bonds" have to be cut. The bulk and surface cohesive energies are therefore of order

$$E_c = E_2 + \sum_{n\geqslant 3} E_n, \quad (3)$$

$$\sigma = \left(1 - \frac{Z_s}{Z_b}\right) E_2 + \sum_{n\geqslant 3} \left[1 - \left(\frac{Z_s}{Z_b}\right)^{n-1}\right] E_n \quad (4)$$

respectively. Since $Z_s/Z_b < 1$, one sees from (4) that the multiplet terms are less reduced than the pair term with respect to their bulk counterpart. Hence, the many-electron correlations, which give rise to the multi-atom interactions, should assume a relatively more important role in the total surface cohesion. This would be particularly true when the convergence of the perturbation series (4) is slow like for the case of very polarizable ions (e.g. in bromide or iodide salts) or when the series does not converge at all which is the case of metals.

In metals, the long-range dispersion energy may be expressed as a zeropoint plasmon oscillation energy and constitutes a substantial part (some 20%) of the total bulk cohesion [7]. One may estimate the long-range correlation energy of a metal surface by calculating the shift in plasmon zero-point energies when the new surface is introduced. All one has to do, as a first approximation, is to extrapolate to zero-separation the Lifshitz theory of adhesion [8] as formulated by van Kampen et al. [9]. To obtain a meaningful result, one must take into account the appropriate sum-rules of the problem. The first sum-rule concerns the frequencies of the characteristic surface plasmons. For a gap of finite width d, in the long wavelength limit, these are given by

$$\omega_\pm(k, d) = \frac{\omega_p}{\sqrt{2}} (1 \pm e^{-kd})^{\frac{1}{2}} \quad (5)$$

and satisfy

$$\omega_+^2(k, d) + \omega_-^2(k, d) = \omega_p^2 \quad (6)$$

for every k and d.

The plasmon energy required to open the gap is

$$\sigma(d) = \frac{1}{(2\pi)^2} \int d^2k \, \tfrac{1}{2} \hbar [\omega_+(k, d) + \omega_-(k, d) - \omega_+(k, 0) - \omega_-(k, 0)]$$

$$= \frac{1}{(2\pi)^2} \int d^2k \, \tfrac{1}{2} \hbar (\omega_+ + \omega_- - \omega_p) \quad (7)$$

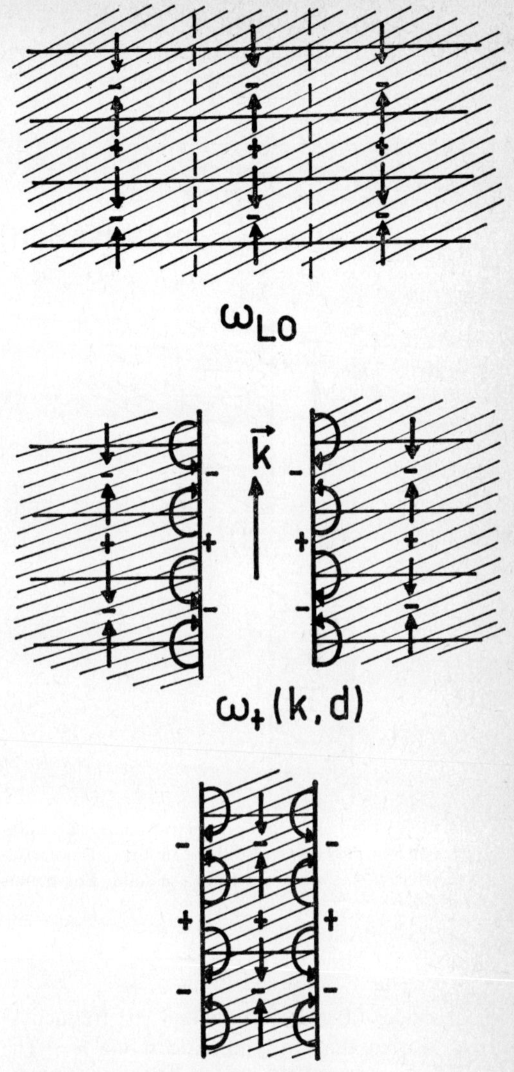

Fig. 3. Generation of the symmetrical surface oscillation mode from the longitudinal bulk mode when a slab is extracted from the bulk solid. For metals ω_{LO} is the bulk plasma frequency.

which is essentially positive due to the sum-rule (6). Notice that the zero-d limit of the surface modes (5) are $\omega_+(0) = \omega_p$ and $\omega_-(0) = 0$. This relates to the second sum-rule which concerns the eigenvectors of surface plasmons. Figs 3 and 4 illustrate the mechanism for the generation of the long wavelength surface plasmons when a slab is extracted from the sample. For each $\boldsymbol{k}$ parallel to the future surfaces the initial longitudinal bulk mode ω_p degenerates into the symmetrical surface mode whereas the initial transverse or

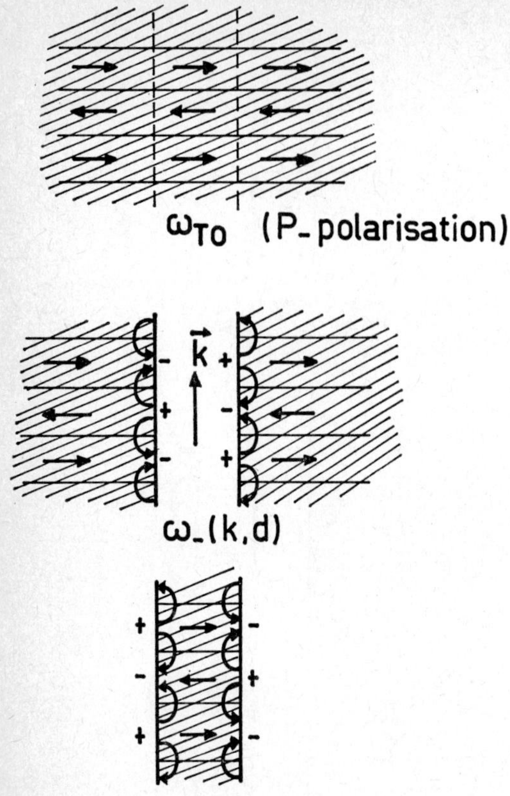

Fig. 4. Generation of the antisymmetrical surface oscillation mode from the transverse shear mode of P-polarisation when a slab is extracted from the solid. For metals ω_{TO} is zero.

shear mode of P-polarization (and zero frequency) gives rise to the antisymmetric mode ω_{-}. The other shear motion, with S-polarisation (flow lines normal to $\boldsymbol{k}$ and parallel to the surfaces) is undisturbed by the cut and contributes nothing to the surface energy. Thus, for long wavelength, the redistribution of collective modes when a gap is produced is such that the newly created surface plasmons represent the vestige of formely bulk modes which have been suppressed by destroying the translational invariance normal to the new surfaces. This has been designated as the "Begrenzung sum-rule" in reference to the so-called Begrenzung effect observed [10] in fast electron spectroscopy of collective oscillations in thin films [11]. Typically, for sufficiently thin slab, the loss peak in the electron transmission spectrum at the characteristic frequency of bulk longitudinal oscillations is nearly cancelled by the appearance of the lower frequency surface oscillation spectrum. This is schematically illustrated in fig. 5.

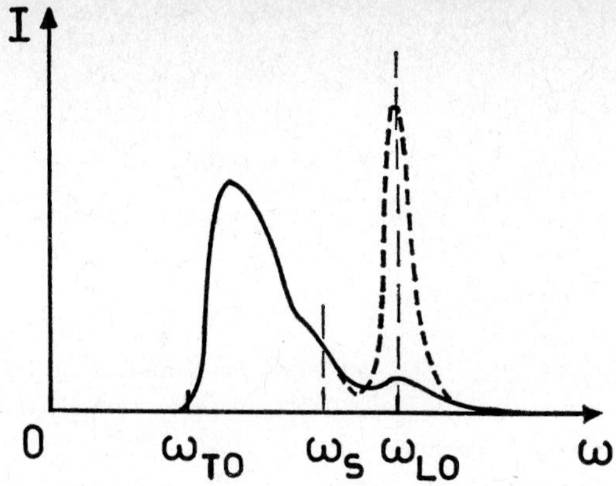

Fig. 5. Schematic illustration of the Begrenzung effect. I, intensity of a fast electron beam transmitted through a thin dielectric film as a function of energy loss $\hbar\omega$. The bulk spectrum of the film (---) is suppressed by the surface phonon loss spectrum.

For short wavelength, the continuum electrostatic approximation breaks down, no clear cut distinction remains between collective and individual excitation spectrum and, within the simple approach discussed here, the only sensible procedure is to extrapolate the Begrenzung sum-rule to some limiting wave-vector k_c. The plasmon

Fig. 6. Choice of cut off wave vector for surface plasmons. The plasmon dispersion is neglected and the k^2-dependence of the edge of the single excitation spectrum is also disregarded. These two approximations tend to compensate each other.

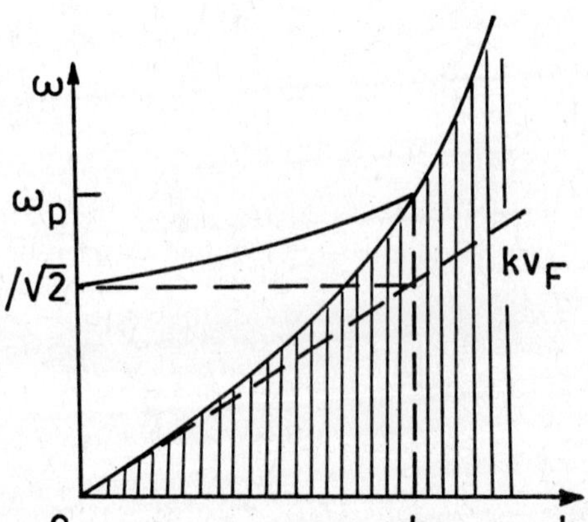

surface energy is then one-half the limit of expression (7) for $d \to \infty$:

$$\sigma_{LR} = \tfrac{1}{2} \frac{1}{(2\pi)^2} \int_0^{k_c} d^2k \, \tfrac{1}{2} \hbar (2\omega_s - \omega_p) \qquad (8)$$

Further neglecting the $\boldsymbol{k}$-dependence in the surface and bulk plasmon dispersion relations and the k^2-dependence of the edge of the individual excitation spectrum (fig. 6), the natural cut off to use is $k_c = \omega_p / \sqrt{2} v_F$ and (8) then reduces to

$$\sigma_{LR} = \frac{\sqrt{2}-1}{16\pi} \hbar \omega_p k_c^2 = C r_s^{-5/2} \qquad (9)$$

where C is a universal constant and r_s the free-electron density parameter. This scaling law is compared with experimental surface energies in fig. 7. The result (9) and the relationship between the plasmon surface cohesion and the Lifshitz theory of adhesion have been discussed elsewhere [6, 12].

An interesting distinctive feature of the empty gap geometry is that, when d is much smaller than the vacuum plasma wavelength λ_p, the symmetric mode $\omega_+(k, d)$ tends towards ω_p when k tends to zero, even when retardation effects are included [13]. This is related to the intuitively obvious fact that if real charges of same sign were accumulated on each face of a narrow gap by some external source and if the latter was suddenly switched off, the surface charges would dissipate into the bulk with a very fast characteristic time, of order ω_p^{-1} (see fig. 8*a*). On the contrary, for a solid slab in vacuum, the mode $\omega_+(k, d)$ tends to zero like kc because the finite amount of material over a given portion of the slab does not permit sidewise evacuation of surface charges faster than the velocity of light (see fig. 8*b*).

Fig. 7. *Abscissa:* r_s (a.u.); *ordinate:* surface energy (erg/cm²).

Comparison between the $r_s^{-5/2}$ scaling law of plasmon cohesion and the observed surface tensions of metals (extrapolated to zero temperature).

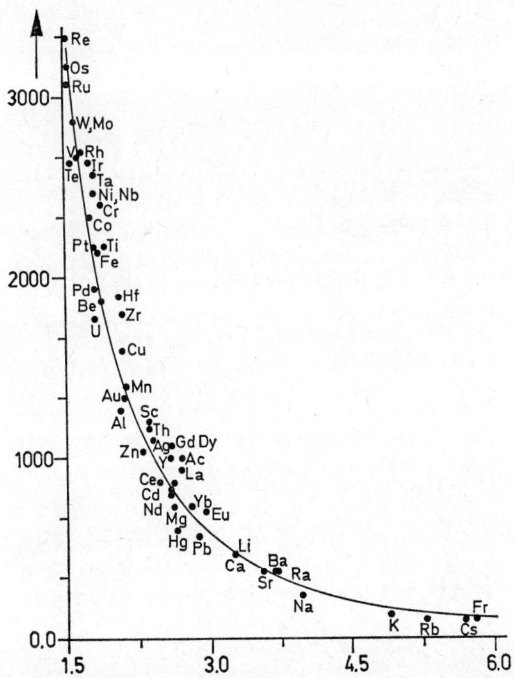

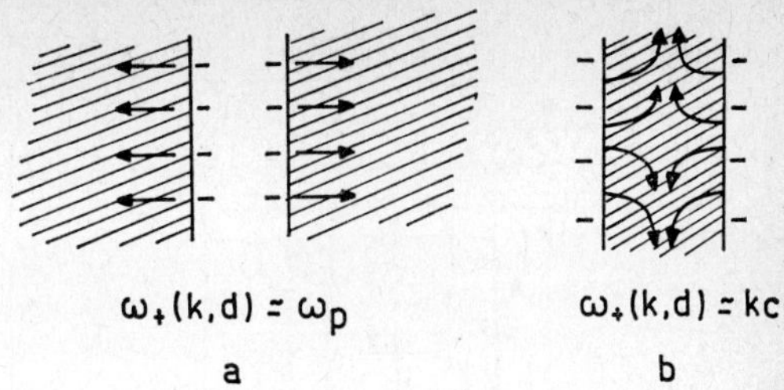

Fig. 8. (*a*) Long wavelength symmetric surface plasmon mode of a gap. The frequency $\omega_+(k,d)$ tends to ω_p when $\boldsymbol{k}$ tends to zero; (*b*) same mode for a slab. The frequency ω_+ goes to zero like kc.

Fig. 9. Spectrum of surface plasmon modes for a cylindrical cavity in a metal (hollow) and for a metal cylinder (solid). $\boldsymbol{k}$ is the wave vector parallel to the cylinder axis and m is the quantum number of the azimuthal eigenfunction $\exp(im\varphi)$.

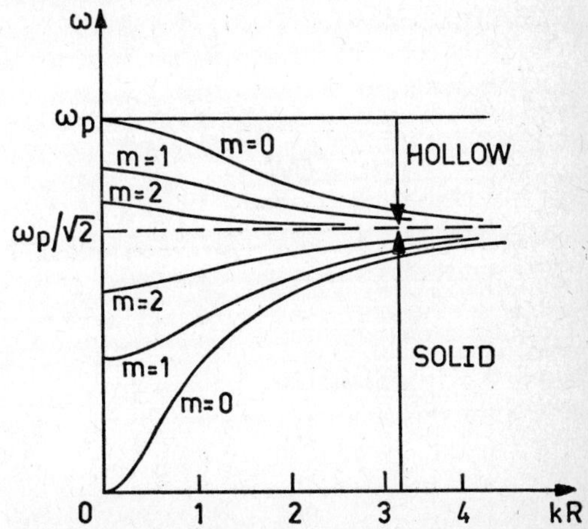

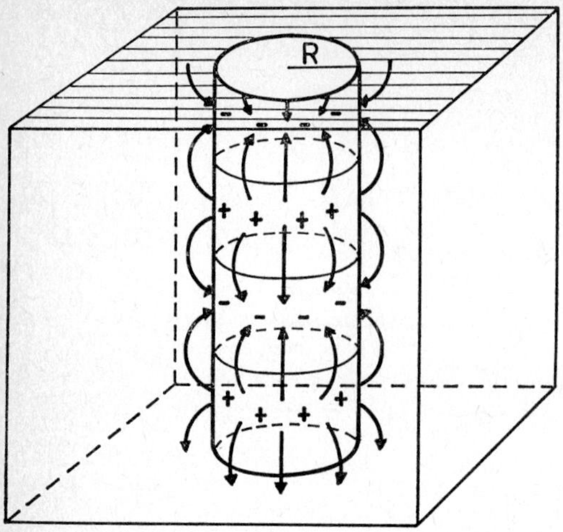

Fig. 10. Eigenvector for the mode (k, $m = 0$) of a cylindrical void.

The equation (9) can also be derived for non planar surfaces. The eigenfrequencies of surface modes in cylindrical geometry have been given by Englman & Ruppin [14] and are illustrated in fig. 9 for both a solid cylinder and an empty cylindrical cavity. The (nonretarded) modes are classified according to their wavevector k parallel to the cylinder axis and the angular quantum number m characteristic of the azimuthal mode patterns. The eigenvector for the (k, $m = 0$) modes of a hollow cylinder is shown in fig. 10. Notice the existence of a totally symmetrical "breathing" mode ($k = 0$, $m = 0$) whose frequency coincide with the bulk plasma frequency ω_p. For a solid cylinder (lower half of the spectrum in fig. 9), the surface mode ($k = 0$, $m = 0$) is softened to zero-frequency as a result of the finite supply of electrons per unit length contained in the cylinder. This is similar to the slab geometry discussed above. For the spherical geometry, the surface plasmon spectrum is shown in fig. 11 [14] and the corresponding eigenvectors in fig. 12. Here, the modes depend on the quantum numbers (l, m) of the spherical harmonics. Again a totally symmetrical breathing mode exists at a spherical void surface but not on a solid sphere. The surface energy in this geometry has been derived elsewhere [15] and coincides with the planar result [9].

All the above cases make it clear that in surface cohesion problems, one cannot disregard characteristic surface collective modes which always appear when new boundaries are introduced. To

Fig. 11. Spectrum of surface plasmon modes for a spherical cavity in a metal (hollow) and for a metal sphere (solid). l is the quantum number of the spherical harmonic Y_l^m. Each mode is $(2l+1)-$ degenerate.

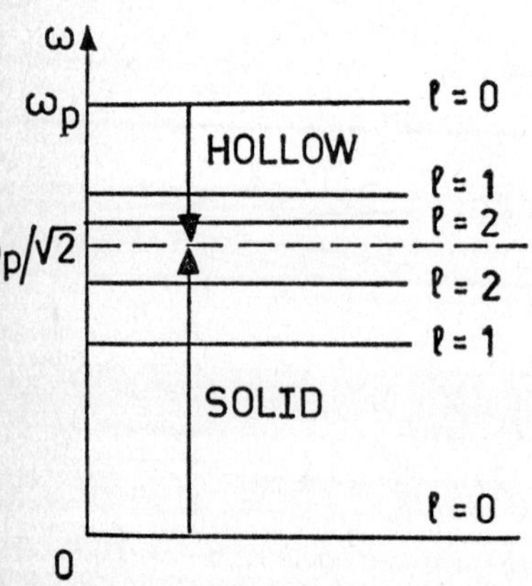

Fig. 12. Eigenvectors for the first few multipolar modes of (*left*) a spherical void; (*right*) a metal sphere.

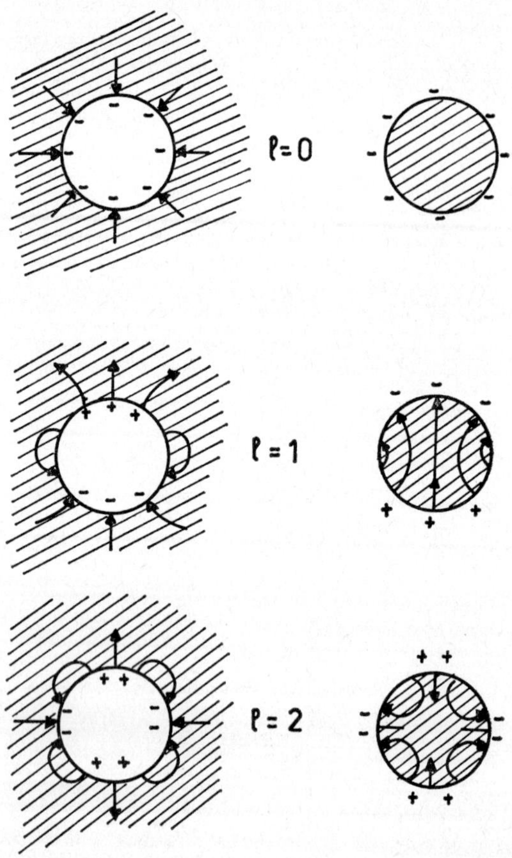

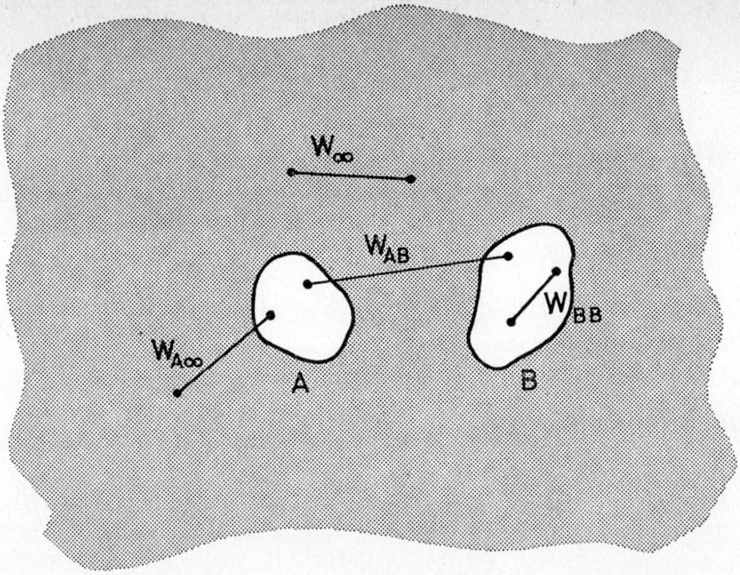

Fig. 13. The various terms contributing to the total cohesion of a solid sample with two neighboring cavities A, B.

illustrate further this point, consider the effective interaction of two cavities in a given large metallic sample. This quantity may be defined as the difference between the total cohesion energies of the sample when the two voids are at a finite distance D of each other and when they are at a large separation. If one has been successful in calculating the bulk binding energy (or some other property) of the perfect crystal on the basis of a pairwise interaction energy W_{ij}, one can think of obtaining the energy of the defect crystal merely by counting all the "broken bonds" with the missing atoms in the cavities. Referring to fig. 13, the total cohesion would be

$$W(D) = \tfrac{1}{2} {\sum_{i,j}}' W_{ij} \tag{10}$$

where the summation $\sum'_{i,j}$ extends over the whole sample, excluding the missing sites in the voids A and B. One has, identically

$$W(D) = \Big(\tfrac{1}{2}\sum_{i,j} - \tfrac{1}{2}\sum_{i,j\in A} - \tfrac{1}{2}\sum_{i,j\in B} - \sum_{i\in A}\sum_{j\ni A} - \sum_{i\in B}\sum_{j\ni B} + \sum_{i\in A}\sum_{j\in B}\Big) W_{ij} \tag{11}$$

where the first summation $\sum_{ij}$ is now unrestricted and where the symbols $\in$ or $\ni$ mean that the summation site are in or out of the indicated void, respectively. Neglecting atomic rearrangements around the defects, the various terms have an obvious interpretation. The first 3 terms (designated as W_∞, W_{AA} and W_{BB} in fig. 13) represent the *bulk* binding energy of the given amount of matter in the perfect crystal. The fourth and fifth terms ($W_{A\infty}$ and $W_{B\infty}$) are the surface energies of each void when isolated from each other. The last term (W_{AB}) represents the required effective interaction energy between the two voids:

$$W_{AB} = W(D) - W(\infty) = \sum_{i\in A}\sum_{j\in B} W_{ij} \tag{12}$$

This last contribution which, in this simple model, is the only D-dependent one, is seen to be identical with the interaction energy of two solid particles of same shape, placed in vacuum at the same distance. The actual functional dependence on D will result from whatever dependence on the atomic distance the pair potential is given. This reasoning and more elaborate ones in which self-consistency between W_{ij} and the defect structure is included [16] overlook the new property of a void to be able to sustain a breathing plasmon mode of monopole type, as distinct from a solid particle whose lowest order surface mode can only be dipolar. The consequence of this difference is the possibility for voids to interact through a dispersion—or van der Waals—type

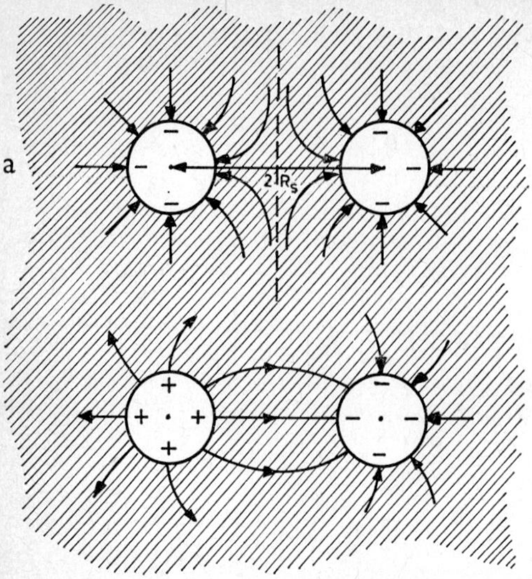

Fig. 14. Basic symmetric (*a*) and antisymmetric (*b*) modes of plasma fluctuations around two spherical voids in a metal. The symmetric mode is antibonding and the antisymmetric one is bonding for the two voids. A net attraction results due to the sum-rule $\omega_+^2 + \omega_-^2 = 2\omega_p^2$ satisfied by the mode frequencies.

force of monopole nature resulting from ground state fluctuations such as illustrated in fig. 14. It has been proposed [15] that such an interaction may be an important element for understanding the phenomena of void nucleation and growth and the occasional spectacular occurrence of void arrays in irradiated metals [17].

Present address: Facultés Universitaires, N.-D. de la Paix, Département de physique II, Namur, Belgium.

References

1. Allen, B C, Liquid metals (ed S Z Beer) p. 161. Dekker, New York, 1972.
2. Tosi, M P, Solid state phys 1964, 16, 1.
3. Lundqvist, S & Sjölander, A, Arkiv fysik 1964, 26, 17.
4. Doniach, S, Phil mag 1963, 8, 129.
5. Lucas, A, Physica 1967, 35, 353.
6. Schmit, J & Lucas, A, Solid state comm 1972, 11, 415.
7. Pines, D, Elementary excitations in solids. Benjamin, New York, 1964.
8. Lifshitz, E M, Soviet phys JETP 1956, 2, 73.
9. van Kampen, N G, Nijboer, B R A & Schram, K, Phys lett 1968, 26A, 307.
10. Boersch, H, Geiger, G & Stikkel, W, Z Phys 1968, 212, 130.
11. Lucas, A A & Šunjić, M, Prog surface sci 1972, 2, 75.
12. Schmit, J & Lucas, A, Collective phenomena 1973, 1, 1.
13. Economu, E N, Phys rev 1966, 182, 539.
14. Englman, R & Ruppin, R, J phys 1968, C1, 614.
15. Lucas, A, Phys rev 1973, 7, 3527.
16. Tewary, V K & Bullough, R, J phys 1972, F2, L69.
17. Evans, J H, Nature 1971, 229, 403.

Discussion

P. W. Anderson: Before examining such suggestions as Dr Lucas has made (which I feel intiutively leads most probably to a short-range force only, and unquestionably to something very weak), one should look into elastic effects. These have an amazing richness—for instance, there is the Friedel potential between two volume defects, which seems to me the most likely candidate for the reason of the swelling phenomena. This is the only potential which is infinitely long range in the sense that it falls off not at all with distance, being of a magnitude inversely proportional to the sample volume at all distances. It is caused by the elastic images in the surfaces, and such image effects will clearly be of importance here.

Lucas: The proposed monopolar dispersion forces are both strong and of infinite range [15]. We are not questioning the classical elasticity theory of swelling phenomena but pointing out a new quantum mechanical interaction between voids which does not seem to be explicitly included in this theory.

Order–Disorder Transitions in Two-Dimensional Overlayers

J. R. Schrieffer

Department of Physics, University of Pennsylvania, Philadelphia, Pa 19174, USA

When atoms or molecules are adsorbed on a clean, perfect crystal face of a crystalline solid, these adsorbates experience interactions not only with the ideal surface, but also lateral interactions between the adsorbates. With the aid of low energy electron diffraction (LEED), one frequently observes the adsorbates to order spacially to form a two-dimensional crystal lattice whose periodicity is commensurate, but not necessarily identical, to that of the substrate [1, 2]. As shown in fig. 1, one finds overlayer structures labelled by (1×1), (2×2), (1×2), $c(2\times2)$, etc. where $(n\times m)$ denotes the period of the overlayer along the principal surface directions being n and m relative to the substrate. $c(2\times2)$ denotes a (2×2) overlayer with a "centered" atom added. At elevated temperature, the overlayer disorders and forms in effect a two-dimensional liquid.

It is important to note that if the total amount of gas allowed to adsorb on the surface is less than that required to completely cover the surface with the overlayer, in many cases "islands" are formed corresponding to a fraction of the surface covered by the ordered overlayer, with regions in between these islands being occupied by a dilute two-dimensional adsorbate gas. The same situation holds above the disordering temperature T_c in a large number of cases. This tendency to form clusters or islands on a partially covered surface indicates that the lateral interactions are attractive for certain spacings between adsorbates, with the occurrence of other than (1×1) structures indicating that the lateral interactions may well be repulsive at nearest neighbor spacing and attractive at next nearest spacing, etc. Experimentally, one finds the stable overlayer structure and T_c vary widely with the substrate and adsorbate species, e.g. W, Mo, Nb, Ni, ... substrates and H, C, N, O, S, CO, ... adsorbates. We restrict the present discussion to adsorbates such as those which are chemically active and bond to the substrate through chemisorption bonds, rather than weak physisorption or van der Waals type of bonds. The overlayer also depends on the particular crystallographic face, and the fractional coverage θ, which varies from $\theta=0$ to $\theta=1$ as a monolayer is formed. For example, H adsorbs on the (100) surface of *W* in a $c(2\times2)$ array for $\theta\leqslant\frac{1}{2}$, and the remaining sites appear to fill in to form a (1×1) array at a full monolayer. LEED measurements of Estrup & Anderson [3] indicates the $c(2\times2)$ structure disorders at $T_c=625$ K.

The physical origin of lateral interactions between adsorbates varies greatly from system to system. If the chemisorption bond is strongly ionic, as in the case of alkali and halide adsorbates in which a fraction of an electron is transferred to the substrate and to the adsorbate, respectively, the interaction between the resulting surface dipoles is repulsive and leads to a spreading out of the adsorbates over the surface, as Andersson will discuss in the case of alkali atoms chemisorbed on Ni (p. 188). In cases like H, C and N which tend to chemisorb with little charge transfer, an indirect interaction appears to dominate the force between adsorbates. As Koutecky proposed 15 years ago [4], an adsorbate, a, will perturb the substrate wave functions through orbital hybridization. The bonding of a second adsorbate, b, to these distorted substrate states is dependent on the spacing R_{ab} between a and b as well as the details of the surface density of states, bond strength, etc. As in the RKKY [5] or the Friedel-Kohn interactions between spins or charges in a metal, this indirect hybridization interaction is expected to be oscillatory and fall off as an inverse power of R_{ab} for large R_{ab}.

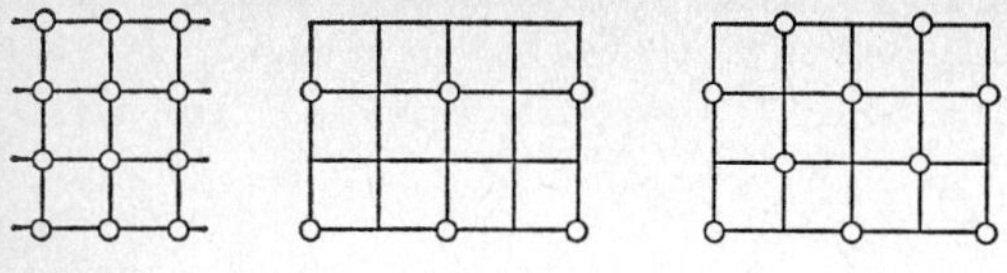

Fig. 1. Overlayer structures on a square surface lattice.

The problem is a two center scattering problem with constructive and destructive interference occurring, depending on the wavelength (i.e. energy) of the incoming wave. By summing over all occupied states, a net repulsion or attraction can occur, as Grimely [6–8], Newns [9], and Einstein and the present author [10] have discussed.

In addition to these dipolar and indirect hybridization effects, in certain systems adsorbate a distorts the geometrical arrangement of the substrate atoms in the vicinity of its adsorption site so that adsorbate b encounters a different conformation at its adsorption site, and vice versa. This "mattress" effect is of considerable importance in enzyme adsorption because of the weak forces maintaining the tertiary structure of enzymes. Because of the high strength of transition metals, the distortion of the surface is less likely to occur in this case, despite the strength of the chemisorption bonds. Further investigation of this question, both experimental and theoretical, is warranted; however, we will assume the substrate conformation to be rigidly fixed in our discussion.

In working out the statistical mechanics of the overlayer system, in addition to the interactions between adsorbates, it is important to take into account the variation of the binding energy of a single adsorbate ΔW as it moves over the clean surface. Generally, the variation of ΔW over a single crystal surface is of the order 10–30% of the mean value of ΔW. We denote the particular binding site in the unit cell by the index X in the following discussion.

LCAO Model

T. L. Einstein and the present author consider systems in which the chemisorption bond is predominantly neutral so that charge transfer effects do not dominate the problem. Furthermore, they assume that electron–electron correlation effects are such that the Hamiltonian can be parameterized in terms of a one-body operator

$$H = H_0 + \hat{V} \tag{1.1a}$$

where

$$H_0 = \sum_{i,j} t_{ij} c_i^+ c_j + \sum_\alpha \varepsilon_\alpha n_\alpha \tag{1.1b}$$

describes the free solid with a perfect clean surface and the free adsorbates, with c_i^+ and c_α^+ creating electrons in the substrate and adsorbate orbitals respectively and the c-operators satisfy anticommutation relations. The interaction $\hat{V}$

$$\hat{V} = \sum_{\alpha,i} (V_{\alpha i} c_i^+ c_\alpha + V_{\alpha i}^* c_\alpha^+ c_i) \tag{1.1c}$$

describes the hybridization interaction between the adsorbate orbital φ_α on the αth adsorbate and the substrate orbital φ_i on the ith substrate atom. More generally, i signifies both the atom and orbital on that atom in the case like transition metals in which the five d-orbitals are of importance, as well as the sp orbitals. Also, several orbitals on the adsorbate may be important, all of which are included in the label α. One can include self-consistent field effects by allowing the parameters t_{ij}, ε_α and $\hat{V}$ to be self-consistently adjusted, if desired. Since we assume that H is diagonal in the electron spin, the spin σ is suppressed in (1.1).

To find the interaction energy due to $\hat{V}$ we denote the eigenvalues of H by E_j and of H_0 by ε_j. Then since the E_j and ε_j's are bounded and finite in number, the function

$$D(E) \equiv \det[E - H_0 - i\delta]^{-1}[E - H - i\delta]$$

$$= \prod_j \left[\frac{E - E_j - i\delta}{E - \varepsilon_j - i\delta}\right] \tag{1.2}$$

$(\delta = 0^+)$

is related to the change of density of one-electron states by

$$\varrho(E) - \varrho_0(E) \equiv \Delta\varrho(E) = \frac{1}{\pi}\frac{d}{dE} \operatorname{Im} \ln D(E) \tag{1.3}$$

as one can see by direct calculation. The determinant is taken with respect to the one-electron basis states $|i\rangle$ and $|\alpha\rangle$. If one defines the unperturbed one-electron Green's function operator as

$$G^0(E) = [E - H_0 - i\delta]^{-1}, \quad (1.4)$$

then from (1.2) and (1.3) one has

$$\Delta\varrho(E) = \frac{1}{\pi}\frac{d}{dE} \operatorname{Im} \ln [1 - G^0(E)\hat{V}]. \quad (1.5)$$

This form is particularly convenient since $\hat{V}$ is zero except for i being in the vicinity of an adsorbate α and therefore the determinant has a dimensionality given by the number of atoms coupled by $\hat{V}$ to adsorbates, plus the number of adsorbates.

In calculating the binding energy, we must insure conservation of electrons

$$2\int_{-\infty}^{E_F} \varrho(E)\,dE = 2\int_{-\infty}^{E_F^0} \varrho_0(E)\,dE \quad (1.6)$$

where E_F and E_F^0 are the Fermi energies of the system with $\hat{V}$ present and $\hat{V}$ absent, respectively. While E_F shifts only by an amount of order ν/N, where ν is the number of adsorbate atoms and N the total number of atoms in the entire substrate, this shift is essential to the analysis, although one need retain only first order terms in ν/N. From (1.6) we have

$$\Delta E_F\, \varrho_0(E_F^0) = -\int_{-\infty}^{E_F^0} \Delta\varrho(E)\,dE. \quad (1.7)$$

On combining this expression with the definition of the total interaction energy

$$\Delta E_{tot} = 2\int_{-\infty}^{E_F} E\varrho(E)\,dE - 2\int_{-\infty}^{E_F^0} E\varrho_0(E)\,dE \quad (1.8)$$

and integrating by parts, one obtains the basic expression giving the binding energy of the system

$$\Delta W = -\frac{2}{\pi}\int_{-\infty}^{E_F^0} \operatorname{Im} \ln \det [1 - G^0(E)\hat{V}]\,dE \quad (1.9)$$

From here on we drop the superscript 0 on G.

Single Adsorbate Problem

For simplicity consider a simple cubic metal substrate having one-orbital per site. We consider only the (100) surface. Let the adsorbate a

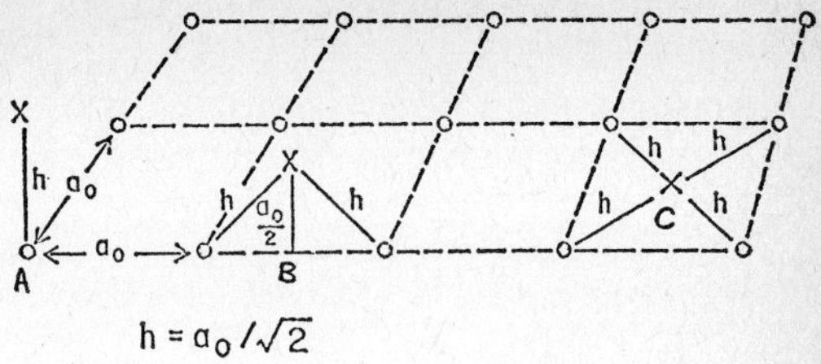

Fig. 2. Atop (A), bridge (B), and centered (C) adsorption sites.

couple only to a single atom, called 1, through $\hat{V}$

$$V_{a1} = V_{1a} = -V$$

$$V_{ai} = V_{ia} = 0, \quad i \neq 1. \quad (2.1)$$

This is appropriate for atop (A) binding, as shown in fig. 2 in which the adsorbate sits on top of a single substrate atom. If the adsorbate sits in the bridge (B) or centered (C) positions and has the symmetry of the site, a single cluster orbital φ_X is involved, where for the bridge and centered sites

$$\varphi_X = \begin{cases} \dfrac{\varphi_1 + \varphi_2}{\sqrt{2}} & X \equiv B \\[2ex] \dfrac{\varphi_1 + \varphi_2 + \varphi_3 + \varphi_4}{\sqrt{4}} & X \equiv C \end{cases} \quad (2.2)$$

with the orbitals $\varphi_1 \ldots \varphi_4$ being the atomic orbitals in the cluster. For the one-adsorbate problem, $D(E)$ is given by

$$D(E) = \begin{vmatrix} 1 & G_{aa}V \\ G_{11}^X V & 1 \end{vmatrix} = 1 - V^2 G_{aa}(E)\, G_{11}^X(E) \quad (2.3)$$

where

$$G_{aa}(E) = \frac{1}{E - \varepsilon_a - i\delta} \quad (2.4)$$

and G_{11}^X is the single-site Green's function for the atom 1 in the clean surface. The surface Green's function $G_{11}(E)$ has been worked out by Kalkstein & Soven [11] for X ≡ A and extended to bridge and centered binding by Einstein. In fig. 3, the surface density of states Im $G_{11}^X(E)$ is plotted for A and C binding, for a nearest neighbor hopping model substrate

$$t_{ij} = \begin{cases} -T & \text{i and j nearest neighbors} \\ 0 & \text{otherwise} \end{cases} \quad (2.5)$$

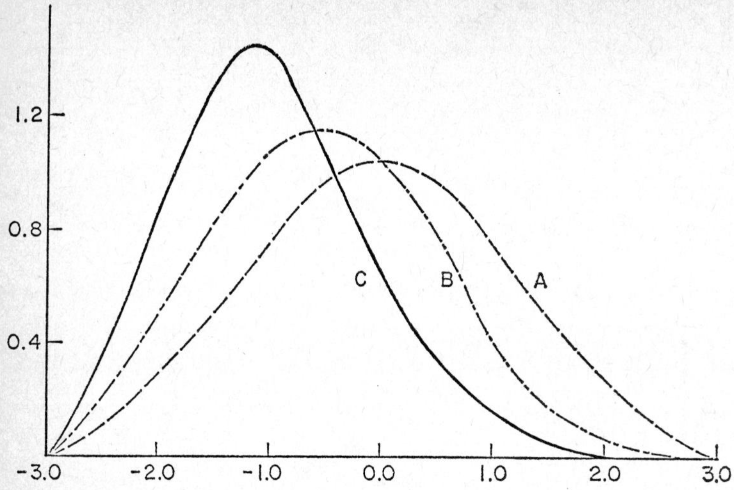

Fig. 3. *Abscissa: E; ordinate:* $\mathrm{Im}\, G^0_{\lambda\lambda}(E)\ \lambda = 1$.
Surface density of states on the (100) surface on a simple cubic s-band solid, for A, B and C binding sites.

Units are chosen so that $2T = 1$, and the bulk band width is 6. Note that for A binding, the density of states peaks at the band center, while for C binding the peak is below the band center since substrate states with roughly the same phase on all four atoms surrounding the C-site is required for a contribution to G^{C}_{11}, that is states of small momentum parallel to the surface, i.e. low energy. We note, states near the band edge are repelled from the surface and the density of states is narrowed relative to that in the bulk.

In fig. 4, the single-adsorbate binding energy derived from (1.9) and (2.3) is plotted as a function of the Fermi energy for $V/T = 3$, $\varepsilon_a = -0.3$, for A, B and C binding sites. This V/T ratio is reasonable for simple adsorbates like H on transition metals, although the model is sufficiently rough that precise values of the parameters are not meaningful. Similar curves hold for different values of V/T and ε_a.

Pair Interaction

Returning to the general expression (1.9) for ΔW, one has in the two adsorbate case

Fig. 4. *Abscissa:* E_F; *ordinate:* ΔW.
Single atom binding energy for $V/T = 3$, $\varepsilon_a = -0.3$ plotted as a function of the Fermi energy for A, B and C binding sites.

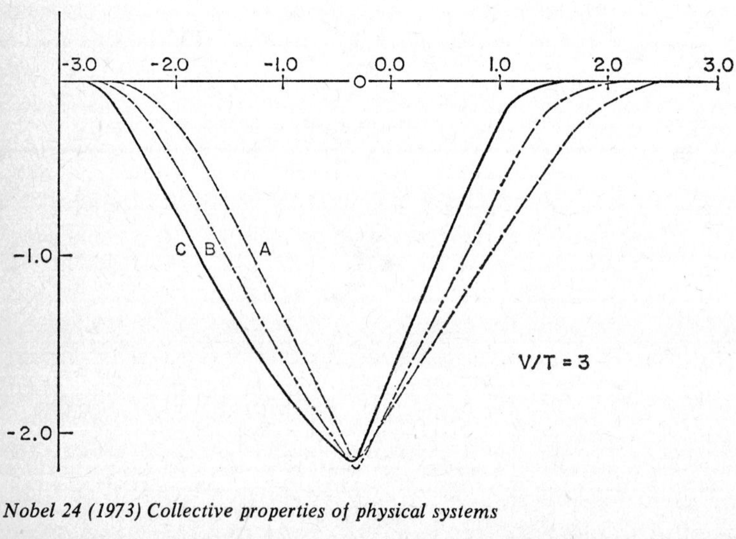

$$\det[1-G^0V]$$

$$= \det \begin{bmatrix} 1 & -G_{11}^X V_{1a} & 0 & -G_{12}^X V_{2b} \\ -G_{aa} V_{a1} & 1 & 0 & 0 \\ 0 & -G_{21}^X V_{1a} & 1 & -G_{22}^X V_{2b} \\ 0 & 0 & -G_{bb} V_{b2} & 1 \end{bmatrix}$$

$$= (1 - G_{aa} G_{11}^X |V_{1a}|^2)(1 - G_{bb} G_{22}^X |V_{2b}|^2) - G_{aa} G_{12}^X G_{bb} G_{21}^X |V_{1a}|^2 |V_{2b}|^2 \quad (3.1)$$

Here a is in contact with surface cell 1, and b is in contact with surface cell 2, where these are at arbitrary locations on the surface. If a and b are at different symmetry sites, the expressions are generalized by using X_a and X_b in G_{11} and G_{22}, respectively, while G_{12} involves X_a in cell 1 and X_b in cell 2. Clearly, the X's need not represent any symmetry other than the requirement that all cluster orbitals orthogonal to φ_{X_a} do not hybridize on to φ_a and similarly for b. Thus, the interaction can be worked out as a continuous function of the positions of a and b, as is required in potential surface calculations. Of course, one must have information on the variation of $\hat{V}$ with the positions of a and b to carry through calculations.

Returning to (1.9), we see that if the sum of the single atom interaction energies $\Delta W_a + \Delta W_b$ is subtracted from ΔW, then since ΔW_a^X is given by

$$\Delta W_a^X = -\frac{2}{\pi}\int_{-\infty}^{E_F} \operatorname{Im}\ln\left(1 - \frac{V^2 G_{11}^X(E)}{E - \varepsilon_a - i\delta}\right) dE \quad (3.2)$$

and similarly for b, one finds the pair interaction is given by

$$\Delta W_{pair} = \Delta W - \Delta W_a - \Delta W_b = -\frac{2}{\pi}\int_{-\infty}^{E_F} \operatorname{Im}\ln(1 - \bar{G}_{aa}^X \bar{G}_{bb}^X \bar{G}_{12}^X G_{21}^X V^4)\, dE. \quad (3.3)$$

The quantities

$$\bar{G}_{aa}^X(E) = \frac{1}{E - \varepsilon_a - V^2 G_{11}^X(E)} \quad (3.4\,a)$$

$$\bar{G}_{bb}^X(E) = \frac{1}{E - \varepsilon_b - V^2 G_{22}^X(E)} \quad (3.4\,b)$$

represent the Green's functions for the adsorbates when bonded to the substrate in the absence of the second adsorbate. For example, as $V \to 0$, $\operatorname{Im} G_{aa}^X(E) \to \delta(E - \varepsilon_a)$ and one has a sharp level. For $V \neq 0$, the ε_a level broadens so long as E is in the region where the surface density of states $(1/\pi)\operatorname{Im} G_{11}^X(E)$ is non zero. In general, $\operatorname{Im} G_{aa}^X(E)$ will have several peaks, and perhaps split off states as well.

The substrate Green's function is given by

$$G_{12}^X(E) = \sum_k \frac{\langle 1_X | k\rangle \langle k | 2_X\rangle}{E - \varepsilon_k - i\delta}, \quad (3.5)$$

where $|k\rangle$ represents a substrate state φ_k of energy ε_k. Physically, G_{12} propagates the disturbance caused by mixing φ_a with the substrate wave functions, thereby influencing the binding of b to the substrate, and vice versa. By expressing $|1_X\rangle$ and $|2_X\rangle$ in terms of a linear combination of atomic orbitals (3.5) can be expressed as a linear combination of off-diagonal G's in the atomic orbital representation. Thus, not only is the single atom binding energy dependent on the adsorption site, but so is the indirect interaction dependent on X, even for a given spacing between adsorbates. Because of the oscillatory nature of the φ_k's, $G_{12}^X(E)$ shows (damped) oscillatory behavior. Except in limiting cases, no simple analytic form of ΔW_{pair} can be derived, with important contributions to the integral in (3.3) coming from a range of E inside the surface band, plus any occupied split-off states.

Using Kalkstein & Soven's form of $G_{ij}(E)$ for the substrate, Einstein calculated ΔW_{pair} for a wide range of parameters V, E_F, ε_a, as well as site symmetry and relative adsorbate locations. Following his notation, ΔW_{ij}^X denotes the interaction energy between two identical adsorbates, each adsorbed on a site having symmetry X (A, B or C were explicitly treated), and with a being in the surface cell with coordinates (1,1) taken along the surface crystal axes and b in the cell with coordinates (i, j), as illustrated in fig. 5. BP denotes "bridge perpendicular", in that there are two inequivalent bridge sites, midway between atoms along the *x*- or *y*-axes, with a in the *x*-like bridge and b in the *y*-like bridge.

In fig. 6, the pair interaction energy ΔW_{12}^X (for adsorbate b in the cell (1, 2), i.e. nearest neighbor cell along the *x*-axis) is plotted as a function of E_F for $V/T = 1$, i.e. $V = \frac{1}{2}$ in the standard units. In this plot, $\varepsilon_a = -0.3$. Qualitatively, as E_F increases the relative phase of the

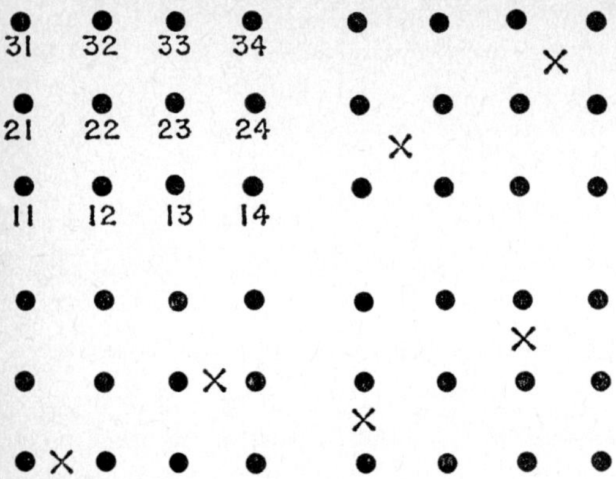

Fig. 5. Pair interaction site notation.

substrate waves φ_k at cell 11 and cell 12 which are added to the integral lead to constructive or destructive interference in the scattering off of the adsorbates. The problem is in a sense analogous to a two center diffraction problem with the proviso that all states of $E < E_F$ contribute to ΔW. As Grimley pointed out, in the asymptotic limit of large spacing R_{ij} between adsorbates, the stationary phase argument dictates that the form of ΔW_{pair} is determined by the upper limit E_F and

$$\Delta W^{X}_{pair} \propto [G^{X}_{ij}(E_F)]^2 \sim \frac{Osc}{R^5_{ij}}, \tag{3.6}$$

where Osc is an oscillating function of R_{ij}. For small E_F, the period of the oscillation is determined by the Fermi momentum, as for the RKKY interaction. In fact, the non-local surface susceptibility χ_{ij} is proportional to ΔW_{pair} for large R_{ij}. The curves in fig. 6 show both attractive and repulsive regions as a function of E_F, with $\Delta W_{pair} \to 0$ as the band fills completely and for $\varepsilon_a < E_F$. This limit is guaranteed by a sum rule which follows from the fact that $\hat{V}$ is purely off-diagonal in our model, or more simply that completey filled levels cannot lower their energy by hybridizing.

In fig. 7, ΔW^{X}_{23} is plotted as a function of E_F for $V/T = 3$, showing the more rapid oscillation of ΔW with E_F since as in most diffraction problems, the "fringes" are more rapidly varying as a function of wavelength for more widely spaced scattering centers, i.e. the phase in a free-electron model of the substrate would behave roughly as $\sqrt{2mE_F}\, R_{ij}$.

In fig. 8, ΔW^{C}_{13}, the pair interaction between atoms adsorbed in the (4-fold) centered site is plotted as a function of E_F, where b is in the next nearest neighbor cell along the x-axis. The curves correspond to values of V/T ranging from 1–5 in integer steps. In practice, V is fixed by setting E_F and W_b from experiment on the bulk, and requiring that the heat of adsorption of a single adsorbate as predicted by this model agree with experiment for an assumed value of ε_a. One finds V/T to be typically in the range 3–5.

We reproduce here in table 1, Einstein's numerical results of ΔW^{X}_{ij} for near neighbor positions. In each case, the number given is 10 plus the common logarithm of the magnitude

Fig. 6. *Abscissa:* E_F; *ordinate:* ΔW^{X}_{12} $(\times 10^{-2})$.
Pair interaction energy W_{23} plotted as a function of E_F for $V/T = 1$.

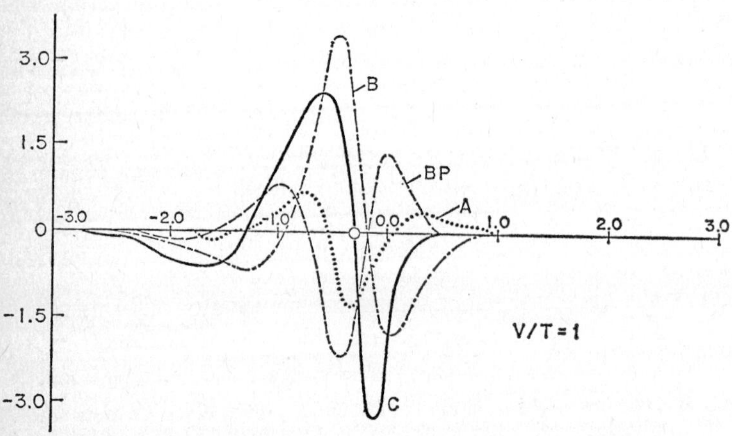

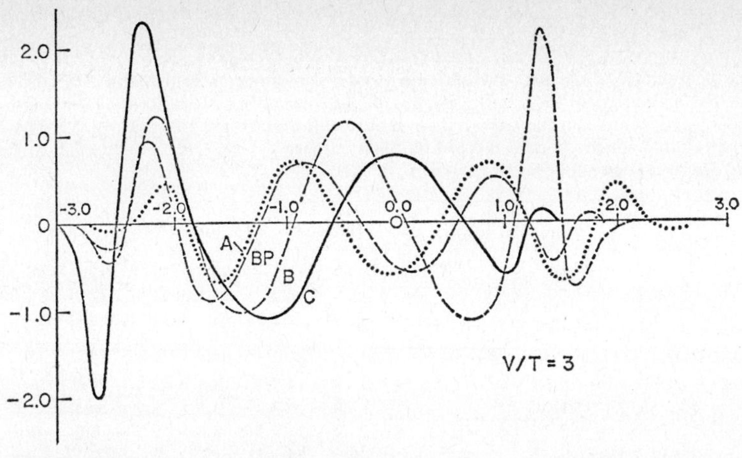

Fig. 7. *Abscissa:* E_F; *ordinate:* $\Delta W_{23}^{X}(\times 10^{-2})$.

Pair interaction for third nearest neighbor site (23) plotted as a function of E_F for $V/T=3$.

of the interactions, with a – sign affixed if ΔW_{ij}^{X} is negative. A nominal surface overlayer structure is assigned in each case [e.g., c(2 × 2), (1 × 1), etc.] corresponding to lowest energy array, in competition with a dilute gas of single adsorbates. Since the interaction drops off numerically roughly exponentially between second and sixth nearest neighbor positions with the fall off being an order of magnitude per cell, the overlayer structure is determined in practice by only the short range interactions. Roughly, aside from the oscillating behavior, the amplitude of ΔW_{ij} varies as

$$|\Delta W_{ij}| \simeq \gamma \exp(-\beta R_{ij}/a_s) \tag{3.7}$$

where $\gamma \simeq 1$ for $V/T=3$ and β is between 5/3 and 2 depending on how one measures the amplitude. a_s is the substrate surface lattice spacing.

The exponential fall off of the amplitude of the pair interaction for near neighbor spacing is to be contrasted with the inverse power law fall-off for $R_{ij} \to \infty$. The asymptotic form, being determined by the behavior of $G_{ij}(E)$ near $E=E_F$, is not sensitive to the strongly interfering contributions from E ranging over the occupied part of the band. Presumably the stationary phase contribution at the Fermi surface, while small in magnitude, wins out when the band-like contributions have nearly completely cancelled each other for large R_{ij}. Fortunately, one finds that the pair interaction is less sensitive to ε_a than to V, E_F and site symmetry, for given R_{ij} so that the results are in essence uniquely determined by the heat of adsorption at low coverage, as mentioned above.

A survey of the experimental literature shows that statistically the frequency of occurrence of a

Fig. 8. *Abscissa:* E_F; *ordinate:* $\Delta W_{13}^{C}(\times 10^{-1})$.

Pair interaction energy W_{13} for V/T ranging from 1–5 in integer steps plotted as a function of E_F.

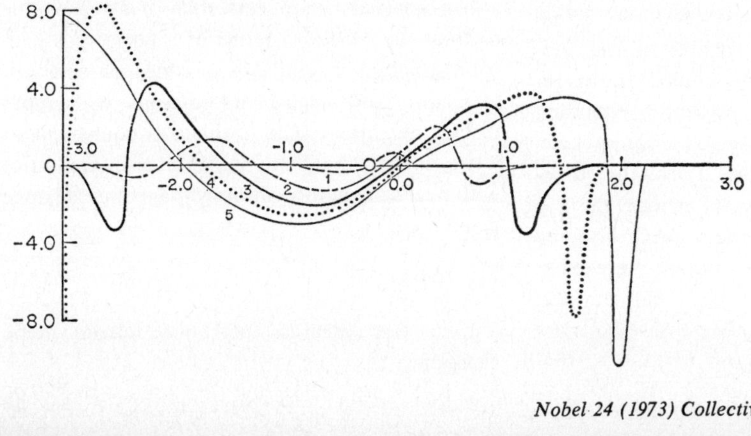

Table 1. *Display of the pair-interaction energy suggesting the sensitivity of adatom arrays to changes in the Fermi level, the hopping potential, the adatom noninteracting level, and the binding type*

One adatom sits at the origin (11) (denoted "0"); the pair energy for a second adatom at each of the nearby sites is indicated by the number at the site. The magnitude of the number given is 10 plus the common logarithm of the magnitude of the interaction. A plus (minus) sign means that the interaction is repulsive (attractive). Thus, $\Delta W = -2.7 \times 10^{-4}$ would be represented by −6.4 in the table. Each chart is labeled by the symmetric surface array predicted. Since ΔW_{ij}, $j \geqslant 4$, is unimportant for this determination, only X = C is shown

ΔW^{X}_{ij} X=	A			C						B			BP		
	31	32	33	31	32	33				31	32	33	31	32	33
ij=	21	22	23	21	22	23				21	22	23	21	22	23
	11	12	13	11	12	13	14	15	16	11	12	13	11	12	13
$E_F = 1.2$	−7.7	−7.5	+7.5	−8.3	+7.1	−7.4				+8.4	+8.3	−6.1	−7.9	−7.4	−6.1
$E_a = -0.3$	+8.9	−8.1	−7.5	−9.0	−8.0	+7.1				−9.3	−8.4	+8.3	+8.2	−8.4	−7.4
$V/T = 3$	0	+8.9	−7.7	0	−9.0	−8.3	−6.6	+6.1	+4.6	0	−9.3	+8.4	0	+8.2	−7.9
		c(2 × 2)			(1 × 1)						(1 × 1)			c(2 × 2)	
$E_F = 1.2$	−7.8	−6.7	+7.4	−8.6	−6.2	−7.8				+8.7	+8.1	−7.6	+7.6	+7.0	−7.6
$\rightarrow E_a = -0.0$	+8.8	−8.4	−6.7	−9.4	−8.6	−6.2				−9.3	−8.3	+8.1	+9.0	−8.3	+7.0
$V/T = 3$	0	+8.8	−7.8	0	−9.4	−8.6	+7.7	−5.8	+5.3	0	−9.3	+8.7	0	+9.0	+7.6
		c(2 × 2)			(1 × 1)						(1 × 1)			c(2 × 2)	
$E_F = 1.2$	−7.7	+7.7	+6.9	+8.5	−7.5	+7.8				+8.5	−7.9	−7.8	−7.8	+6.9	−7.8
$E_a = -0.3$	+8.6	−8.6	+7.7	−9.3	−8.4	−7.5				−9.1	+7.5	−7.9	+9.1	+7.5	+6.9
$\rightarrow V/T = 4$	0	+8.6	−7.7	0	−9.3	+8.5	+7.6	−6.1	+4.9	0	−9.1	+8.5	0	+9.1	−7.8
		c(2 × 2)			(1 × 1)						(1 × 1)			(2 × 2)	
$\rightarrow E_F = 0.9$	−7.5	+7.7	−7.0	+8.1	−7.9	+7.7				+8.4	−7.7	−7.7	+6.3	+7.8	−7.7
$E_a = -0.3$	+8.5	−8.5	+7.7	−9.3	−8.4	−7.9				−9.2	−7.1	−7.7	+9.0	−7.1	+7.8
$V/T = 3$	0	+8.5	−7.5	0	−9.3	+8.1	+7.5	−6.3	+4.7	0	−9.2	+8.4	0	+9.0	+6.3
		c(2 × 2)			(1 × 1)						(1 × 1)			c(2 × 2)	
$E_F = 1.5$	−6.7	−7.8	+4.4	−7.3	+5.8	−6.8				−8.4	−7.7	+7.0	+6.7	−7.0	+7.0
$E_a = -0.3$	+8.7	+8.3	−7.8	+6.2	−7.2	+5.8				−8.9	+6.2	−7.7	−8.7	+6.2	−7.0
$V/T = 3$	0	+8.7	−6.7	0	+6.2	−7.3	−6.3	+5.5	+4.5	0	−8.9	−8.4	0	−8.7	+6.7
		c(4 × 2)			c(2 × 2)						(1 × 1)			(1 × 1)	

given overlayer array corresponds qualitatively with the distribution of arrays predicted by the above calculations, if one averages smoothly over the parameters. This is no more than a statement that the theory is not qualitatively wrong in this regard, and is certainly not a substantive check of the theory; nevertheless, it is a significant result. To predict the actual overlayer structure for a given chemisorption system, the actual geometric and electronic structure of the substrate and adsorbate must be included. In this regard, J. W. Davenport is presently extending Kalkstein & Soven's analysis by treating f.c.c. and b.c.c. lattices, and the five d-orbitals. Hopefully, it will prove possible to include the s–p bands as well. By fitting the LCAO hopping and overlap parameters to obtain agreement with spectroscopic data on the clean surface, one will have a realistic starting point to consider indirect interaction effects.

Overlayer Statistical Mechanics

As mentioned above, hydrogen on the (100) surface of tungsten forms a c(2 × 2) structure for coverage $\theta \leqslant \frac{1}{2}$ for $T < T_c$, above which the overlayer disorders, entering a two-dimensional liquid phase. Estrup & Anderson observed the transition to occur at $T_c = 625$ K. From the above analysis, it appears that it is reasonable to treat the model in terms of a lattice gas with nearest neighbor repulsion, next nearest neighbor attraction and neglect interactions with more distant neighbors. The problem is equivalent to a two-dimensional Ising model on a square lattice with nearest neighbor antiferromagnetic exchange given by

$$w_1 \equiv \Delta W_{12} = -4J \tag{4.1}$$

and a next nearest neighbor ferromagnetic exchange,

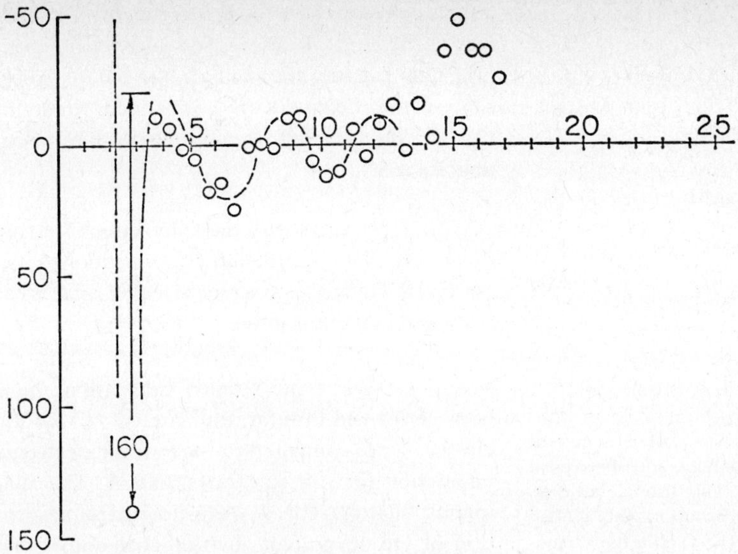

Fig. 9. *Abscissa: R*, distance (Å); *ordinate: V(R)*, cohesive energy (meV).
Effective pair interaction potential determined by Tsong from low density two-dimensional gas data.

$$w_2 \equiv -\Delta W_{22} = 4J'. \tag{4.2}$$

Since spin up is identified with an occupied adsorption site and spin down corresponds to an empty site, the low energy state corresponds to an antiferromagnetic spin array. The approximate calculations of Fan & Wu [12] for the two-dimensional Ising model with first and second neighbor interactions lead to the relation determining the melting temperature T_c,

$$\exp\left(\frac{\beta w_1}{2}\right) = 2\exp\left(\frac{-\beta w_2}{2}\right) + e^{-\beta w_2}, \beta = \frac{1}{kT_c} \tag{4.3}$$

which is accurate to within a few percent. Most of the cases calculated above have w_2 ranging from w_1 to an order of magnitude smaller than w_1. For $w_1 = w_2$, $w_1 = 0.747\ kT_c$ while for $w_1 = 7.83\ w_2$, $w_1 = 1.495\ kT_c$. Since $kT_c \sim 1/20$ eV for H on (100) W, w_1 will range between 0.037 eV and 0.075 eV. This range of values is in reasonable agreement with the calculated values of w_1 for $V/T = 3$ and ε_a near the band center, with W_b of order 10 eV.

Recently, Tsong [13] has observed the effect of interactions between rhenium atoms adsorbed on a $W(110)$ plane, using field ion microscopy. He observed the position of five rhenium atoms at 34 successive times separated by periods of 3 min, during which time the temperature was maintained at 320 K. From these observations, the pair distribution function $p(R)$ was determined. If $p_0(R)$ is the pair distribution for $V(R) \equiv \Delta W(R_{ij}) = 0$, then in the low coverage (dilute two-dimensional gas) limit

$$V(R) = -kT \ln\left[\frac{p(R)}{p_0(R)}\right]. \tag{4.4}$$

This quantity is plotted in fig. 9. Presumably, at nearest neighbor sites, the direct interaction between Re atoms bonds them strongly, while the indirect interaction dominates at larger spacings. The oscillating interaction has a period of roughly twice the lattice spacing for $W(110)$, which is 2.74 Å. While Tsong used for $p_0(R)$, the correlation function for a dilute two-dimensional gas on a translationally invariant substrate, the site variation of ΔW_{single} leads to a variation of $V(R)$ with the period of the lattice spacing and these effects should not qualitatively change the results.

Thus, the observed periodic overlayer structures, their melting and the pair distribution function in the dilute gas all support the picture of significant indirect interactions between adsorbates. While we have neglected the effect of substrate lattice distortion induced by the adsorbates in treating the indirect interaction, it is difficult to see how this could have an oscillatory behavior, although such effects should be investigated. Clearly much remains to be done in determining the interaction potentials and in relating these interactions to the observed overlayer phase diagrams.

Work supported in part by NSF and the Laboratory for Research on the Structure of matter, University of Pennsylvania, Philadelphia, Pa 19174, USA.

References

1. Somorjai, G A & Szalkowski F Z, J chem phys 1971, 54, 389.
   This work contains references to most of the current experimental observations of patterns and offer a lucid discussion of nomenclature. This listing, however, suggests that the (1 × 1) pattern occurs roughly as often as c(2 × 2) or (2 × 2). In fact, (1 × 1) is the most frequently achieved pattern, by a substantial margin. It is, however, far more difficult to detect from LEED patterns and also less interesting to deal with.
2. Estrup, P J. Private communication.
3. For example, Estrup, P J, in The structure and chemistry of solid surfaces (ed G A Somorjai) chap. 28. Wiley, New York, 1969.
4. Koutecky, J, Trans Faraday soc 1958, 54, 1038.
5. Kittel, C, Quantum theory of solids, p. 360. Wiley, New York, 1963. (Who references the pioneering RKKY papers.)
6. Grimley, T B, Proc phys soc 1967, 90, 751.
7. — Ibid 1967, 92, 776.
8. Grimley, T B & Walker, S M, Surface sci 1969, 14, 395.
9. Newns, D M, Phys rev 1969, 178, 1123.
10. Einstein, T L & Schrieffer, J R, Phys rev 1973, B7, 3629.
11. Kalkstein, D & Soven, P, Surface sci 1971, 26, 85.
12. Fan, C & Wu, F Y, Phys rev 1969, 179, 560.
13. Tsong, T T, Proc 33rd ann conf on phys electronics. Berkeley, Calif, March 1973.

Discussion

S. Andersson: Is there any possibility to estimate the interaction strength V/T from first principles theory?

Schrieffer: One can estimate V/T from a number of approaches, for example, by using Hartree–Fock atomic orbitals and a superposition of atomic charge distributions. Presumably, more reliable numbers can be obtained by treating a cluster containing the adsorbate and a group of substrate atoms through the density variable approach described by Kohn, through a scattered wave scheme, as employed by Johnson & Connolly. By treating the same cluster by an LCAO scheme, the effective V will be obtained. No doubt other techniques will be developed in the forseeable future.

Fisher: I presume that the interactions via the "mattress effect" of the lattice, as mentioned earlier by De Gennes, play a much smaller role? What is their relative magnitude?

Schrieffer: While no detailed calculations have been performed thus far, estimates show that the effect is weak compared to the indirect electronic interaction for stiff substrates, like W. The analogous "mattress effect" was estimated for adsorption of He on graphite by Campbell and Schick who also concluded it was a weak effect. Clearly, this will break down in easily deformed substrates, as occurs for enzymes, or for very reactive chemisorption on metals in which reconstruction of the surface occurs.

Suhl: You stated that at short distances, the adatom–adatom interaction falls off exponentially, but at large distances it looks like RKKY interaction. Can you comment on this changeover.

Schrieffer: The asymptotic form is determined by the integrand near the stationary phase point in the energy integral, that is E_F. As in the three-dimensional (RKKY) interaction, the asymptotic form of the envelope of the oscillating interaction is an inverse power law. For small interadsorbate spacing R, all states below E_F contribute to the integral, with relative phases which change rapidly as R increases. This band contribution to ΔW_{1j} dies away roughly exponentially with increasing R, while the stationary phase term remains for large R.

Kohn: As I recall two atoms, or ions adsorbed on a metal surface have an r^{-3} interaction due to the electric dipoles which are induced. Is this effect included in your work?

Schrieffer: We have not included the electric dipolar interaction between adsorbates. For neutral chemisorption these effects are likely to be small, although this is most likely not the case for ionic chemisorption, as occurs for alkali atoms adsorbed on transition metals, about which Dr Andersson will be speaking. In this case, a dipolar repulsion occurs between the adsorbates and no ordered overlayers occur. For partially ionic adsorption, a competition between the dipolar repulsion and the indirect interaction effects we treated will occur.

Halperin: How general is this oscillating $1/R^5$ interaction? If there is a band of surface states overlapping the Fermi energy, I would expect there to be an asymptotic oscillatory interaction which decreases as $1/R^2$.

In your tight binding model, of course, there is no surface band because cutting the surface bands will always tend to narrow the energy band for any given k_x, k_y. I do not know how common surface bands would be in real metals.

Schrieffer: In the model we treated, surface states have not occurred for the range of parameters we considered. If parameters were chosen so that surface states occur, it is possible that a less rapid drop off of the oscillation envelope would occur for large R although we have not investigated this situation at present. In practice, the interaction which dominates the statistical mechanics of the overlayer are no doubt the short range, rather than the long range forces. Thus, the asymptotic behavior of the interaction for large R may be difficult to observe experimentally.

March: Is it possible to estimate quantitatively the lack of sharpness of the Fermi surface from your exponential decay of the interaction?

Schrieffer: Since the surface spoils momentum conservation in the direction normal to the surface, presumably only the momentum parallel to the surface is important in determining the fall-off of the interaction at large spacings between adsorbates. Since the interaction is predicted to drop very rapidly even without disorder scattering, it would appear unlikely that this is a good method to experimentally determine the sharpness of the "Fermi surface" at the surface of the crystal.

Mahan: The formula you used to compute the interaction contains the term Im log (Det). Is this not equivalent to evaluating all of the electron energy eigenvalues with and without the impurities, and then substracting their zero point energies?

Schrieffer: The expression is simply the sum of the energy shift of the occupied single particle levels.

Structure and Ordering of Sodium on Nickel (100)

S. Andersson

Department of Physics, Chalmers University of Technology, S-402 20 Göteborg 5, Sweden

Summary

The dynamical LEED theory has been used to analyse the crystallography of the $c(2\times2)$ sodium structure on nickel (100). The sodium atoms are found to occupy the fourfold symmetrical hollows in the nickel surface and the spacing between the sodium and topmost nickel layer is 2.9 ± 0.1 Å. The $c(2\times2)$ sodium structure exhibits an order–disorder transition at $T_c = 535$ K, and it is shown that the ordering closely follows Onsager's solution for the Ising problem of a square lattice gas.

Lack of structure knowledge has been an intriguing problem in most attempts to understand the chemisorption phenomena in a quantitative way. In this respect adsorption of alkali metal atoms on metal surfaces has offered relatively simple systems since all available information indicate that the adlayer is composed of only alkali metal atoms. Furthermore, the electronic properties of the alkali metal atoms are simple enough and any charge transfer has been found to manifest itself in a dramatic lowering of the work function. A wealth of experimental and theoretical work that considers this phenomenon is available [1]. More sophisticated experiments have utilized single crystal surfaces of the substrate metal and some experiments have also monitored desorption energies and two-dimensional periodicity of the alkali metal layer. The classical dipole picture of the adsorption mechanism [2] has been replaced by quantum mechanical models that treat the charge transfer in a self-consistent way. The charge transfer itself cannot be found from the experimental data but is expressed in terms of a dipole moment. Exact knowledge of the adsorption site is required in order to get a separation of charge and geometry. This is a frustrating problem and it effectively inhibits a comparison of the status of available theories. The purpose of this paper is to describe the application of low-energy electron diffraction (LEED) to a determination of the $c(2\times2)$ sodium structure on nickel (100). The ordering properties of this system will also be considered in some detail.

It is interesting to notice that for this system, dipole moments as well as desorption energies are known experimentally for a wide range of alkali metal coverage [3]. In conjunction with the present knowledge of structure and ordering one realizes that this system is far less intriguing than e.g. the adsorption of hydrogen on tungsten.

(1) Crystallography of the $c(2\times2)$ Sodium Structure

At the moment a structure analysis within the framework of the dynamical LEED theory has to be performed in a trial-and-error manner. There is simply no explicit Fourier transform of the diffraction intensities into a structure, but intensities for various trial structures are calculated and compared with experimental data. Recent work has for simple metals yielded diffraction intensities that agree well with experimental data [4–7], and this knowledge is the basis for any structural investigation of chemisorption layers on such metals. Certainly there is no shortage of simple systems that can be tackled in this way and as yet there is no viable alternative.

The method has been applied to the $c(2\times2)$ sodium structure on nickel (100) [8]. The adopted structural picture regards the overlayer as being just constituted of alkali metal atoms. The $c(2\times2)$ sodium structure is thus close to a physical monolayer of sodium (adatom–adatom separation on nickel (100) is 3.52 Å) and the atomic diameter of sodium (3.70 Å) thus closely specifies

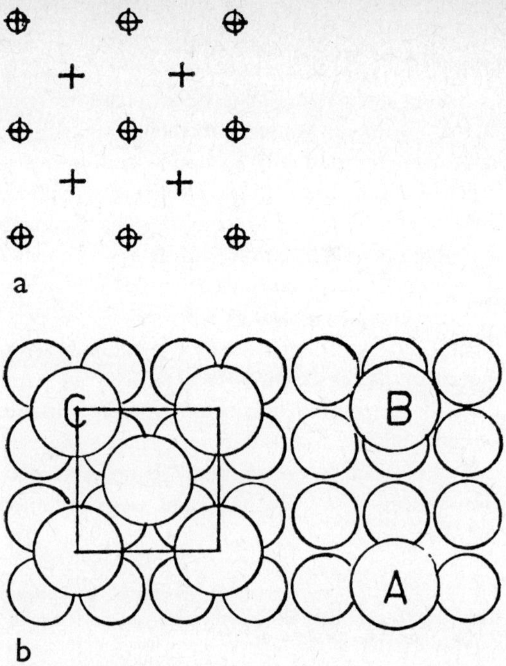

Fig. 1. (*a*) Diffraction pattern from c(2 × 2) sodium on nickel (100); (*b*) model structure for c(2 × 2), three different adsorption sites are indicated, onsite A, bridge B and fourfold hollow C. (*a*) ○, Ni beams; + c(2 × 2) beams; (*b*) small ring, Ni; large ring, Na.

the dimension of the adsorbed species. This model is a reasonable starting point for the trial-and-error analysis.

The overlayer introduces new diffracted beams (see fig. 1) due to the change in the two-dimensional symmetry brought about by the layer. The extra beams yield the size and shape of the new unit mesh as sketched in fig. 1. Positions of atoms within the unit mesh has to be found from an intensity analysis and it is thus of prime importance to understand how the overlayer affects intensities of beams. The scattering problem to be handled is illustrated in fig. 2. The amplitudes of beams incident on and reflected by the substrate are denoted by b_g^+ and b_g^-, respectively,

$$b_g^- = \sum_{g'} R_{gg'} b_{g'}^+ \tag{1}$$

where **g** are the appropriate two-dimensional reciprocal vectors and $R_{gg'}$ denotes the scattering matrix of the substrate. With the overlayer present the following equations relate the various amplitudes

$$b_g^+ = \sum_{g'} T_{gg'}^{++} a_{g'}^+ + R_{gg'}^{+-} b_{g'}^- \tag{2}$$

$$a_g^- = \sum_{g'} T_{gg'}^{--} b_{g'}^- + R_{gg'}^{-+} a_{g'}^+ \tag{3}$$

T^{++}, T^{--} are respectively the forward and backward transmissivities of the overlayer, R^{-+} and R^{+-} the forward to backward, and backward to forward reflectivities. Solving eqs 1–3 gives

$$a_g^- = \sum_{g'} [T^{--}(1 - R^{+-})^{-1} R T^{++} + R^{-+}]_{gg'} a_{g'}^+ \tag{4}$$

a_g^+ are amplitudes of waves incident on the system, and a_g^- are the amplitudes of the reflected waves we wish to find. From eq. 4 one can see how various interferences come about. The term including R^{-+} represents waves reflected directly by the overlayer. The other term involves transmission through the overlayer (T^{++}) followed by multiple scattering between the overlayer and substrate adding to give $(1 - RR^{+-})^{-1}$ R, then transmission through the overlayer (T^{--}) to emerge into vacuum.

T and R are preferably calculated for each layer within a self-consistent multiple scattering framework including the ion core scattering. Other important parameters entering the calculations are the real, V_{or}, and imaginary, V_{oi}, parts of the constant potential between the ion cores. Of certain importance are also the surface potential barrier and the thermal vibrations of the atoms. In the particular case of sodium on nickel, the thermal vibrations were neglected assuming that they exert a small influence at the

Fig. 2. Scattering problem for an overlayer on a substrate. The a_g's and b_g's denote amplitudes of waves outside the system and between overlayer and substrate.

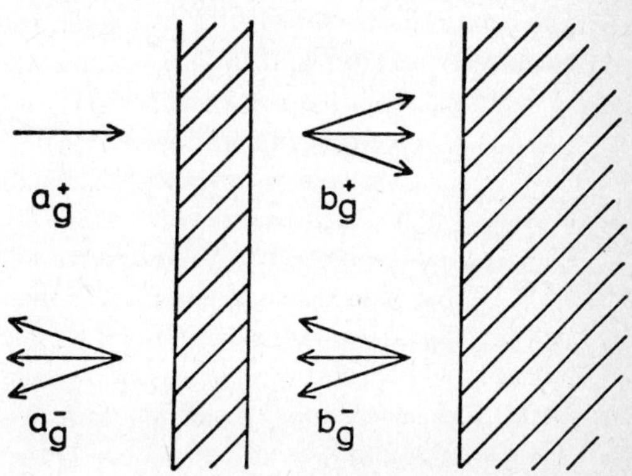

low electron energies considered (0–40 eV). There are other factors that emphasize the use of low energy intensity/energy spectra in structural determinations. These spectra are most sensitive to the immediate surface region. Multiple scattering processes are particularly strong at low energies which increases the sensitivity to the parameters we wish to determine (analogous to optical interferometry). Since peaks in the spectra originate mainly in Bragg reflections within the substrate the influence of the overlayer will mainly be to modify a basic structure dictated by the substrate. The overlayer can affect peak positions by, for example, changing the amplitude of excitation of a Bloch wave as the energy increases through a peak in reflectivity. The apparent position of a peak can be shifted in this fashion by the order of the width of the original peak (5–10 eV). It will be appreciated that it is important to know the properties of the substrate to a high degree of accuracy. The intensity/energy spectra calculated for the clean nickel (100) surface reproduced experimental peak positions to within 1–2 eV and peak intensities to within 50% [8].

Fig. 3 shows the calculated intensity/energy spectra for the specular, 00, beam from the c(2 × 2) sodium structure on nickel (100). The sodium atoms are situated above the fourfold symmetric hollows in the nickel substrate (see fig. 1) and spectra for various values of the spacing, *d*, between the centre of the sodium layer and the centre of the underlying nickel layer are presented. It is obvious that the 00 beam contains several peaks that are sensitive to *d*, some peaks shift by about 10–20 eV/Å. With an accuracy of 1–2 eV in peak positions these yields a sensitivity of about 0.1 Å in a determination of *d*—about the same error as is introduced by approximations made in the model calculation. The 00 beam is rather insensitive to lateral displacements of the layer and the lateral position has to be found from the non-specular $\{\frac{1}{2}\frac{1}{2}\}$ and $\{10\}$ beams. The calculated spectra for fourfold positioning are compared with the experimental ones in fig. 4. It is conclusive that the curves for $d = 2.86$ Å gives the best fit to the experimental data, i.e. a separation that is slightly larger than $d = 2.55$ Å obtained from considerations of atomic radii based on bulk lattice spacings of sodium and nickel. The conclusion is based on a comparison with a substantial set of calculated spectra for a range of *d*-values and different lateral position. It is of course crucial for an analysis of this kind that the experimental data are as accurate as those calculated. Thus peak positions and intensities must be reproducible and without systematic errors due to contamination and disorder for example (effects of disorder will be illustrated in section 2.) It is most valuable to use a simple set of diffraction parameters, as normal incidence of the primary electron beam and low electron energies (< 100 eV). In this way experimental diffraction intensities were obtained from the clean nickel (100) surface as well as from the c(2 × 2) sodium structure on this surface. Surface purity and order were monitored by means of electron energy distributions and electron reflectivities. The best data were obtained

Fig. 3. Calculated intensity/energy spectra of the 00 beam from a c(2 × 2) sodium structure on nickel (100) for various spacings of the sodium layer. The sodium atoms are above the fourfold hollows in the nickel layer below.

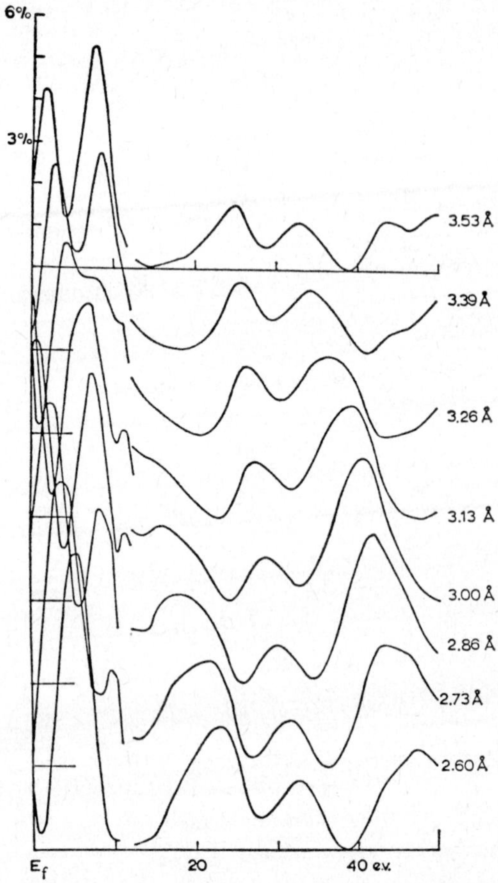

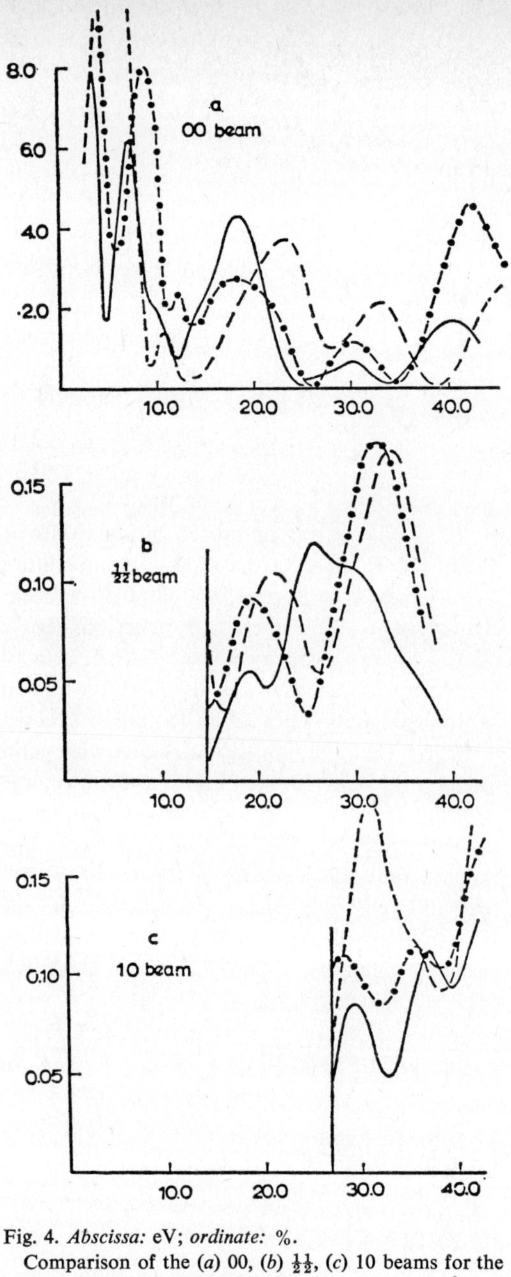

Fig. 4. *Abscissa:* eV; *ordinate:* %.
Comparison of the (*a*) 00, (*b*) ½½, (*c*) 10 beams for the best c(2 × 2) sodium trial structure (–·–). $d = 2.87$ Å and for $d = 2.55$ Å (– – –) with experiment (——). Experimental 00 beam × 3 above 25 eV; theoretical ½½ beam ÷ 4; theoretical 10 beam ÷ 2.

from a nickel surface cleaned by argon ion bombardment while kept at 400°C. Sodium was then deposited from a zeolite source onto the nickel surface (kept at room temperature) and the c(2 × 2) sodium structure formed was brought to optimum order at 450 K. The experimental spectra shown in fig. 4 are averages of several runs and peak positions were reproducible to within 0.5 eV and peak intensities to within 30 %.

(2) Order–Disorder Transition of the c(2 × 2) Sodium Structure

The nature of the interaction between pairs of chemisorbed atoms has been discussed by several authors [9, 10]. The first attempt to evaluate theoretically the two-adatom indirect, interaction at nearby binding sites has been given by Einstein–Schrieffer (E–S) [11]. They dealt with adsorption on the (100) surface of a simple cubic semi-infinite lattice of a tightbinding solid. They found that the pair interaction energy oscillates in space, i.e. relative to a given adatom, certain positions became attractive and others repulsive. For a reasonable range of parameters describing the substrate metal, the adatom electron level and the adatom-substrate interaction strength the model predicts as stable overlayers a number with symmetries that have been observed in LEED experiments. The essential limitation of the L–S model concerning its description of the pair interaction is that the direct adatom–adatom interaction is not included. This is unfortunate since many of the simplest chemisorption systems approach a physical monolayer of adatoms. This is for example the case with dense alkali metal layers like the c(2 × 2) sodium structure on nickel (100) where the adatom–adatom charge overlap must be appreciable. Within the E–S model a c(2 × 2) structure is caused by repulsive nearest neighbour interaction and attractive next nearest neighbour interaction. Such a system can reasonably well be considered as a two-dimensional square lattice gas and its ordering properties can be treated by the equivalent Ising model for an AB-alloy. The ordered lattice critical temperature, T_c, can be found from Fan & Wu's [12] results for the Ising model with second-neighbour interactions. The approximate solution for T_c is

$$e^{\beta W_1/2} = \sqrt{2}\, e^{-\beta W_2/2} + e^{-\beta W_2} \tag{5}$$

where $\beta = 1/kT_c$ and W_1 and W_2 are the pair interaction energies for nearest and next nearest neighbours, respectively. Neglecting W_2 (i.e. $W_2 = 0$) gives Onsager's exact solution to the

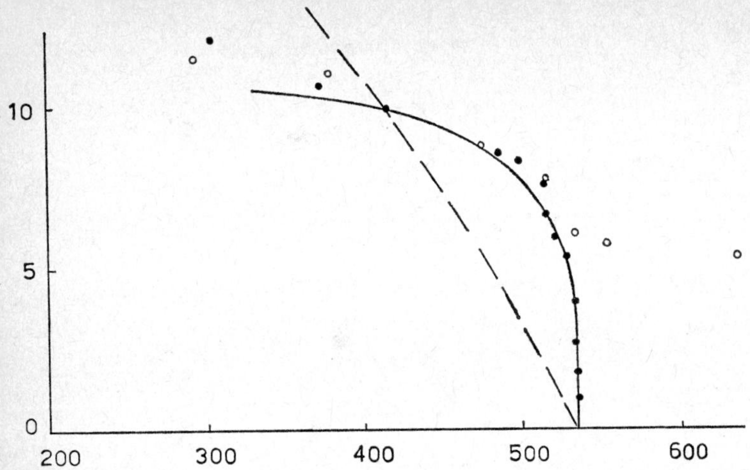

Fig. 5. *Abscissa:* T(K); *ordinate:* I (arb) ●, $\frac{1}{2}\frac{1}{2}$ beam; ○, 10 beam.

Intensity/temperature dependence of the $\frac{1}{2}\frac{1}{2}$ and 10 beams for the c(2 × 2) sodium structure on nickel (100). The solid curve is the theoretical prediction from Onsager's exact solution to the Ising problem for $T_c = 535$ K and the dashed cruve is due to the Bragg–Williams approximation for $T_c = 535$ K. Theory and experiment are matched at 416 K.

Ising problem for a two-dimensional square lattice

$$kT_c = \frac{W_1}{2\ln(1+\sqrt{2})} \tag{6}$$

The ordering properties of the c(2 × 2) sodium structure was investigated in a LEED experiment. One can easily prove generally that the intensity of the diffraction beams characteristic for the overlayer depends on the long range order parameter S as

$$I \propto S^2 \tag{7}$$

S is defined in such a way that

$$S = 2r - 1 \tag{8}$$

where r denotes the probability that an atom occupies the "right" site (i.e. its equilibrium site in the ordered layer).

The experimentally determined temperature dependence of the $\frac{1}{2}\frac{1}{2}$ and 10 beams from the c(2 × 2) sodium structure is shown in fig. 5. The diffraction conditions were: 25 eV electron beam normally incident onto the surface. The ordering of the overlayer was optimized by annealing at about 450 K prior to the intensity/temperature measurements. It is obvious that the $\frac{1}{2}\frac{1}{2}$ beam characteristic for the c(2 × 2) structure decays rapidly to zero intensity around 535 K. The 10 beam does not vanish since the overlayer as well as the substrate contribute to the diffracted intensity. There is, however, a sudden change in intensity around 535 K followed by a slower intensity reduction due to thermal vibrations in the substrate (the Debye–Waller effect). The experimental I/T dependence is compared with two different theoretical predictions for the Ising problem of a two-dimensional square lattice gas. The solid curve is calculated from Onsager's exact solution according to which [13]

$$I \propto \left\{1 - \left[\sinh\left(\frac{T_c}{T}\ln(1+\sqrt{2})\right)\right]^{-4}\right\}^{1/4} \tag{9}$$

The broken curve is calculated from the result of the Bragg–Williams approximation

$$S = \tanh\left(S\frac{T_c}{T}\right) \tag{10}$$

Experiment and theory were matched at 416 K and it is reasonable to assume that thermal vibrations will not affect the comparison strongly in the range 416–535 K. It is most striking that Onsager's result gives a very good fit to the experimental data.

From a critical temperature of $T_c = 535$ K we find from eq. 6 that

$$W_1 = 8.2 \times 10^{-2} \text{ eV}.$$

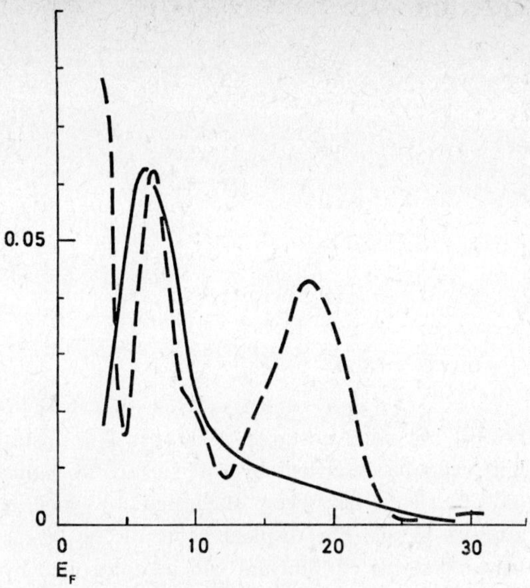

Fig. 6. *Abscissa:* E(eV); *ordinate:* I_{00}/I_0. ——, Ni (001) − Na; ---, Ni (001) c(2 × 2) − Na.

00 beam intensity/energy spectra from the ordered c(2 × 2) sodium layer (dashed) and from the disordered sodium layer (solid).

This is of the same order of magnitude as found by E–S [11] for an interaction strength of $V \approx 1$ eV which is a reasonable figure with respect to the binding energy of sodium on nickel [3].

The transition is reversible but ordering occurs very slowly. At room temperature several hours are required for the c(2 × 2) structure to crystallize again.

There is an interesting point to make about the disordered sodium overlayer. Although the long range lateral periodicity is destroyed one can from intensity spectra conclude that the separation to the top-most nickel layer is practically unchanged. This can be concluded from fig. 6 where the 00 beam intensity for the ordered c(2 × 2) as well as the disordered sodium layer is shown. The prominent difference between the two spectra is the disappearance of a peak at about 18 eV which is due to an interference effect involving the $\frac{1}{2}\frac{1}{2}$ beams. These beams vanish and so does the interference. The peak around 8 eV is caused by an interference in the 00 beam scattered by the overlayer and the substrate and the fact that its position is practically unchanged means that the separation is unchanged (see fig. 3).

The author wishes to express his thanks to B. Kasemo and J. B. Pendry for cooperation in the work reported here.

Note added in proof: There are certain indications that some contaminant could be responsible for the observed order-disorder behavior.

References

1. For a review see Lang, N D, Solid state phys (ed H Ehrenreich, F Seitz & D Turnbull) New York, London, 1973. To be published.
2. Langmuir, I, J Am chem soc 1932, 54, 2798.
3. Gerlach, R L & Rhodin, T N, Surface sci 1970, 19, 403.
4. Jepsen, D W & Marcus, P M, Phys rev 1972, B5, 3933.
5. Strozier, J A & Jones, R O, Phys rev lett 1970, 25, 516.
6. Tong, S Y & Rhodin, T N, Phys rev lett 1972, 28, 553.
7. Pendry, J B, J phys C 1971, 4, 2501 & 2514.
8. Andersson, S & Pendry, J B, J phys C 1973, 6, 601.
9. Koutecky, J, Trans Faraday soc 1958, 54, 1038.
10. Grimley, T B, Proc phys soc (London) 1967, 90, 751.
11. Einstein, T L & Schrieffer, J R., Phys rev 1973, B7, 3629.
12. Fan, C & Wu, F Y, Phys rev 1969, 179, 560.
13. Onsager, L, Nuovo cim, suppl. 1949, 6, 261.

Discussion

Fisher: Can you actually fit to Onsager's solution for the order? My recollection is that Onsager's result saturates much more closely to T_c than your data appear to.

S. Andersson: The order, *S*, saturates certainly quite closely to T_c but the diffraction intensity I which depends on S^2 has a smoother behaviour. In fact the fit depends somewhat on the matching point to the experimental data but is in general very good.

Kohn: I would like to emphasize that the estimation of the distance of the adsorbed ions is from the last lattice plane. This is an important new, independent piece of information which presents a significant test for theoretical models. Lang and I have used these data to verify with reasonable accuracy our estimation of where, relative to the last lattice plane, the image plane is situated.

Remarks on Superstructures and Reconstruction

A. Blandin

Laboratoire de Physique des Solides associé au CNRS,
Université Paris-Sud, Centre d'Orsay, 91405-Orsay, France

Summary

We give a discussion of the various effects which may occur on the surface of a crystal: relaxation, superstructures, reconstruction. In the case of superstructures, which are linked with the existence of soft surface phonons, we discuss the possibility of phase transitions and their characteristics. For magnetic surfaces, soft surface magnons have similar effects; possible phase transitions are discussed.

Phase transitions have been widely studied in three dimensions. The effects of finite size, "reduced" dimensionality are of current interest. True two-dimension transitions may occur on surfaces: surface superconductivity is well known; cooperative phenomena of adsorbed layers begin to be well understood (p. 188). We concentrate in this contribution on the case of clean surfaces of pure crystals, where good experimental information is now available through the development of Low Energy Electron Diffraction (LEED) experiments and of the interpretation of LEED intensities [1]. Magnetic surfaces will then be discussed briefly.

1. Crystallography of Surfaces

When a crystal is cut by a plane, the atoms may keep the same positions as in the infinite crystal. But different other possibilities may occur to move the atoms out of these "normal positions":

(*a*) The atoms of the first plane or planes near the surface may experience non zero forces. In this non equilibrium situation the distance between the first planes will change in order to restore equilibrium. This will lead to surface relaxation [2].

(*b*) The atoms near the surface are in equilibrium, but with unstable positions. This instability can be described by soft surface phonons ($\omega_s^2 < 0$). This has been suggested by various authors [3, 4]. The displacements of the atoms towards stable equilibrium will be controlled by anharmonic forces. In this situation each atom can be labelled by its original (normal) position and the new surface structure is specified by finite (and usually small) displacements from the normal positions. In this case we shall speak of superstructures.

(*c*) The atoms in normal positions are in a stable configuration. But there exist other situations of lower energy.

A simple case is a finite displacement of the first plane leading to a fault in a stacking sequence. For example a face-centered cubic lattice, can be described by the sequence of (111) planes—ABC ABC ABC—. With a fault for the first plane at the surface, this would give CBC ABC ABC—. For the observation of such faults, one needs the comparison of intensities of LEED spots. It has not yet been observed.

One can think of other cases where the first plane keeps the same symmetry as in the bulk crystal, but has a different lattice parameter; a simplified model has been discussed [5]; it is related to an old problem, the movement of dislocations in crystals [6–8]. Another possibility exists: the first plane has a different symmetry than in the bulk crystal. In these cases, we shall speak of reconstruction. In both cases, one cannot label each atom of the first plane by a "normal" position and a small displacement. The surface will not be a perfectly flat two-dimensional lattice but will be rather rough. If the average lattice parameter of the first plane has a relation of rationality with the parameter of the planes in the bulk crystal, the situation will appear as if the surface has a structure (note that the number of atoms per unit area is not the same for the first plane and other planes). Here again the distinction between the cases of reconstruction and superstructure can be done only through the investigation of LEED intensities. This is the difficulty with LEED experiments, where multiple scattering takes place.

Such reconstructed surfaces have been observed. (100) surfaces of gold, for example gives LEED patterns which correspond to a hexagonal first layer [9, 10]. Compared to a hexagonal plane (direction (111) in the bulk crystal) there is a contraction of 4.4%. This seems to be an example where the first plane has a different symmetry than the bulk crystal. (111) surfaces of gold exhibit anomalous spots which are interpreted as a first layer being contracted by the same amount [11].

2. Interactions near a Surface

In the case of adsorbed layers, one has three types of interactions, between (1) adsorbed atoms; (2) atoms of the substrate; (3) adsorbed atoms and substrate atoms. Depending on the relative strength of these interactions, a variety of possibilities are observed [1]. For a clean surface of a pure crystal, interactions cannot be very different and we expect a priori less interesting surface structures. This is not completely true and many interesting surface structures are observed [1]. The reason is that even if the interactions between atoms near the surface are the same as in the bulk crystal, instabilities may occur and superstructures or reconstruction may be observed. We give here a simple example of this idea [12].

Let us consider, for simplicity, a monoatomic crystal with central forces from first to n'th nearest neighbour. The potential energy is for "normal" positions:

$$V = \tfrac{1}{2} \sum_{i} \sum_{j=1}^{n} v(\boldsymbol{R}_i + \boldsymbol{a}_j) \qquad (1)$$

where $\boldsymbol{R}_i$ is the position of the i atom, $\boldsymbol{a}_j$ the distance between the atom at site i and one of its j'th nearest neighbours. One can scale the distances with the first neighbour distance $a_1 = a : a_j = \alpha_j a$. At equilibrium in the bulk, $dV/da = 0$ which gives:

$$\sum_{j=1}^{n} z_j \alpha_j \, dv/d\alpha_j = 0 \qquad (2)$$

where z_j is the number of j'th nearest neighbour. Eq. (2) defines the equilibrium value a_{eq} of a in the bulk crystal. For a crystal with a free surface, one can make the following qualitative remarks: let us consider the first plane; if it were alone (two dimensional crystal) it would have, with the same potential energy $v(\boldsymbol{r})$, an equilibrium value a'_{eq}, which is defined by an equation similar to (2) (the numbers z_j being different). In general $a'_{eq} \neq a_{eq}$.

If $a'_{eq} > a_{eq}$ the first plane would like to expand due to the forces within the plane; if a'_{eq} is not too different from a_{eq}, the interaction with the other planes will determine the number of atoms per unit area. The best solution is a "buckled" surface. The condition $a'_{eq} > a_{eq}$ is certainly not sufficient and one must take into account the relaxation of the bulk half-crystal, calculate the phonon spectrum; the instability will be manifested by the existence of a soft surface phonon and this will lead to a superstructure. If $a'_{eq} < a_{eq}$, one expects stability of the normal positions. But, if a'_{eq} and a_{eq} are noticeably different, there is a possibility of reconstruction. Qualitatively, one may draw the following diagram:

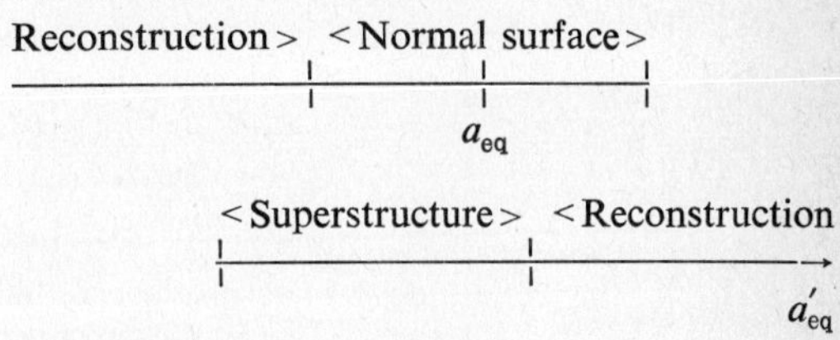

The relative value of a'_{eq} in comparison with a_{eq} cannot be discussed a priori: it depends on the crystal structure and on the direction of the surface: for example, for a simple cubic lattice, a'_{eq} can be larger than a_{eq} for a (111) surface and smaller than a_{eq} for a (001) surface. Simple examples have been discussed with first and second neighbour interactions and simple lattices, which support this qualitative discussion [12].

The interactions between atoms on the surface can be different and realistic models are needed in order to discuss real crystals and surfaces. A general method can be used for the discussion of stability and the occurrence of superstructures:

the first step is to deduce from the experimental bulk phonon spectrum the best parameters (force constants) which can describe the spectrum; an analysis of surface phonons and of their stability can be made, keeping the same force constants on the surface. If the predicted results agree with the experimental results (stability, or instability leading to superstructures), the forces on the surface are the same as in the crystal. If there is a discrepancy (predicted stability and observed superstructure or vice versa), this method enables us to judge how to modify the forces on the first plane or between the first planes in order to explain the experiment. This method can, in principle, give some idea about the change of surfaces forces and validate the introduction of more sophisticated effects [13, 14] or long range dipolar forces [4]. Various applications can be made to the case of rare gas crystals where the interactions are known (without or with adsorbed layers) and to the case of metals and semiconductors.

3. Order and Characteristics of Surface Transitions

Let us make some remarks about the effects of temperature and the existence of phase transitions. For the cases discussed in section 1c, p. 194, if there is a transition towards the normal structure below the melting point, it will evidently be a first order transition.

The cases of superstructures is more interesting [15]. At $T=0$, when the normal structure is unstable, the amplitudes of the displacements away from the normal positions are controlled by anharmonic forces. At $T\neq 0$, it is possible to make a mean field theory: one can define a temperature dependent effective harmonic Hamiltonian, which is the best harmonic Hamiltonian minimizing the energy. This method is analogous to the Hartree–Fock theory and has been used for the discussion of displacive transition [16]. At high temperature, the surface structure is normal. Superstructure will occur below a temperature T_s where one surface phonon becomes soft.

Let us discuss the situation at T_s. It is well known [17] that the lowering of symmetry at T_s is an essential feature of the transition. From a symmetry point of view, a crystal limited by a surface is equivalent (isomorphic) with the two-dimensional crystal of the surface (with different number of non equivalent atoms per unit cell). For example, a simple cubic lattice with a (001) surface is equivalent with a monoatomic square lattice; a face-centered cubic lattice with a (111) surface is equivalent with a hexagonal (triangular) lattice with three non identical atoms per unit cell. A soft surface phonon is characterized by a wave vector $\boldsymbol{q}$ which is parallel to the surface and within the first Brillouin zone of the surface. To a given value of $\boldsymbol{q}$ are associated the values which can be deduced from the symmetry transformations of the two-dimensional reciprocal lattice. This defines an "irreducible star" of wave vectors $\boldsymbol{q}_1$. In general, at T_s, there is only one soft mode for a given $\boldsymbol{q}$ with a polarization unit vector $\boldsymbol{\alpha}$ and one can deduce the polarization α_1 of the soft phonons of the star with wave vectors $\boldsymbol{q}_1$ by symmetry considerations (α_1^z being independent of l). The position of an atom in normal position is $\boldsymbol{R}_i := \boldsymbol{\varrho}_i + na\boldsymbol{k}$; $\boldsymbol{\varrho}_i$ is the position in the n'th plane off the surface, a the distance between plane and $\boldsymbol{k}$ the unit vector perpendicular to the surface. Near T_s, the average displacements $\boldsymbol{u}_i$ are given by their components u_i^σ:

$$u_i^\sigma = \lambda_n \sum_1 \eta_1 \alpha_1^\sigma \exp(i\boldsymbol{q}_1 \cdot \boldsymbol{\varrho}_i), \tag{3}$$

$\sigma = x, y$ or z; λ_n defines the amplitude away from the surface; it decreases exponentially with n for large n. η_1 is the component of the order parameter relative to the wave vector $\boldsymbol{q}_1$. Note that $\eta_1 = \eta_{-1}^*$ (η_{-1} being associated with the vector $-\boldsymbol{q}_1$ of the irreducible star).

Let us now discuss the order of the transition. For a one-dimensional parameter of order, the free energy can be written near T_s:

$$F - F(T_s) = A\eta^2 + B\eta^3 + C\eta^4. \tag{4}$$

In order to have a transition at T_s, $A(T_s)$ must vanish and in a mean field theory $A \simeq \lambda(T - T_s)$ near T_s. Now, if $B \neq 0$, the transition is a first order transition. If $B=0$, the transition may be a second or a first order. The order of the transition will depend on the sign of C. Let us generalize these results to the surface transition problem. Near T_s, one can write the free energy:

$$F(T)-F(T_s)=A\sum_{l}|\eta_l|^2+\sum_{\sigma\sigma'\sigma''}\sum_{ijk}b(ijk;\sigma\sigma'\sigma'')$$

$$\times u_i^{\sigma}u_j^{\sigma'}u_k^{\sigma''}$$

$$+\text{fourth order terms.} \qquad (5)$$

In (5) the coefficients b are invariant by simultaneous translations of i, j and k (the translations are the two-dimensional translations leaving invariant the crystal and its surface). It follows that it appears in the third order term of (5) factors of the type:

$$\sum_{i}\exp\left(i(\boldsymbol{q}_1+\boldsymbol{q}_2+\boldsymbol{q}_3)\cdot\boldsymbol{\varrho}_i\right),$$

where $\boldsymbol{q}_1$, $\boldsymbol{q}_2$ and $\boldsymbol{q}_3$ belong to the same irreducible star. These factors are zero unless

$$\boldsymbol{q}_1+\boldsymbol{q}_2+\boldsymbol{q}_3=0 \text{ (or } \boldsymbol{\tau}) \qquad (6)$$

$\boldsymbol{\tau}$ being a vector of the two dimensional reciprocal lattice. If the irreducible star is such that (6) is obeyed, third order terms in F do not vanish; the free energy F becomes

$$F-F(T_s)=A\sum_{l}|\eta_l|^2+B\sum_{l_1l_2l_3}\eta_{l_1}\eta_{l_2}\eta_{l_3}$$

$$+\text{fourth order terms.}$$

The transition will be a first order transition at a temperature $T_c>T_s$. It will be observed if T_c is smaller than the melting temperature T_f.

If the third order terms vanish and if the lower symmetry phases involves only two vectors $\boldsymbol{q}_1$ and $-\boldsymbol{q}_1$ (surface plane of low symmetry), the parameter η has one component. The free energy is given by (4) with $B=0$. If $C<0$ the transition is a first order transition; if $C>0$, it is a second order transition characterized by $n=1$, $d=2$. Near T_s the system will behave as a two-dimensional Ising model. If the low symmetry phase involves four vectors $\boldsymbol{q}_1$, $-\boldsymbol{q}_1$, $\boldsymbol{q}_2$, $-\boldsymbol{q}_2$ the parameter of order may be a two-dimensional vector. This is the case for surfaces with square lattices, $\boldsymbol{q}_i$ being along a direction of high symmetry. Particular cases appear for special values of $\boldsymbol{q}$: for example, for a monoatomic simple cubic lattice with a (001) surface, $q_x=q_y=\pi/a$, two surface phonons with polarization parallel to the surface are degenerate and become soft together. The same situation occurs for a diatomic simple cubic lattice, a (001) surface and optical phonons with $\boldsymbol{q}=0$ and polarization parallel to the surface. In such cases, the free energy near T_s can be written:

$$F-F(T_s)=A(\eta_1^2+\eta_2^2)+C(\eta_1^2+\eta_2^2)^2+D\eta_1^2\eta_2^2 \qquad (7)$$

(7) describes the $x-y$ model ($n=2$) in a two dimensional case ($d=2$) with anisotropy (if $D\neq0$). This is the two-dimensional analogy of the problem discussed in this conference by Dzyaloskinsky. When $D=0$, one knows that no long range exists for the $x-y$ model [18]. With $D\neq0$ the question is open. Intuitively, one might expect that anisotropy brings back the system to the case $n=1$.

Applications to simple cases give interesting results. If the half crystal (surface) has a threefold symmetry axis (point) of symmetry, (6) is always satisfied. This is the case for hexagonal (triangular) lattices where only first order transitions should be observed. On the contrary, for square lattices, (001) surfaces of cubic crystals for example, second order transitions can be observed.

The observation of surface transitions is related with the measurements of LEED intensities. Few experiments have been done. The transformation of the (7×7) structure to the (1×1) structure of (111) surfaces of silicon has been studied [12]: the variation is smooth and was interpreted as a second order transition in contradiction with the above results. It might be a first order transition with a small jump of the parameter at the critical temperature. Hysteresis may be much less important on a surface, where atoms can take advantage of the free space to follow equilibrium conditions.

Similar symmetry considerations can be made for the case of transition within adsorbed layers.

4. Magnetic Surfaces

Let us come to the case of magnetic surfaces. We consider a magnetic crystal with a surface. The atoms keep their normal positions. Each atom has a spin $\boldsymbol{S}_i$. The spins are coupled by exchange. For simplicity, we shall assume classical spins.

One can discuss the various possibilities which occur at $T=0$ in a manner similar to that in section 1 [15].

The spins of the first plane (or planes) may experience non zero torques

$\boldsymbol{C}_i=-2\sum_j J_{ij}\boldsymbol{S}_i\times\boldsymbol{S}_j$.

The vectors $\boldsymbol{C}_i$ may be different from zero for helical structures: this will give something similar to a relaxation.

The spins at the surface are in equilibrium but with unstable directions. This instability can be described by soft surface magnons ($\omega_s < 0$). The spins are in stable equilibrium but there exists another spin configuration of lower energy: the first plane has a different magnetic structure. This is equivalent to the case of reconstruction of section 1.

Soft surface magnons have been discussed when the exchange constants are different for spins on the surface and in the crystal ($J < 0$ for spins on the surface $J > 0$ otherwise) [21]. But, this is not necessary. Even, if the exchange constants are the same, instability may occur [20]. Let us consider, for example, a crystal with nearest neighbour ($J_1 > 0$) and next nearest neighbour interactions ($J_1 > 0 J_2 < 0$). The three-dimensional crystal is ferromagnetic when $|J_2|$ is smaller than a critical value J_c. Now let us consider the first plane. If it was alone (two-dimensional crystal), it would be ferromagnetic when $|J_2| < J_c'$ (J_c' depending on the two-dimensional structure) and antiferromagnetic if $|J_2| > J_c'$. If $J_c' < J_c$ and if $J_c' < |J_2| < J_c$, the first plane would like to have antiferromagnetic order (due to the exchange within the plane), but the interactions with the other planes have the opposite effect. We are in a favourable position to have soft magnon modes. For given structures and surfaces, one can calculate the magnon spectrum and one finds that there exists a domain of values of $|J_2|$ near J_c (smaller than J_c) where surface magnons are unstable. Simple examples have been discussed [20]. At $T = 0$, the magnetic surface should reconstruct, as in the crystallographic case; the amplitude of the spin deviations are controlled by the fourth order terms (in the spin deviations) which are included in the Heisenberg Hamiltonian.

One can extend these consideration at finite temperature [20]. Within mean field theory, when the exchange constants are the same for the atoms in the crystal and on the surface, surface ordering cannot occur at a temperature above the bulk Curie temperature T_c. But there exists a temperature $T_s < T_c$, where one surface magnon becomes soft. This surface magnon is characterized by a vector $\boldsymbol{q}$ parallel to the surface within the first Brillouin zone. At T_s, the antiferromagnetic component is perpendicular to the bulk magnetization $\boldsymbol{M}$. It can rotate freely in the plane perpendicular to $\boldsymbol{M}$. We are in the case where $n = 2$, $d = 2$, ($x - y$ model in two dimensions). In that case, one knows that no long range order exists [18]; only "quasi long range order". This is characterized by zero spontaneous magnetization, though the magnetic susceptibility diverges. Thus, below T_s, strictly there will not exist any superstructure, only this kind of quasi order giving rise to large spin fluctuations.

When the exchange constants on the surface and in the bulk are different, one could obtain, within mean field theory, a critical temperature T_s larger than T_c. Depending on the values of exchange constants, the order could be the same as in the bulk [22] (ferro- or antiferromagnetic surface order, the bulk crystal being ferro- or antiferromagnetic) or different (antiferromagnetic surface order, the bulk crystal being ferromagnetic below T_c), we are in a case where $n = 3$ and $d = 2$. True long range order should not exist in this case. In fact one expects that the surface exchange constants (indirect exchange constants) to be equal or smaller than the bulk exchange constants. The case $T_s > T_c$ is unlikely to occur for pure crystals; it might occur with adsorbed magnetic layers.

No direct observation of surface structures has been reported though interesting results can be inferred from LEED data [23] or photoemission data [24]. Possible "superstructures", as discussed above ($T_s < T_c$) could occur for the ferromagnetic Europium chalcogenides EuO, EuS, EuSe: in these compounds J_1 is positive and decreases in the series, J_2 is negative.

Let us end this contribution by a last remark: When there are cooperative phenomena in the bulk crystal, it is always possible that a surface structure occurs. As an example, long range order on the (001) surface of Cu_3Au has been reported within a range of temperature above the critical temperature for atomic order of this alloy [25]. The structure that is observed appears to coincide with the (001) plane of ordered Cu_3Au. This is a case where $T_s > T_c$.

Part of the work which is reported here has been done with D. Castiel and L. Dobrzinski. The author wishes to thank J. L. Domange, J. Friedel, P. G. de Gennes and

G. Toulouse for stimulating discussions. Thanks are also due to J. Perdereau and G. E. Rhead for communicating their results before publication.

References

1. Somerjai, G A, Principles of surface chemistry. Prentice Hall, Englewood Cliffs, N.J., 1971.
2. Dobrzynski, L & Maradudin, A A, Phys rev 1973, B7, 1207.
3. Feuchtwang, T E, Phys rev 1967, 155, 715.
4. Trullinger, S & Cunningham, S, Phys rev lett 1973, 30, 913.
5. Ying, S C, Phys rev 1971, B3, 4160.
6. Dehlinger, U, Ann phys 1929, 9, 749.
7. Kontorova, T & Frenkel, Y I, Zh eksp teor, fiz 8, 1938, 89, 1340, 1349.
8. Nabarro, F R N, Theory of crystal dislocations, p. 130. Clarendon Press, Oxford, 1967.
9. Rhead, G E, J phys 1973, F3, L53.
10. Grønlund, F & Højlund Nielsen, P E, J appl phys 1972, 43, 3919.
11. Perdereau, J & Rhead, G, E. Private communication.
12. Blandin, A, Castiel, D & Dobrzynski, L, Solid state comm (and in press.)
13. Ducros, P, Surface sci 1968, 10, 295.
14. Dobrzynski, L & Mills, D L, Phys rev 1973, B7, 2367.
15. Blandin, A, Solid state comm. In press.
16. Boccara, N & Sarma, G, Physics 1965, 1, 219.
17. Landau, L D & Lifshitz, E M, Statistical physics, Pergamon Press, New York, 1959.
18. Jancovici, B, J phys (Paris) 1971, 32, suppl. 10, 185.
19. Florio, J V & Robertson, W D, Surface sci 1970, 22, 459.
20. Blandin, A, Phys lett A. In press.
21. Trullinger, S E & Mills, D L, Solid state comm 1973, 12, 819.
22. Mills, D L, Phys rev 1971, B3, 3887.
23. de Wames, R E & Wolfram, T, Phys rev lett 1969, 22, 137.
24. Sattler, K & Siegman, H C, Phys rev lett 1972, 29, 1565.
25. Højlund Nielsen, P E, Phys lett 1973, 42A, 468.

Reaction Kinetics at Solid Surfaces

E. G. d'Agliano, W. L. Schaich, P. Kumar and H. Suhl[1]

Department of Physics, University of California, San Diego, La Jolla, Calif. 92037, USA

The work I am going to describe here was inspired by a steadily growing body of evidence that catalytic reaction rates on metals and certain insulators show enhancements or other anomalies whenever certain types of fluctuations in the substrate degrees of freedom are unusually large, either as the result of a phase transition, or as the result of alloying, or for other reasons. Thus it is at least conceivable that solid state physics has something to contribute to catalysis, beyond the purely local steric and energetic considerations regarding the character of the chemi- or physisorbed states of the participating chemicals. Our aim will be to write, as far as possible, the various quantities appearing in the semi-phenomenological expression for the reaction rate in terms of physical quantities pertaining to the substrate.

First we briefly summarize the theory of reaction rates in general. With few exceptions, chemical reactions are thermally activated rather than tunneling processes, and since the inert cores of the participating atoms are heavy and the temperature is rather high, a classical description is generally considered adequate. The coordinates of the cores, and their momenta specify a point in phase space, and the reaction may be considered as a Brownian motion of this representative point which we shall sometimes call the "particle". The forces activating this motion are of two kinds: (1) systematic, coming from the direct interaction between the participants and (especially in catalysis) from the averaged coupling to the "bath" or substrate; (2) random, due to the thermal fluctuations of the coupling to the bath.

Usually the systematic forces are calculated in Born–Oppenheimer (BO) approximation, that is to say they are the gradient of the ground state energy of all the outer electrons calculated as a function of the (fixed) core positions $\boldsymbol{R}$. This energy $E_g(\boldsymbol{R})$ represents a hypersurface with various minima or valleys corresponding to different compounds (or dissociated fragments). By differential geometry, any two minima are separated by a saddlepoint. Next, the representative point is allowed to move, and a chemical reaction corresponds to motion of that point from one minimum (or valley) to the neighboring one; for obvious energetic reasons it takes place along the steepest line over the saddlepoint or along a path within the thermal energy kT of that line. The formula for the rate of transfer of particles from one minimum to the other, i.e., the reaction rate $\varkappa$, basically has the form first proposed by Arrhenius

$$\varkappa = \nu e^{-\beta V_b} \tag{1}$$

where V_b is the height of the saddlepoint above the minimum representing the initial situation, and the prefactor ν may loosely be interpreted as the number of "attempts" to surmount the barrier per unit time. β is $1/kT$, (T the temperature and k Boltzmann's constant).

A general derivation of (1) in all possible physical conditions is still not available. The most extensive study is probably that of Kramers [1]. The Brownian motion of the phasepoint is described by a transport equation for the distribution function f, $f(\boldsymbol{R}, \boldsymbol{P})$, in the phase space $\boldsymbol{R}, \boldsymbol{P}$. Inasmuch as the momentum changes of the "particle" in a collision with the bath are small, this transport equation plausibly takes Fokker–Planck form. Furthermore, since the motion, as already mentioned, is essentially one-dimensional along the line of steepest ascent and descent over the saddlepoint a one dimensional form of that

[1] Read by H. Suhl.

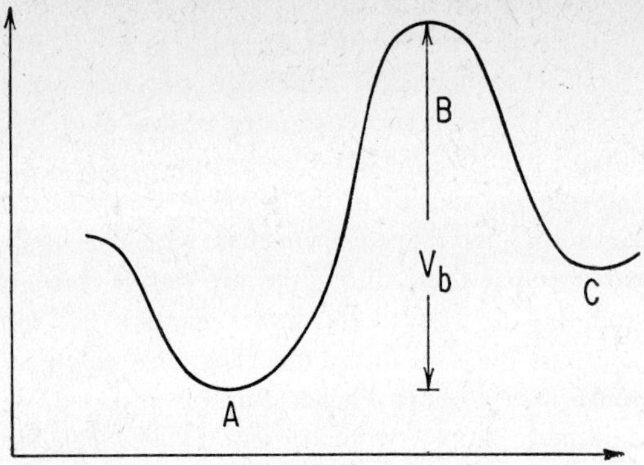

Fig. 1. *Abscissa*; reaction coordinate; *ordinate*: E_g. Plot of typical E_g versus reaction coordinate. The activation energy, V_b for a reaction from A to C is marked.

equation,

$$\frac{\partial f}{\partial t}+p\frac{\partial f}{\partial q}+F\frac{\partial f}{\partial p}=\eta\frac{\partial}{\partial p}\left(pf+MkT\frac{\partial f}{\partial p}\right) \qquad (2)$$

suffices. Here η is a friction coefficient, $F = -\partial E_g/\partial q$, in BO approximation, and q is the position along the reaction coordinate and p the momentum. We do not pause here to amplify on the meaning of M. Evidently in the reduction of the equation from the full dependence on the $6N$ coordinates and momenta of all the N cores, to the one-dimensional form (2) (with the full distribution factoring into a Boltzmann term in the degrees of freedom "normal" to q, p, and the f of eq. (2)) will lead to an "effective" mass M, which, moreover will generally be q dependent.

Kramers discusses eq. (2) in 3 cases: η much smaller than, comparable with and much larger than certain characteristic frequencies measured by the curvature of E_g in the appropriate places. For very large η, so large that the distance a thermally drifting particle traverses in time η^{-1} is small on the scale of variation of F, eq. (2) reduces the Smoluchowski equation for the density $\varrho = \int f du$ alone:

$$\frac{\partial \varrho}{\partial t} = -\frac{1}{\beta M\eta}\frac{\partial}{\partial q}\left(\frac{\partial \varrho}{\partial q}-\beta F\varrho\right) \qquad (3)$$

Kramers solves this for a steady state reaction (from $A \rightarrow C$ in fig. 1) as follows: Since $\partial\varrho/\partial t = 0$, the current $-(\beta M\eta)^{-1}(\partial/\partial q - \beta F)\varrho$ must be constant, $= j$, say. This relation may be solved for ϱ, and the rate is then defined as

$$\varkappa = j/\text{total no. of particles near } A$$

$$= j\Big/\int \varrho\, dq$$

The result is $\varkappa = (\omega_A\omega_B/2\pi\eta)\exp(-\beta V_b)$, where

$$\omega_A^2 = \frac{2}{M}\left(\frac{\partial^2 E_g}{\partial q^2}\right)_{q=A} : \omega_B^2 = -\frac{2}{M}\left(\frac{\partial^2 E_g}{\partial q^2}\right)_{q=B},$$

that is to say, the rate decreases with increasing η in this regime.

When $\eta = 0$ one may be tempted to make a purely kinematic argument; no friction or thermal agitation being explicitly present, one may take f to be a Boltzmann distribution well to the left of the barrier, and simply "count" up the number of particles traveling towards the barrier and being energetic enough to surmount it. This is a crude form of Eyrings argument leading to his so-called absolute rate theory. It leads to the result $\varkappa \sim \omega_A \exp(-\beta V_b)$, in the situation described by fig. 1.

At very small η, Kramers points out, this argument must fail. For η is not only the friction, but also a measure of the thermal "nudging" of the particle by the bath. In the absence of the η the tail of the Maxwell distribution, with $p > \sqrt{2V_bM}$ will rapidly deplete by escape over the barrier, and the reaction will cease. Kramers notes that for small η the particle drifts through a large distance before colliding, its energy remaining constant between collisions. It is then more appropriate to regard the contact with the bath as a cause for diffusion in energy rather than configuration space. A transformation of (2) to angle and action (I, φ) or equally well to energy and epoch (E, τ) variables is then appropriate. Kramers shows that in the extreme small η limit the τ-variable (or the φ variable) may be averaged out, and the resulting rate is

$$\varkappa = \eta\frac{V_b}{kT}\exp(-V_b/kT)$$

A finite rate thus requires replenishment of the tail by thermal agitation of which η is a measure. However, Kramers goes on to show that for intermediate values of η, large enough to ensure this replenishment, but not so large as to inhibit the motion altogether by friction, the Eyring absolute rate theory applies. For an overwhelming number of homogeneous phase reactions, the latter theory can indeed be relied on.

In regard to catalysis, the evidence cited in appendix I strongly suggests that one is in a regime in which the prefactor is in fact η dependent. In these cases, when systematic progressive changes are made in the substrate (for example by changing T across a phase transition or by varying degree of alloying) V_b in many cases does not change significantly, while $\varkappa$ changes considerably, suggesting a non-constant prefactor. (Strictly speaking V_b and η are not totally independent. The part of E_g due to interaction with the substrate, generalized to include retardation effects is connected by causality principles to the similarly generalized friction. However, since only the "spot value" V_b of E_g and the zero frequency friction η enters the formula for $\varkappa$ the relation of V_b and η is tenous except in special cases.) For these reasons we have extended the Kramers theory to the next order in $1/\eta$ and η, respectively, and the results are summarized in appendix II.

This brings us to the problem of reactions over solid substrates. Here the heat bath, the solid, serves a double function: not only does it provide the fluctuation–dissipation mechanism of which η is a measure, but it also contributes crucially to the systematic force occurring in the drift terms of the Fokker–Planck equation: The potential well responsible for physisorption or chemisorption arises out of adatom-solid interactions. The contribution to catalysis that solid state physics may be able to make lies in the possibility of finding at least approximate expressions for the various coefficients (like F and η) occurring in the Fokker–Planck equation, in terms of quantities involving the substrate alone.

A satisfactory way of doing this from the standpoint of rigor would be, for example, to begin with the Liouville equation for the density matrix of the whole system adparticle + solid, and to extract from it, by suitable tracing operations over the bath degrees of freedom an equation of motion for the "reduced" density matrix of the adparticles alone. Finally, since quantum effects in the motion of the latter are known to be small, a Wigner (or Husimi) transform should reduce that equation to the classical Fokker–Planck form.

Such (and, so far as we can see, similar) procedures run into formal contradictions among themselves. The conflicts are of the same kind that all semiclassical expansion procedures in statistical mechanics encounter at low temperatures.

From the success of the classical viewpoint of reaction kinetics one can conclude with some confidence that these difficulties are purely formal, and that a classical transport equation for the adatoms does, in fact, hold. That this equation must take Fokker–Planck form is dictated by extremely simple symmetry and detailed balance considerations, once it is agreed that the unsystematic part of the bath-adparticle interaction causes a succession of very small momentum changes in the adparticles. For simplicity we take the substrate to be translationally invariant parallel to its surface (as would be the case for the jellium model of a metal) and consider the case of only one adparticle. The equation is then (in terms of velocities U rather than momenta p):

$$\frac{\partial f}{\partial t} + \boldsymbol{U}\cdot\frac{\partial f}{\partial \boldsymbol{R}} + \frac{\boldsymbol{F}_T}{M_T}\cdot\frac{\partial f}{\partial \boldsymbol{U}_T} + \frac{F_z}{M_z}\frac{\partial f}{\partial U_z} = \eta_T\frac{\partial}{\partial \boldsymbol{U}_T}\cdot\left(\boldsymbol{U}_T f + \frac{kT}{M_T}\frac{\partial f}{\partial \boldsymbol{U}_T}\right) + \eta_z\frac{\partial}{\partial U_z}\left(U_z f + \frac{kT}{M_z}\frac{\partial f}{\partial U_z}\right) \tag{4}$$

$\boldsymbol{F}_T = -\partial V/\partial \boldsymbol{R}_T$; $F_z = -\partial V/\partial z$, the subscript T and z referring to dimensions in and normal to the surface, and the problem is to find expressions for the effective masses M_T, M_z, the effective potential V and the friction coefficients η_T, η_z.

One expects the M's and V to follow from equilibrium considerations alone. In equilibrium

$$f \sim \exp(-\beta[\tfrac{1}{2}(M_T U_T^2 + M_z U_z^2) + V(R)])$$

where $\beta = 1/kT$. Evidently, with $\boldsymbol{p}_T = M_T U_T$ etc., we have $\langle \boldsymbol{p}_T^2\rangle = 2M_T/\beta$, $\langle p_z^2\rangle = M_z/\beta$ (equipartition) on the other hand, these averages may be found from the equilibrium density matrix exp $(-\beta H)$ of the total sytem with Hamiltonian H.[1] Thus

---

[1] The effect of zero-point motion, which makes $\langle p^2\rangle$ finite even at $T=0$ is here disregarded [14]. This and related matters will be discussed in a forthcoming publication.

$$M_z = \beta \frac{\text{Trace}\{p_z^2 \exp - \beta H\}}{\text{Trace} \exp - \beta H},$$

$$M_T = \frac{\beta}{2} \frac{\text{Trace}\{\boldsymbol{p}_T^2 \exp - \beta H\}}{\text{Trace} \exp - \beta H} \qquad (5)$$

where $p_z = -i\hbar\, \partial/\partial z$, z being the z-coordinate of the adparticle. In the same way, we have for the density of adparticles at $\boldsymbol{R}_0$,

$$\varrho(\boldsymbol{R}_0) = \int f\, du \Big/ \int f\, du\, dx = A \exp - \beta V(\boldsymbol{R}_0)$$

where A is a constant. On the other hand $\varrho(\boldsymbol{R}_0) = \text{Tr}[\hat{\delta}(\boldsymbol{R}_0)\exp(-\beta H)]/\text{Tr}[\exp(-\beta H)]$ where $\hat{\delta}(\boldsymbol{R}_0)$ is the density operator for the adparticle ($\delta(\hat{\boldsymbol{R}} - \boldsymbol{R}_0)$ in position representation). Thus

$$F_z(\boldsymbol{R}_0) = -\partial V/\partial Z_0 = \frac{1}{\beta} \frac{\partial}{\partial Z_0} \log \text{Tr}[\hat{\delta}(\boldsymbol{R}_0) \exp(-\beta H)] \qquad (6)$$

and similarly for F_T.

Note that M_z and M_T differ from the bare mass M only as the result of quantum effects. We shall presently return to this matter.

The computation of the η's cannot be achieved on the basis of equilibrium considerations alone. However, linear response theory may be used in the similar dual way as was used for M and V:

A small external disturbance applied to the adparticles evokes current and heat current responses that are given by the well-known Kubo formalism. On the other hand, if the Fokker–Planck equation is assumed to hold, these responses are linked together by an equation for various moments of the velocity. The link involves η, which thus can be expressed in terms of these responses. Inasmuch as the F.P. equation is expected to describe only slowly varying phenomena, the external disturbance is taken to be independent of space and time. For simplicity let us take the Fokker–Planck in one dimensional form (i.e., along the reaction coordinate x only). Writing $\varrho_1 = \int f_1\, du$, $j_1 = \int u f_1\, du$, $K_1 = M/2 \int u^2 f_1\, du$ for the density, current, and kinetic energy density responses induced by a small uniform external field F_1 we find, multiplying the F.P. equation by u, integrating and linearizing

$$M\eta(x) = [F_1\varrho_0(x) + F_0\varrho_1(x) - 2\partial K_1(x)/\partial x]/j_1(x) \qquad (7)$$

Expressions 5, 6 and 7 (or the obvious generalization of 7 for η_T and η_z) are based solely upon the hypothesis that a Fokker–Planck equation holds, but make no other assumptions. (In particular, they do not depend on the validity of the Born–Oppenheimer approximation.) One test of the validity of the basic hypothesis would be to compute the analogue of (7),

$$M(\eta(x, \omega, k) - i\omega) = [F_1\varrho_0(x)e^{+ikx} + F_0\varrho_1(x, \omega, k)$$
$$-2\partial K_1(x, \omega, k)/\partial x]/j_1(x, \omega, k) \qquad (8)$$

for disturbance $F_1 \exp(-i(\omega t - kx))$ rather than a constant one. If it turns out that $\eta(x, \omega, k)$ strongly depends on ω and k for small ω and k, the F.P. hypothesis must be rejected. (For large values of ω, for example, the hypothesis certainly fails, since the F.P. equation gives a response that is essentially Lorentzian, whereas it is known (e.g. [2]) that at very high ω, the exact response falls off faster than any inverse power of ω.)

Another test that the above procedure must meet is that in the limit of large adparticle mass, $\eta(x)$ should become proportional to the integrated autocorrelation function of the fluctuation part of the force at x, a classical result established, for example, by Kim & Oppenheim [3]. That this is the case follows rather simply in the limit in which M tends to infinity first and ω to zero afterwards. (The opposite order of limits gives the same result.)

The proof may be sketched as follows: With $\hat{\varrho}$ denoting the full density matrix of the system and $\hat{\delta}$ the density operator for the adparticle (in position representation $\hat{\delta}(\boldsymbol{R}_0) = \delta(\boldsymbol{R} - \boldsymbol{R}_0)$, the quantities

$$\bar{\varrho} = \text{Tr}(\hat{\varrho}\hat{\delta}); \bar{j} = \tfrac{1}{2} \text{Tr}\left[\hat{\varrho}\left(\frac{\hat{P}}{M}\hat{\delta} + \hat{\delta}\frac{\hat{P}}{M}\right)\right]$$

$$\overline{K} = \frac{1}{8M} \text{Tr}[\hat{\varrho}(\hat{P}^2\hat{\delta} + 2\hat{P}\hat{\delta}\hat{P} + \hat{\delta}\hat{P}^2)]$$

$$\overline{F\varrho} = \text{Tr}(\hat{\varrho}\hat{F}\hat{\delta})$$

satisfy an equation similar to (8):

$$M\frac{\partial \bar{j}}{\partial t} + 2\frac{\partial \overline{K}}{\partial x} - \overline{F\varrho} = 0 \qquad (9)$$

and the same equation holds for the disturbances $\bar{j}_1$, $\overline{K}_1$, $(\overline{F\varrho})_1$ due to a small applied field. Making the identifications $\bar{j}_1 = j_1 \bar{\varrho}_1 = \varrho_1$ $\bar{K}_1 = K_1$ and eliminating K_1 between eqs (8) and (9), we find

$$M\eta(x) = -\frac{1}{j_1}((\overline{F\varrho})_1 - F^* \varrho_1)$$

where we have written F^* for $-\partial V/\partial x$.

Now the Kubo response of any observable A due to a force $-x \cdot F_1 e^{-i\omega t}$ is

$$\langle \hat{A}_1 \rangle - \langle \hat{A} \rangle_0 = \frac{1}{M}\int_0^\infty \mathrm{d}\tau\, e^{i\omega\tau} \times \int_0^\beta \mathrm{d}\lambda \langle \hat{P}(-\tau - i\lambda\hbar)\, \hat{A} \rangle F_1$$

where the bracket $\langle ... \rangle$ means $\mathrm{Tr}\, e^{-\beta H}...$, and $\hat{P}(\tau)$ is $\hat{P}$ in the interaction representation.

Thus we find

$$M\eta(x) = -\int_0^\infty \mathrm{d}\tau\, e^{i\omega\tau} \int_0^\beta \frac{\mathrm{d}\lambda}{\beta} \langle \hat{P}(-\tau - i\lambda\hbar) \times (\hat{F}\hat{\delta} - F^*\hat{\delta}) \rangle \Big/ \int_0^\infty \mathrm{d}\tau\, e^{i\omega\tau} \int_0^\beta \frac{\mathrm{d}\lambda}{\beta} \times \langle \hat{P}(-\tau - i\lambda\hbar) \cdot \frac{1}{2M}(\hat{P}\hat{\delta} + \hat{\delta}\hat{P}) \rangle$$

Integration by parts (with respect to τ) in the numerator gives (since $\mathrm{d}\hat{P}/\mathrm{d}\tau = F(\tau)$)

$$M\eta(x) = \frac{\int_0^\infty \mathrm{d}\tau\, e^{i\omega\tau} \int_0^\beta \frac{\mathrm{d}\lambda}{\beta} \langle \hat{F}(-\tau - i\lambda\hbar)(\hat{F} - F^*)\hat{\delta} \rangle}{-i\omega \int_0^\infty \mathrm{d}\tau\, e^{i\omega\tau} \int_0^\beta \frac{\mathrm{d}\lambda}{\beta} \langle \hat{P}(-\tau - i\lambda\hbar) \frac{1}{2M}(\hat{P}\hat{\delta} + \hat{\delta}\hat{P}) \rangle} \qquad (10)$$

Now let M become large while ω is finite. Then $\hat{F}(\tau)$ may be replaced by $F_0(\tau)$, the force on the immobile particle. The integration over the λ-variable may be omitted if the real parts are taken in the limit of zero frequency. Writing

$$\hat{\mathfrak{F}} = \hat{F} - \langle \hat{F} \rangle_0$$

we then get $\int_0^\infty \mathrm{d}\tau \langle \hat{\mathfrak{F}}_0 \hat{\mathfrak{F}}_0(\tau) \hat{\delta} \rangle$ in the numerator of eq. (10).

To process the denominator, we note that P is of order $\sqrt{MkT}$, on the other hand $\mathrm{d}P/\mathrm{d}t$ is F, which is mass independent. Thus if we introduce a "scaled" P by writing $P = \sqrt{M/m}\tilde{P}$, where $\tilde{P}$ is of order unity, we see that the scale of time variation is $\sqrt{m/M}\tau$. The integral in the denominator of (10), for large ω thus has the form

$$-i\omega \int_0^\infty e^{i\omega\tau} g\left(\sqrt{\frac{m}{M}}\tau\right) \mathrm{d}\tau$$

$$= -i\omega \sqrt{\frac{M}{m}} \int_0^\infty \mathrm{d}\tau\, e^{i\omega\sqrt{\frac{M}{m}}\tau} \sum \frac{\tau^n}{n!} g^{(n)}(0)$$

$$= \sum_{n=0}^{\infty} g^{(n)}(0) \Big/ \left(i\omega\sqrt{\frac{M}{m}}\right)^n$$

$$\to g(0) \quad \text{for large } M/m$$

But

$$\int_0^\beta \frac{\mathrm{d}\lambda}{\beta} \langle \hat{P}(-i\lambda\hbar) \frac{1}{2M}(\hat{P}\hat{\delta} + \hat{\delta}\hat{P}) \rangle = kT\langle \hat{\delta} \rangle_0$$

where $\langle \hat{\delta} \rangle_0$ is the mean density.

A close look at the averaging operations then yields the well-known result

$$M\eta(x) = \frac{1}{kT} \lim_{\omega \to 0} \mathrm{Re} \int_0^\infty e^{i\omega\tau} \langle \hat{\mathfrak{F}}_0 \hat{\mathfrak{F}}_0(\tau) \rangle_0 \,\mathrm{d}\tau \qquad (11)$$

where, as above, $\langle ... \rangle_0$ denotes the bath average.

As a simple example of the large mass limit results, we discuss a one dimensional model of a metal + adatom potential consisting of a semi-infinite well of depth V_0 extending from $x = 0$ to ∞, to the left of which there is a well of depth V extending from $x = -X - b/2$ to $x = -X + b/2$ (see fig. 2).

Coulomb interactions were neglected, the main objective being to pinpoint the contribution to η of the covalency effect which causes an electron in the outside well to spread into the metal. (This effect also causes a bonding energy $E_g(X)$ which serves as potential $V(X)$ in the Fokker–Planck equation.) The relevant formulae are

Fig. 2. Electronic potential energy versus position for a one-dimensional model of metal plus adatom. The position of the Fermi level is marked.

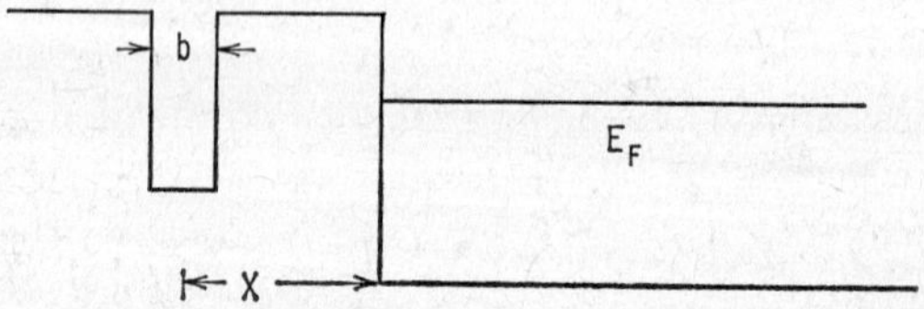

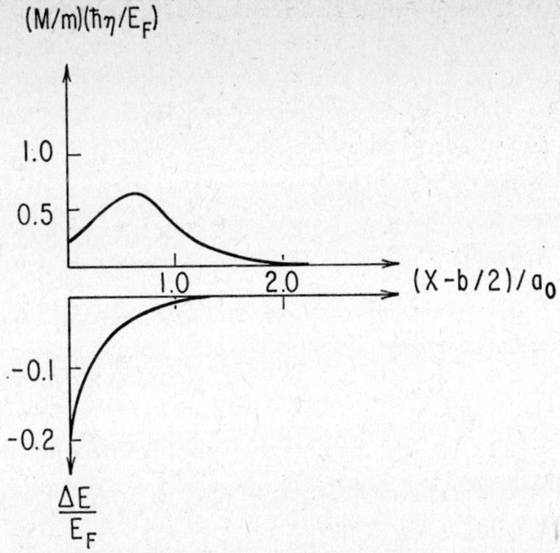

Fig. 3. Friction and binding energy of one-dimensional model versus adatom position.

$$\Delta E(X) = \int_0^{E_F} (E - E_F)\, \Delta\varrho(E)\, dE$$

where $\Delta\varrho$ is the incremental change

$$\frac{d}{dE}\frac{1}{\pi}[\delta(-X, E) - \delta(-\infty, E)]$$

in state density and

$$\eta = \frac{\pi\hbar}{M}\int dE\left(-\frac{\partial f}{\partial E}\right)\lim_{\omega\to 0}\sum|\langle n|F|m\rangle|^2 \times \delta(E - E_n)\,\delta(E - E_m - \hbar\omega). \qquad (12)$$

Here $\delta(X, E)$ is the one-particle phaseshift of an electron in the metal when the outside well is centered at $-X$, E_F is the Fermi energy, f the Fermi function, and

$$F = -\frac{\partial H}{\partial X}$$

where H is the Hamiltonian. A typical pair of curves of ΔE and η is shown in fig. 3 and the qualitative trend as a function of E_F is as follows, for the case of a well with one bound state E_b: Beginning with E_F above E_b, the peak in η diminishes in height and moves to larger values of X as E_F falls towards E_b. At the same time $|\Delta E(X)|$ becomes larger for all X.

At $E_F = E_b$, $|\Delta E|$ reaches a maximum (presumably due to the special sensitivity of electrons at Fermi-level to this resonant condition). At the same time η falls to zero for all X (alternatively the peak in η has moved off to infinity). As E_F falls below E_b, $|\Delta E|$ diminishes again, and the peak in η moves back towards $X=0$. The formal reason for the vanishing of η at $E_F = E_b$ is the fact that, for a square well, the force F is the difference of two δ-functions, so that any one of the diagonal elements of F in eq. (12) becomes the difference of the electron densities at the two edges and this difference vanishes at "resonance". This disappearance of η may however simply be an artifact of a one-dimensional calculation.

Returning to the general case, we see from eq. (11) how the substrate correlations, or alternatively substrate responses, enter the formula for η. For example, if the substrate is a metal in Coulomb interaction with the ad-ion,

$$\mathfrak{F}(\boldsymbol{R}) = -\int d\boldsymbol{r}\,\frac{\partial V(\boldsymbol{R}, \boldsymbol{r})}{\partial \boldsymbol{R}}\,\delta\hat{n}(\boldsymbol{r})$$

where $\delta\hat{n}(\boldsymbol{r}) = \hat{n}(\boldsymbol{r}) - \langle\hat{n}(\boldsymbol{r})\rangle$ is the density operator of the electron gas. Then, from (11)

$$M\eta(\boldsymbol{R}) = \frac{1}{kT}\lim_{\omega\to 0}\mathrm{Re}\int_0^\infty e^{i\omega\tau}d\tau\iint\frac{\partial V(\boldsymbol{R}, \boldsymbol{r})}{\partial \boldsymbol{R}} \times \frac{\partial V(\boldsymbol{R}\boldsymbol{r}')}{\partial \boldsymbol{R}}\,c_1(\boldsymbol{r}, \boldsymbol{r}', \tau)\,d\boldsymbol{r}\,d\boldsymbol{r}'$$

where

$$c_1(\boldsymbol{r}, \boldsymbol{r}', \tau) = \langle\delta\hat{n}(\boldsymbol{r}, 0)\,\delta\hat{n}(\boldsymbol{r}', \tau)\rangle \qquad (13)$$

In the same way, if we suppose that an electron remains more or less localized on the adatom (at least at certain distances from the surface) which then enters into exchange interaction $\int J(\boldsymbol{R}, \boldsymbol{r})\,\hat{\boldsymbol{S}}\cdot\hat{\boldsymbol{s}}(\boldsymbol{r})\,d\boldsymbol{r}$ with the substrate spin density $\hat{\boldsymbol{s}}(\boldsymbol{r})$, one finds (assuming $\langle\boldsymbol{S}\rangle = \langle\boldsymbol{s}\rangle = 0$ for simplicity)

$$M\eta(\boldsymbol{R}) = \frac{1}{kT}\lim_{\omega\to 0}\mathrm{Re}\int_0^\infty e^{i\omega\tau}\iint\frac{\partial J(\boldsymbol{R}, \boldsymbol{r})}{\partial \boldsymbol{R}} \times \frac{\partial J(\boldsymbol{R}, \boldsymbol{r}')}{\partial \boldsymbol{R}}\,c_2(\boldsymbol{r}\boldsymbol{r}', \tau)\,d\boldsymbol{r}\,d\boldsymbol{r}'\,S(S+1)$$

where

$$c_2(\boldsymbol{r}, \boldsymbol{r}', \tau) = \langle\hat{s}_z(\boldsymbol{r}, 0)\,\hat{s}_z(\boldsymbol{r}', \tau)\rangle \qquad (14)$$

The correlation functions in (13) and (14) may be reworked into the form of charge and spin response functions of the substrate. Because one takes the real part only the "lossy" part of the responses survive. To illustrate these remarks, we consider the evaluation of η in a uniform medium, where it will be independent of $\boldsymbol{R}$ and where a Fourier representation of (13) and (14) is appropriate, if we make the further approximation of neglecting the presence of the heavy particle in the calculation of c_1 and c_2. This last approximation amounts to evaluating η to lowest order in the forces coupling the heavy particle motion to the electron bath. Hence we expect to obtain expressions similar to those previously derived from the "golden rule" [4]. We find from eq. (13)

$$M\eta = \frac{4\pi e^2}{3}\int \frac{d^3q}{(2\pi)^3} \lim_{\omega\to 0} \mathrm{Im}\left[-\frac{1}{\omega}\frac{1}{\varepsilon(\boldsymbol{q},\omega)}\right] \tag{15}$$

where the dielectric constant $\varepsilon(\boldsymbol{q},\omega)$ is given by

$$\varepsilon(\boldsymbol{q},\omega) = \left[1 - V(\boldsymbol{q})\frac{i}{\hbar\Omega}\int_0^\infty d\tau\, e^{i\omega\tau}\langle[\hat{n}(\boldsymbol{q},\tau), \hat{n}(-\boldsymbol{q},0)]\rangle\right]^{-1},$$

and $V(\boldsymbol{q}) = 4\pi e^2/q^2$ and Ω is the volume of the system. Similarly, from eq. (14) we derive

$$M\eta = \frac{S(S+1)}{3}\int \frac{d^3q}{(2\pi)^3} q^2 |J(\boldsymbol{q})|^2 \times \lim_{\omega\to 0} \mathrm{Im}\left[-\frac{1}{\omega}\chi(\boldsymbol{q},\omega)\right] \tag{16}$$

where the susceptibility $\chi(\boldsymbol{q},\omega)$ is defined by

$$\chi(\boldsymbol{q},\omega) = -\frac{i}{\hbar\Omega}\int_0^\infty d\tau\, e^{i\omega\tau}\langle[\hat{s}_z(\boldsymbol{q},\tau).\, \hat{s}_z(-\boldsymbol{q},0)]\rangle.$$

A particularly simple demonstration of the relation between friction and response may be given for the case of an ion embedded in a uniform medium. In that case a formula of essentially the form 10 has been derived by Edwards [5] and Rousseau et al. [6] for the resistivity ϱ of the substrate. The results are

$$\varrho = \frac{n_1}{n_e}\frac{m}{e^2}\frac{1}{\tau}$$

$$\eta = n_e\frac{m}{M}\frac{1}{\tau}$$

where $1/\tau$ is proportional to the integrated $\hat{\mathfrak{J}}\hat{\mathfrak{J}}$ correlation function. It is intriguing to speculate whether such a relation might also apply in an inhomogeneous system, thereby implying that η is related to the additional resistance of a film due to adsorbed particles.

Appendix I

Experimental information regarding the role of the substrate

(1) *Metals*

(1.1) To the extent that the desorption of hydrogen is a bottleneck in some hydrogenation reactions, the desorption rate of hydrogen from nickel thin films has been observed to be smaller by a factor of 1/2 in the vicinity of 526 K, the temperature corresponding to resistivity anomaly for the sample [7]. The width of the minimum is approx. 100 K.

(1.2) By sandwiching Al_2O_3 between 2 gold layers [8], the decomposition rate of formic acid was found to increase as a tunneling current was passed between the layers. The rate of increase being roughly 2 orders of magnitude for an effective power of 0.1 W/cm².

(1.3) Couper & Eley [9] have observed the dramatic change in the activation energy of orthopara conversion at 60% Au and 40% Pd alloy. While attempts have been made to explain this effect by the filling of the d band [10], the prefactor of the reaction rate also shows a sharp dip at that concentration.

(4) Dowden & Reynolds [11] have observed the correlation between the activity for the dehydrogenation of styrene, methanol and formic acid and the specific heat of the substrate. The substrate consists of transition metal alloys.

(5) The log of hydrogenation rate of benzene on paramagnetic Ni–Cu alloys is found to be proportional to the effective number of Bohr magnetons of the substrate [12]. The activation energy of the reaction remains unchanged.

(2) *Insulators*

(2.1) The ratio of ortho to para hydrogen over Cr_2O_3 changes by a factor of 10 at the Neel temperature of the sample [13]. Such variation is not seen on an amorphous substrate. On a substrate of small particles, the jump broadens.

(2.2) On some ferroelectrics, the oxydation rate of CO is found to be large near the transition temperature [14].

Appendix II

Contractions of the Fokker–Planck equation for extreme values of the friction coefficient

(1) *Large η regime and the generalization of the Smoluchowski eq.* The F.P. eq., in one dimension is

$$\frac{\partial f}{\partial t}+u\frac{\partial f}{\partial x}+\frac{F}{M}\frac{\partial f}{\partial u}=\eta\frac{\partial}{\partial u}\left(uf+\frac{kT}{M}\frac{\partial f}{\partial u}\right)$$

The substitution $f=\exp(-\frac{1}{2}\mu^2)\,\Phi(\mu,x)$, with $\mu=u\sqrt{M/(2kT)}$ reduces it to

$$\frac{\eta}{2}\left[\frac{\partial^2}{\partial\mu^2}+(1-\mu^2)\right]\Phi$$

$$=\sqrt{\frac{2kT}{M}}\,\mu\frac{\partial\Phi}{\partial x}+\frac{F}{M}\sqrt{\frac{M}{2kT}}\left(\frac{\partial}{\partial\mu}-\mu\right)\Phi+\frac{\partial\Phi}{\partial t} \qquad \text{(A 1.1)}$$

The eigenfunctions of the left hand side are $\Phi_n=H_n(\mu)\,e^{-\mu^2/2}$ with eigenvalues $-\eta n$. (Here H_n is the Hermite polynomial of order n). The substitution

$$\Phi=\sum_{n=0}^{\infty}\Phi_n(\mu)\,\Psi_n(x,t)$$

gives a set of differential difference equations for the Ψ_n:

$$\left(\frac{\partial}{\partial t}+\eta n\right)\Psi_n=\sqrt{\frac{kTn}{M}}\left(\frac{\partial}{\partial x}-\frac{F}{kT}\right)\Psi_{n-1}$$

$$+\sqrt{\frac{kT(n+1)}{M}}\frac{\partial\Psi_{n+1}}{\partial x} \qquad \text{(A 1.2)}$$

with $\Psi_n=0$ for $n<0$.

From the orthogonality relations it follows that Ψ_0 is essentially density, $\int f\,du$ and Ψ_1 the current (indeed for $n=0$, the last equation is the continuity equation).

Smoluchowski's equation is obtained by keeping only Ψ_0 and Ψ_1 in the hierarchy, setting all higher Ψ_n's equal to zero. Also, $\partial/\partial t$ is neglected compared with η. Elimination of Ψ_1 gives Smoluchowski's eq.

$$\frac{\partial\Psi_0}{\partial t}=\frac{kT}{M}\frac{\partial}{\partial x}\frac{1}{\eta}\frac{D\Psi_0}{Dx}$$

where

$$\frac{D}{Dx}=\frac{\partial}{\partial x}-\frac{F}{kT}=e^{-\beta V}\frac{\partial}{\partial x}e^{\beta V}$$

Successive generalizations, that proceed in inverse powers of η, are obtained by retaining Ψ_n's up to successively higher values of n. For example, keeping Ψ_0, Ψ_1 and Ψ_2 and eliminating Ψ_1 and Ψ_2 one obtains

$$\frac{\partial\Psi_0}{\partial t}=\frac{kT}{M}\frac{\partial}{\partial x}\frac{1}{\eta}\left(1-\frac{kT}{M}\frac{\partial}{\partial x}\frac{1}{\eta}\frac{D}{Dx}\frac{1}{\eta}\right)\frac{D\Psi_0}{Dx} \qquad \text{(A 1.3)}$$

Closer examination shows that the expansion parameter is essentially $\sqrt{kT/M}\,|\partial(\log F)/\partial x|/\eta$. Also, the assumption $\partial/\partial t \ll \eta$ must be made, otherwise an equation non-local in time is obtained. Successively higher equations give more and more accurate values for density, current, kinetic energy density, heat current, etc. In this respect the procedure resembles that of Chapman–Enskog. Also the price that must be paid here is similar: More and more boundary conditions on Ψ_0 must be specified as higher derivatives occur in the successive equations. The resulting reaction rate due to the generalization (A 1.3) can easily be calculated to be

$$\varkappa=\varkappa_{\text{Kramers}}\left(1-\frac{\omega_B^2}{\eta^2}\right)=\frac{\omega_A\omega_B}{2\pi\eta}\left(1-\frac{\omega_B^2}{\eta^2}\right)e^{-\beta V_b}$$

(2) *Small η*: In this case Kramers envisages that the phase point oscillates many times on a curve of constant energy before undergoing a collision. Therefore, he considers the Brownian motion as a diffusion in energy space. To generalize his procedure in this regime we make a contact transformation from position q and momentum p to energy and epoch E and τ as follows

$$\lambda\frac{\partial E}{\partial p}=\frac{\partial\tau}{\partial q};\quad -\lambda\frac{\partial E}{\partial q}=\frac{\partial\tau}{\partial p} \qquad \text{(A 2.1)}$$

where E is the Hamiltonian function. The curves $\tau=$ constant are thus orthogonal trajectories to

E = const. We determined λ by imposing the condition that this transformation be canonical:

$$\frac{\partial\tau}{\partial q}\frac{\partial E}{\partial p} - \frac{\partial\tau}{\partial p}\frac{\partial E}{\partial q} = 1$$

whence

$$\lambda = 1 \Big/ \left[\left(\frac{\partial E}{\partial q}\right)^2 + \left(\frac{\partial E}{\partial p}\right)^2\right]$$

(q and p are here considered scaled to both have dimensions of $\sqrt{\text{action}}$.)

The drift term in the F.P. eq. is the Poisson bracket $[E, f]$, and thus transforms into $\partial f/\partial\tau$. The F.P. eq. thus becomes

$$\frac{\partial f}{\partial t} + \frac{\partial f}{\partial \tau} = \eta \mathfrak{F} f \qquad \text{(A 2.2)}$$

where

$$\mathfrak{F} = \eta \frac{\partial}{\partial u}\left(u + \frac{kT}{M}\frac{\partial}{\partial u}\right)$$

transformed into E, τ variables.

Kramers assumes that the density $\int f \mathrm{d}\tau$ in energy space is zero when the phase point gets to that energy contour in p, q space that passes through the top of the barrier due to irrevocable escape. On this assumption only the motion in well A (fig. 1) is of interest. There it is simply periodic, were it not for the $\mathfrak{F}$ operator, and thus it is reasonable to assume a discrete Fourier decomposition

$$f = \sum_{\mathrm{n}} e^{\mathrm{i}n\tau} \Psi_{\mathrm{n}}(E, t) \qquad \text{(A 2.3)}$$

with respect to τ (for contours outside the well, a Fourier integral would have to be used).

Successive recurrence relations are then obtained by substituting (A 2.3) into (A 2.2) multiplying by $e^{-\mathrm{i}n\tau}$ and integrating:

$$\frac{\partial \Psi_{\mathrm{n}}}{\partial t} + \mathrm{i}n\, \Psi_{\mathrm{n}} = \frac{\eta}{2\pi}\int_0^{2\pi} \sum_{\mathrm{m}} (e^{-\mathrm{i}n\tau} \mathfrak{F} \Psi_{\mathrm{m}} e^{\mathrm{i}m\tau})\, \mathrm{d}\tau$$

For a simple harmonic well $\mathfrak{F}$ in the new coordinates is relatively tractable. Kramers keeps only $\mathrm{n} = 0$.

It turns out that even and odd sets Ψ_{n} do not couple. If Ψ_0 and Ψ_2 are kept, the result (which always has the form of a continuity eq.) is

$$\frac{\partial \Psi_0}{\partial t} = \frac{\partial j}{\partial E}$$

where

$$j = \eta \hat{K} \Psi_0 - \frac{\eta^3}{8\omega_A^3} E\left(1 + 2kT\frac{\partial}{\partial E}\right) \times \left(\frac{\partial}{\partial E}\hat{K} - \frac{kT}{E}\right) E \frac{\partial}{\partial E}\left(1 + kT\frac{\partial}{\partial E}\right)\Psi_0$$

and where

$$\hat{K} = E\left(1 + kT\frac{\partial}{\partial E}\right).$$

The correction to the Kramers rate can be calculated, in principle, from this relation, but with much more difficulty than for the large η case.

References

1. Kramers, H A, Physica 1940, 7, 284.
2. Martin, P C, Problème à N Corps (ed C DeWitt & R Balian). Gordon & Breach, New York, 1968.
3. Kim, S & Oppenheim, I, Physica 1972, 57, 469.
4. Suhl, H, Smith, J H & Kumar, P, Phys rev lett 1970, 25, 1442.
5. Edwards, S F, Proc phys soc 1965, 86, 977.
6. Rousseau, J, Stoddart, J C & March, N H, J phys 1972, C5, 175.
7. Shanabarger, M. Private communication.
8. Haidinger, W & Figar, J, Chem phys lett 1971, 11, 545.
9. Couper, A & Eley, D D, Disc Faraday soc 1950, 8, 172.
10. Newns, D M, Phys rev 1969, 178, 1144.
11. Dowden, D A & Reynolds, P W, Disc Faraday soc 1950, 8, 184.
12. Evzerikhin, E I & Lyubarskii, G D, Scientific selection of catalysts (ed A Blandin et al.). Israel Program of Scientific Research, Jerusalem, 1968.
13. Selwood, P W. Result communicated by M. Shanabarger.
14. Paravano, G, J chem phys 1952, 20, 342.

High-Energy Excitations in Solids

Theory of Plasmon Effects in High Energy Spectroscopy

D. C. Langreth

Department of Physics, Rutgers University, New Brunswick, N.J. 08903, USA

In recent years the whole area of high energy excitations in solids has begun to flourish, both experimentally and theoretically. Their spectroscopy and theoretical understanding will have important implications for our knowledge of the structure and properties of materials, both in the bulk and near the surface. Even the limits indicated by the title of this paper are much broader than can be covered in a work of this size. The choice of topics has been rather personal, with some review of earlier work, and some presentation of new results.

The theory discussed here relates to the plasmon satellites in various experiments where a deep electron is excited. I discuss the effects in the spectra intrinsic to the rapid creation of a deep hole as well as the extrinsic effects which arise in some experiments from the energy loss of a fast electron. The understanding of the intrinsic spectra will be important, aside from its theoretical interest, for interpretation of experimental results in terms of electronic state densities, especially when the theory is extended to apply to valence electrons. Understanding the extrinsic effect has an important bearing on our knowledge of electronic mean free paths and inelastic escape depths, which in turn is a fundamental question involved in the spectroscopy of solid surfaces. The old question of whether one is measuring a bulk or surface property comes up again and again. The understanding of plasmon interactions will be crucial in providing the answer in each case, as they are generally the most strongly coupled excitation.

In this paper the theory for the most part will be couched in the language of x-ray photoemission. Nevertheless, many of the results and methods used can be taken over to other forms of spectroscopy.

The arrangement of material in this work is more chronological than logical. I have done this purposefully, to emphasize the importance of the earlier work [1–5] on the intrinsic effect. In my view the import of this work has not been until recently sufficiently recognized, especially by experimental practicioners who have, for example, used supposedly "structureless" deep spectral lines to rid their valence band data of unwanted "background," which was assumed to come from energy loss effects alone. Section I (pp 211–212) discusses the intrinsic plasmon satellites occurring on the spectral density of a rapidly excited deep-hole. In section II (pp 212–214) we review a sum rule for the spectral density, as well as presenting some new work showing that a sum rule recently discussed by Doniach is simply another form of this one. It should be emphasized of course that these sum rules are completely general and are not limited to the case of plasmon interactions. In section III (pp 214–215) is a general discussion of intrinsic and extrinsic plasmon effects, of coupling constants, and of the types of experiments in which the respective effects might be expected to be strong. Section IV (pp 215–217) contains a review of the extrinsic effect, as well as the interference term between fast and slow electrons. It is emphasized that extrinsic and intrinsic satellites have different strength dependences which might be used to separate them experimentally. Section V (pp 217–219) generalizes the results of the preceding section to include surface effects. Finally in section VI (pp 219–222) are some results for thin films which I present here for the first time, and which might offer another method for separating the intrinsic and extrinsic effects.

I. The Deep-Hole Spectral Density

The spectral density, or the analogue of the single particle density of states in a many particle system, has a direct relevance to almost all experiments and in some, such as x-ray photoemission is almost measured directiy [5] at least when the "extrinsic" [6–8] effects discussed in section IV are accounted for. Here we consider the spectral density of a hole state sufficiently deeply bound in an atom of a solid to be considered essentially a recoil-free spectral line, but not so deep as to be lifetime broadened to any significant degree. In the absence of interaction then this spectral density would be simply a delta function. In this section we consider the spectral density appropriate for the excitation of the deep hole, and simply note that the de-excitation case is similar [9–11] although its relevance is sometimes complicated by incomplete relaxation effects [12].

Much of the important early work in this area was done by Hedin, Lundqvist & Lundqvist [1–5] and indeed it was B. I. Lundqvist [3] who recognized and stressed the importance of the coupling of the deep hole to the plasmon-like excitations of the valence band. These excitations are characterized by $S(q, \omega)$, the Fourier transform of the density–density correlation function, and consist of screened single particle excitations (particle-hole pairs) as well as plasmons. Lundqvist approximately solved for the deep-hole spectral density $A(\omega)$ by 2nd order perturbation theory in the self energy:

$$\Sigma(\omega) = \int d\nu \sum_q |V_q|^2 S(q, \nu) G_0(\omega - \nu) \tag{1}$$

and

$$A = -2 \operatorname{Im} G \equiv -2 \operatorname{Im} [\omega - \Delta - \Sigma(\omega)] \tag{2}$$

as illustrated in fig. 1. Here Δ is the position of the core level in the Hartree Fock approximation, and $V_q \simeq 4\pi e^2/q^2$. The above result showed (in addition to the original core "line" broadened by electron-hole excitations and shifted) a broad "plasmaron satellite" beginning at the plasma energy $\hbar\omega_p$ below the main peak. Although subsequent calculations predicted a difference in the shape and number of satellites, the importance of the above calculation lay in the demonstration that plasmon satellites could

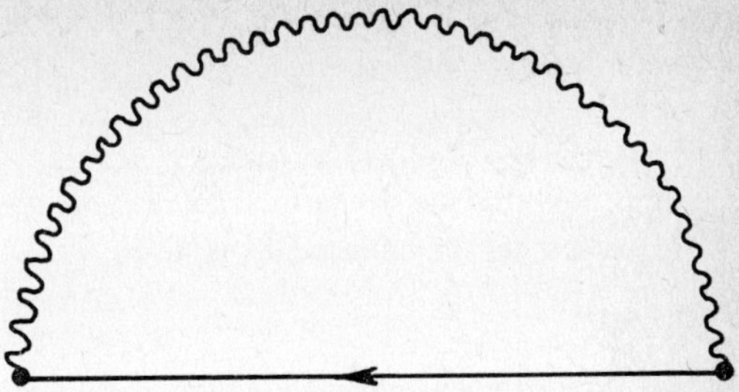

Fig. 1. The diagram for the self-energy of the deep-hole leading to the approximation of eq. (1). ——, (in this figure only), bare deep hole of Green's function; ∿∿∿, propagator for density fluctuation excitations of valence electrons.

carry a substantial fraction of the spectral weight—around 50% in the case of the lower density metals.

The next development was the realization [11] that within the above assumptions plus an assumed independence of the valence band excitations, there existed an exact solution. The simplest model treated only the plasmons, in which case the "plasmaron" problem reduced to a set of displaced harmonic oscillators. The more sophisticated approach applied the same method as used by Nozières & de Dominicis [9] to solve the threshold problem [13]. It was shown that

$$A(\omega) = \int dt \exp[i(\omega - \Delta) t - C(t)], \tag{3}$$

where C is an integral over linked diagrams as illustrated in fig. 2. The most important terms is fig. 2*a* which treats the plasmons as non-interacting excitation and yields

$$C(t) = \sum_q |V_q|^2 \int_{-\infty}^{\infty} d\omega S(q, \omega)(1 + i\omega t - e^{i\omega t})/\omega^2 \tag{4}$$

It can then be shown [5, 11] that instead of a broad plasmon peak, eq. (4) yields a series of fairly sharp plasmon peaks with their most singular parts at multiples of the plasmon energy. Eq. (4) also predicts that plasmon peaks as well as the main threshold are softened by the Anderson orthogonality block [9, 14] which is caused by the multiple emission of soft electron-hole pairs. In any case, the mean spread of any low order plasmon satellite due to electron-hole emission can be shown to be considerably less than the plasmon energy $\hbar\omega_p$, so that the emission of e-h pairs serves mainly to redistribute the spectral weight within a given satellite. It follows from (4) then that the strength of the nth satellite (n = o represents the main peak) is given by

$$e^{-\beta}\beta^n/n! \tag{5}$$

where β is a dimensionless coupling parameter characterizing the plasmon deep-hole interaction, which in simple systems is of the order of magnitude $r_s/6$ (where r_s is the usual interparticle spacing parameter in Bohr units); therefore, for say Al, we would expect about a 30% first plasmon satellite on the deep-hole spectral density. That (5) is simply a Poisson distribution is a reflection of the physical fact that the virtual plasmons in the cloud around the deep hole cannot correlate with each other through the hole (because it is recoil-free), plus the approximation that the plasmon in a given range of wave vector is a good excitation which does not interact either with other plasmons or with electron-hole pairs.

Fig. 2. Examples of diagrams for the exponent *C*. (*a*) approximation of eq. (4); (*b*) next order correction to *C*. ——, propagator for valence electrons.

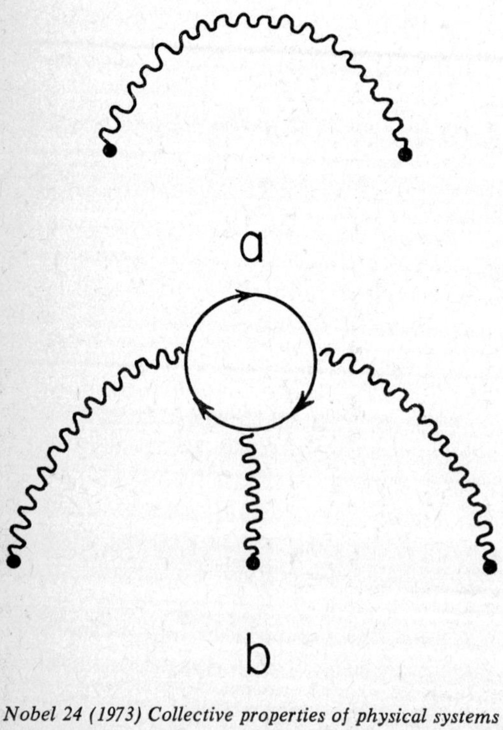

The validity of this latter approximation was recently checked by Chang & Langreth [8] by calculation of the leading correction to it. This correction is pictured in fig. 2*b*. The contribution to the strength of the second plasmon satellite is most easily represented by the diagrams of fig. 3 where the dotted line represents where the imaginary part is to be taken; such diagrams represent a correlated emission of two plasmons, and therefore are the leading terms producing a deviation from the Poisson distribution. The result of the approximate calculation is that the sum of all diagrams of the form of fig. 3 give a contribution to the strength of the second plasmon satellite which is only about equal to 0.06 ($\beta^2/2$), for $r_s \sim 2$, that is the leading terms in the deviation from the Poisson distribution is only about 6%. The tentative conclusion is that (4) and hence (5) represent a good approximation indeed, at least so far as the strength of plasmon satellites goes.

II. Sum Rule

Another important contribution of Lundqvist [3] arose from his observation that the fairly large energy shift of the deep-hole quasiparticle line predicted by (1) was in apparent contradiction to our usual notion that correlations among the valence electrons should not have a large effect on the binding energy of an electron deep in an atomic core. Lundqvist also supplied the resolution by noting that the average energy of the deep hole state was unchanged by the valence electron correlations; this was done by the numerical integration of the approximate form (1) and finding that to within computational error

$$\int \frac{d\omega}{2\pi}\,\omega A(\omega) = \Delta. \tag{6}$$

It was later shown [11] that the expression (6) which Lundqvist inferred to be at least approximately true on physical grounds, and verified numerically in the approximation (1), was an exact consequence of the assumption that one has a hole state with no internal degrees of freedom and which cannot decay. Thus the sum rule (6) follows if one neglects the term in the crystal (or atomic) Hamiltonian which cause the

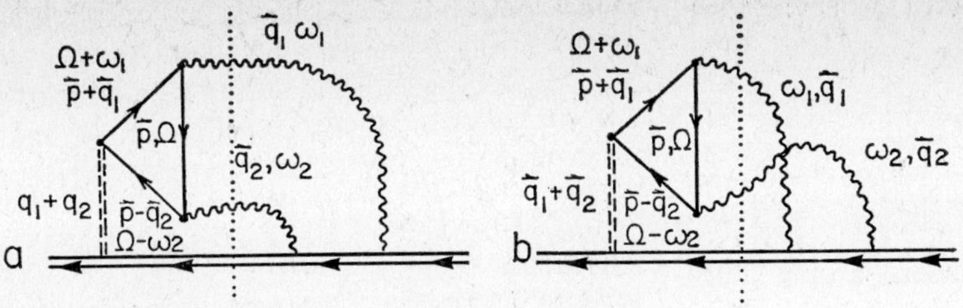

Fig. 3. Contributions to the strength of the second plasmon satellite resulting from specific time orderings of fig. 2. ═══, bare Coulomb interaction which is a permissible approximate way to estimate the dynamically screened interaction at frequency transfers of twice the plasma frequency.

hole to decay or to have a width, and is hence valid to the extent these widths are much smaller than say (in our case) the plasma energy.

To the previously stated physical arguments for the correctness of the sum rule, let me add the following. The quantity Δ is by definition the energy required to remove the deep electron with the valence electrons frozen in their normal states. Now if the instantaneous removal of the core electron is viewed with such poor energy resolution as not to get into trouble with the uncertainty principle, one should on one hand view a "line" at the "mean" position and on the other hand, one should be perceiving the fact that the transition was fast so that the position of this "mean line" is to be calculated using frozen wave functions—hence the sum rule.

But if one looks at the position of the beginning of the quasiparticle peak in the spectrum, one is of necessity looking with very fine energy resolution, implying a large time uncertainty, so that the valence electrons will have had time to relax, and at the threshold they will have relaxed to their ground state in the presence of the hole. If we denote the position of this threshold by Δ_r (the relaxed energy of the core electron as opposed to the frozen energy Δ) and write (3) as

$$A(\omega) = \int \mathrm{d}t \exp[\mathrm{i}(\omega - \Delta_r)t - B(t)], \tag{7}$$

we are in a position to show the equivalence between (6) and a sum rule recently proved by Doniach [15, 16]. Note that (7) serves a definition of the quantity B which might be termed the "satellite generator" [16] in the sense that it contains a constant term which subtracts weight from the threshold plus an oscillatory term which when the exponential is expanded adds this weight in the form of multiple satellites. From (6) one can write

$$\Delta = \mathrm{i}\frac{\partial}{\partial t}\int_{-\infty}^{\infty}\frac{\mathrm{d}\omega}{2\pi}A(\omega)\,\mathrm{e}^{-\mathrm{i}\omega t}\Big|_{t=0}. \tag{8}$$

Substitution of (7) into (8) gives

$$\Delta = \left(\Delta_r - \mathrm{i}\frac{\partial B}{\partial t}\right)\Big|_{t=0}\mathrm{e}^{-B(0)}. \tag{9}$$

Now $B(0)=0$, a fact which follows from the normalization integral of $A(\omega)$:

$$\int \frac{\mathrm{d}\omega}{2\pi}A(\omega) = 1. \tag{10}$$

Therefore eq. (9) may be written as

$$\Delta - \Delta_r = \int \frac{\mathrm{d}\omega}{2\pi}\,\omega B(\omega), \tag{11}$$

where $B(\omega)$ is the Fourier transform of $B(t)$. This is Doniach's form of the sum rule, and it states simply that the first moment of the satellite generator is equal to the difference between the "frozen" and "relaxed" energies of the core hole. The advantage of the form (11), is that it has a phase shift expansion in a model problem described in [14], and if this can be generalized, then it may prove very useful in the estimation of relaxation shifts in real materials [6].

Finally we note that this relaxation shift is of course just the most divergent part of $C(t)$—which oscillates linearly with time, that is

$$\Delta_r - \Delta = \lim_{t\to\infty} C(t)/\mathrm{i}t \tag{12}$$

For the particular approximation (4) this is just

$$\Delta_r - \Delta = \sum_q |V_q|^2 \int_{-\infty}^{\infty}\frac{\mathrm{d}\omega}{\omega}S(q,\omega) \tag{13}$$

This may be written in a more transparent form by writing $S(q, \omega)$ in terms of the dielectric function $\varepsilon(q, \omega)$:

$$S(q, \omega) = \frac{q^2}{4\pi^2 e^2} \operatorname{Im}\left[1 - \frac{1}{\varepsilon(q, \omega)}\right] \theta(\omega) \tag{14}$$

and then applying the Kramers–Kronig transform to yield

$$\Delta - \Delta_r = \tfrac{1}{2} \sum_q |V_q|^2 \frac{q^2}{4\pi e^2} \left[\frac{1}{\varepsilon(q, 0)} - 1\right]. \tag{15}$$

This is physical because if V_q is the Coulomb interaction this is just the work involved in moving a charge into a neutral polarizable material calculated according to classical electrostatics. The generalization to an arbitrary potential V_q is trivial.

III. Intrinsic vs Extrinsic Plasmon Effects

We term plasmon effects like that considered in section I, "intrinsic" effects as opposed to the energy loss of an incident or escaping electron which we would term an "extrinsic" effect. In experiments involving the excitation of a core hole, the electrons involved in intrinsic effects are generally either bound in a core state, or else have energies close to the Fermi level; in either case, we term such electrons "slow". On the other hand the energy of say an escaping photoelectron can be either high or low depending on the photon energy and which core level is involved; therefore electrons involved in extrinsic effects can be either "fast" or "slow". In this work we discuss only the case in which they are "fast" in the sense defined below.

There can be a considerable difference in the size of the coupling between a "slow" electron and plasmons and that of a "fast" electron. As mentioned earlier the "slow" electron coupling is of order

$$\beta \sim e^2 q_c/\pi\hbar\omega_p \sim r_s/6 \tag{15}$$

where q_c is the plasmon cutoff wave vector. Eq. (15) is of course model and material dependent, but can be of order unity for lower density materials; it would be expected to be only about 0.3 for Al, Si, or Ge, although many-body renormalizations may change estimates somewhat. On the other hand the coupling of a fast electron of velocity v and energy E to a plasmon is

$$\sim e^2/\hbar v = (R_\infty/E)^{\frac{1}{2}} \tag{16}$$

where R_∞ is the Rydberg constant. This will serve to define what I mean by "fast" because in this work I always assume $e^2/\hbar v$ to be a small parameter; for example for a 1.5 keV electron it is about 0.1.

There exist a class of experiments in which the intrinsic effects tend to be small [6]; this is when the number of "slow" electrons is conserved in the transition. Physically this is because the monopole moment does not change in the transition, and it is such a change which is most effective in shaking up the valence electrons [the satellites on the deep-hole spectral density for example reflect the response of the valence electrons to the creation of this deep hole]. Mathematically what happens is that the plasmon satellites on the various spectral functions of the electrons involved tend to be cancelled by the vertex corrections (interference terms). Of course the above just represents a more general statement of the cancellation noticed and explicitly calculated [17, 18] a long while ago for X-ray absorption and emission.

A second class of experiment is characterized by a change in the number of "slow" electrons during the transition. In this case the monopole moment of the "slow" electrons changes during the transition, and the contribution to this monopole by the fast electrons can be neglected due to the assumption that $e^2/\hbar v$ is small. Therefore the net cancellation is incomplete, and there is a chance of having relatively strong intrinsic satellites if the coupling constant β is sufficiently large.

Unfortunately, in experiments where "slow" electrons are not conserved, one always must worry about the extrinsic effect, which also produces plasmon satellites. Merely the fact that the fast electron coupling constant may be quite small is not sufficient to insure that the extrinsic effect is small. This is because the depth to which a fast electron penetrates is itself limited by this extrinsic coupling, and reducing it simply enables the electron to penetrate further, giving it a greater chance to interact. As it has been pointed out previously [6–8], the parameter α which characterizes the strength of the extrinsic effect is given by

$$\alpha = \Gamma_1/\Gamma \tag{17}$$

where Γ_1 is the decay rate of the fast electron via plasmon production and Γ is the total decay rate. Since both Γ_1 and Γ are of the order $e^2/\hbar v$ at high velocities, the quantity α does not approach zero. It is clear that if one is to separate the intrinsic and extrinsic effects experimentally, one is going to need a theory of the extrinsic effect as well. The beginnings of such a theory have been developed [6–8] and are reviewed in the next two sections.

IV. Extrinsic and Interference Effects in X-ray Photoemission

Here we discuss X-ray photoemission[1] as perhaps the simplest form (theoretically) of core excitation spectroscopy in which extrinsic effects are important. The theory of the extrinsic effect can of course be easily applied to other forms of spectroscopy, and we give an example of this later. For the interference effects this can usually only be done simply in a certain approximation [6, 7] which is probably not sufficiently accurate for detailed calculations of the strength of the fast–slow interference term, but which in any case is a useful starting point to fix our ideas about the order of magnitude of the various effects.[2]

For photoemission one must calculate to second order the current induced by an oscillating electric field representing the photons [21, 22, 23]. This is most easily done by calculating the electronic Green's function to second order in the photon's field, because the perturbation theoretic methods for such a non-equilibrium expansions are standard [24, 25], although

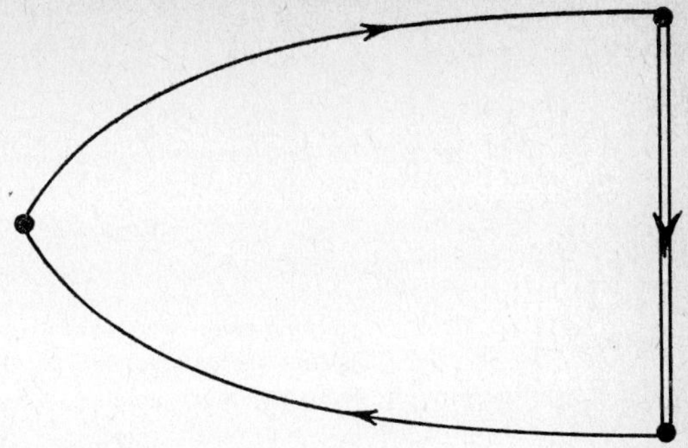

Fig. 4. The lowest order approximation to the photo current.

perhaps not so widely known as they should be. The basic diagram is shown in fig. 4. The double line represents the renormalized deep-hole propagator discussed in section I, while the single lines are the renormalized fast electron propagator. The most important extrinsic effects are those shown in fig. 5, and these represent multiple energy losses by the photoelectron on its way out. Without these it is clear that one would be measuring simply the deep hole spectral density. The sum of diagrams in fig. 5 is the convolution of the deep-hole spectral density and what can be termed the escape function $E(\boldsymbol{k}, \omega)$ for an electron of energy ω and wave vector $\boldsymbol{k}$; it satisfies the equation

$$E(\boldsymbol{k}, \omega) = |G(\boldsymbol{k}, \omega)|^2 + \left\{ \int \mathrm{d}\nu \left[\sum_{q} |(4\pi e^2/q^2)|^2 S(\boldsymbol{q}, \nu)\right] |G(\boldsymbol{k}, \omega)|^2 \right\} \times E(\boldsymbol{k}-\boldsymbol{q}, \omega-\nu). \tag{18}$$

The essential mathematical point to notice is that fig. 5 represents the only class of diagrams to which the additional excitation line also adds a pair of propagators G and G^* which are forced to have the same energy and momentum. Therefore even though the square bracketed quantity is of the order Γ where

$$\Gamma = \sum_{q} |(4\pi e^2/q^2)|^2 S(q, \omega) \backsim \omega_\mathrm{p} \frac{e^2}{\hbar v} \ln\left(\frac{mv^2}{\hbar\omega_\mathrm{p}}\right) \ll \omega_\mathrm{p}. \tag{19}$$

Neverthless

$$|G|^2 \sim [(\omega-\varepsilon_k)^2 + (\Gamma/2)^2]^{-1} \backsim \Gamma^{-1}\delta(\omega-\varepsilon_k), \tag{20}$$

[1] For a recent review of this and other methods of electron spectroscopy see D. A. Shirley [19] and references contained therein.

[2] In calculations in which the recoil of the slow electron is accounted for, it is extremely important to include plasmon dispersion and the plasmon-Coulomb collision lifetime [27] or else the theory will be riddled with false divergences or anomalously small energy denominators. It is just these lifetime effects which make the simple theory applicable to threshold quasi-singularities on plasmon satellites.

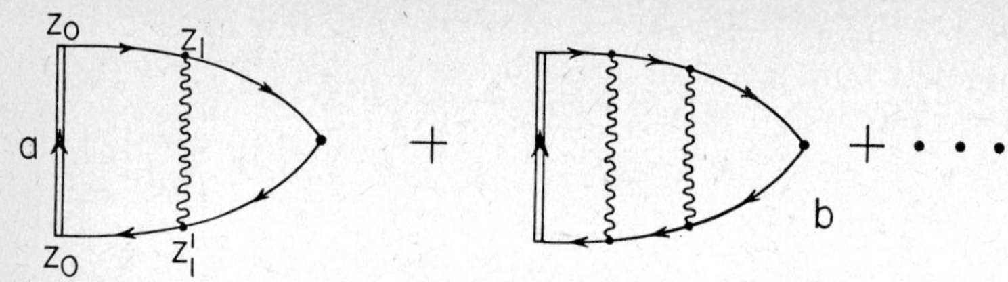

Fig. 5. Extrinsic effects in photoemission which survive the limit $e^2/\hbar v \to 0$ (assuming that the sample thickness and the x-ray penetration depth are much greater than the inelastic mean-free path).

so that the curly bracketed quantity in (18) is of order unity rather than $e^2/\hbar v$. Therefore this class of terms survives the limit $e^2/\hbar v \to 0$.

Aside from simplifications which come from the assumption that $E \gg \hbar\omega_p$ and $k \gg q_c$, the important approximate fact which enabled the analytic solution [6, 7] of (18), is that once an electron loses energy by emitting an electron-hole pair, its energy is smeared out over a range of 20 eV or more, and its contribution to the photocurrent will be deposited somewhere in the background, and it can no longer contribute to plasmon satellites, which are quite sharp. On making this approximation it is easy to show [6, 7] from (18) that the extrinsic contribution to the nth plasmon satellite relative to a main peak normalized to unity is

$$\alpha^n \tag{21}$$

where α is defined by (17). Of all the various corrections to (21) involving only the fast electron, the leading ones have all been calculated [7]. They are not only of order $\alpha \cdot e^2/\hbar v$ but also have rather small coefficients, and are therefore quite negligible, at least if v is large enough for any of this to make sense.

One must also consider the "fast–slow" interference terms of the type depicted in fig. 6. It is this sort of term that produces the cancellation when "slow" charge is conserved. These have been calculated [6, 7] to order $e^2/\hbar v$ and the results may be summarized by noting that their effect is to replace β in (5) by

$$\beta' = \beta - (e^2/\hbar v)F \tag{22}$$

where F is a slowly varying but model dependent function of the velocity, and whose magnitude is about unity. Therefore the effect of the interference term is indeed to weaken the intrinsic satellites by an amount that gets small for large velocities, in agreement with our earlier arguments. On the other hand if β is not very large to begin with (say ~ 0.3 as for Al), then the weakening (~ 0.1 for 1.5 keV electron) is still not completely negligible, since it would effectively reduce β from ~ 0.3 to ~ 0.2.

To find the combined effect of intrinsic and extrinsic terms, we must convolute the results implied by (5) and (21) with account taken of (19). The result is that strength of the nth plasmon satellite (normalized to unit threshold strength) is simply the coefficient of x^n in the Taylor series expansion of

$$\frac{e^{\beta' x}}{1-\alpha x}. \tag{23}$$

Note that (23) predicts, for example, that the strength ratio of the second peak to the first is smaller than the strength ratio of the first peak to the main threshold. On the other hand, if

Fig. 6. The lowest order "fast-slow" interference term or vertex correction. Such a term would weaken greatly any intrinsic plasmon effects in low energy photoemission, but this weakening effect becomes increasingly less important and goes to zero as $e^2/\hbar v$ at high photoelectron energies.

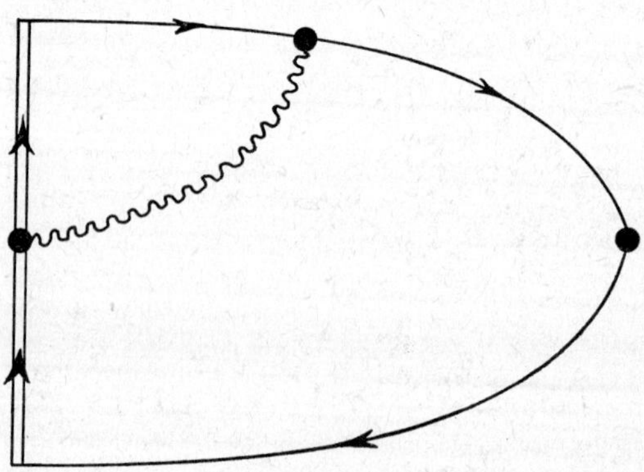

Table 1. *Recent experimental data on the strength of plasmon satellites in X-ray photoemission*

Material	Core State	(1st/ Main peak)	(2nd/ 1st)	Reference
Al	2s	0.74	0.76	[26]
Al	2s	0.61	0.40	[27]
Al	2p	0.50	0.30	[27, 28]
Si	2s	0.47	0.34	[29]
Ge	3d	0.38	0.27	[29]
Ge	3s	0.37	0.24	[29]
Ge	$2p_{\frac{3}{2}}$	0.37	0.18	[29]

only extrinsic effects were important, then this ratio would be constant according to (21). Therefore an experimental verification of this falloff would constitute evidence for the existence of the extrinsic effect.

In table 1, I summarize the data of several different experiments on several different materials. Although probably neither the accuracy of the data, nor the internal consistency of the table is as good as one might wish, there is an unmistakable tendency for the strength ratio to decrease in a way consistent with the existence of an intrinsic effect; it would probably be premature, however, to use the data to determine α and β'.

The fact that there is no well defined way to separate "peak" from "background" is a serious drawback to this type of analysis. It would be desirable to have a complete iterative solution of (18) which explicitly includes the excitation of electron-hole pairs, even though this would surely introduce some model dependence into the prescription.

V. Surface Effects

Even at energies of 1.5 keV, the inelastic mean free path of an electron in matter is only around 15 or 20 Å. Therefore one might expect the surface to have a noticeable effect on experiments such as X-ray photoemission. This has recently been investigated [8] using the model Hamiltonian

$$H_{\mathrm{e-pl}} = \sum_q g_{q\mathrm{B}}\, \varrho(\boldsymbol{r})\, a_q \sin q_\perp z\, \mathrm{e}^{\mathrm{i}\boldsymbol{q}\cdot\boldsymbol{r}_\parallel}\, \theta(-z) + \sum_q g_{q\mathrm{S}}\, \varrho(\boldsymbol{r})\, b_{q_\parallel}\, \mathrm{e}^{\mathrm{i}\boldsymbol{q}\cdot\boldsymbol{r}_\parallel}\, \mathrm{e}^{-|q_\parallel z|} + \text{c.c.}, \tag{24}$$

with

$$g_{q\mathrm{B}}^2 = \frac{4\pi e^2}{q^2}\,\omega_\mathrm{B}\left(1-\frac{1}{\varepsilon}\right), \tag{25}$$

and

$$g_{q\mathrm{S}}^2 = \frac{\pi e^2}{q}\,\omega_\mathrm{S}\left(\frac{\varepsilon-1}{\varepsilon}\right), \tag{26}$$

where ω_B and ω_S are the bulk and surface plasmon frequencies respectively, a_q and $b_{q_\parallel}$ are the respective plasmon destruction operators, and $\varrho(\boldsymbol{r})$ is the electronic density. Here ε is the static dielectric constant of the material (infinity for a metal). The Hamiltonian (24) contains both the surface modification of the bulk charge density and the existence of surface plasmons; nevertheless the description of these is rather crude, so that although these calculations should serve as good estimates of the magnitudes of the various effects, the results should not be considered quantitatively (except for asymptotic forms a long way from the surface).

The first effect considered was the modification of the deep-hole spectral density. The method of solution used in ref. [11] works here as well and it was found that

$$A(\omega) = \exp\left(-[\beta_\mathrm{S}(z)+\beta_\mathrm{B}(z)]\right) \sum_{\mathrm{nm}} \frac{\beta_\mathrm{S}^\mathrm{n}(z)\,\beta_\mathrm{B}^\mathrm{m}(z)}{\mathrm{n}!\,\mathrm{m}!} \times \delta(\omega-\Delta+\mathrm{n}\omega_\mathrm{S}+\mathrm{m}\omega_\mathrm{B}-\Delta E_\mathrm{S}(z)-\Delta E_\mathrm{B}(z)) \tag{27}$$

Note that the surface and bulk plasmon satellites still have an uncorrelated Poisson-like distribution; the satellites are delta functions in this model because of our neglect of plasmon dispersion as well as the smearing due to electron-hole excitations. The quantities $\beta_\mathrm{S}(z)$ and $\beta_\mathrm{B}(z)$ control the respective weightings of the bulk and surface plasmon satellites as a function of the distance $|z|$ inside the surface. For $|z|\to\infty$ we have of course $\beta_\mathrm{B}\to\beta$ inside and $\beta_\mathrm{S}\to 0$, while on the other hand $\beta_\mathrm{B}\to 0$ as $z\to 0$. Another interesting feature of the calculation is the energy shift of the spectrum $\Delta E_\mathrm{S}(z)+\Delta E_\mathrm{B}(z)$. At large z (inside the solid), these take the form

$$\Delta E_\mathrm{B}(z) \to \beta\omega_\mathrm{B} - \frac{e^2}{4|z|}\left(1-\frac{1}{\varepsilon}\right), \tag{28}$$

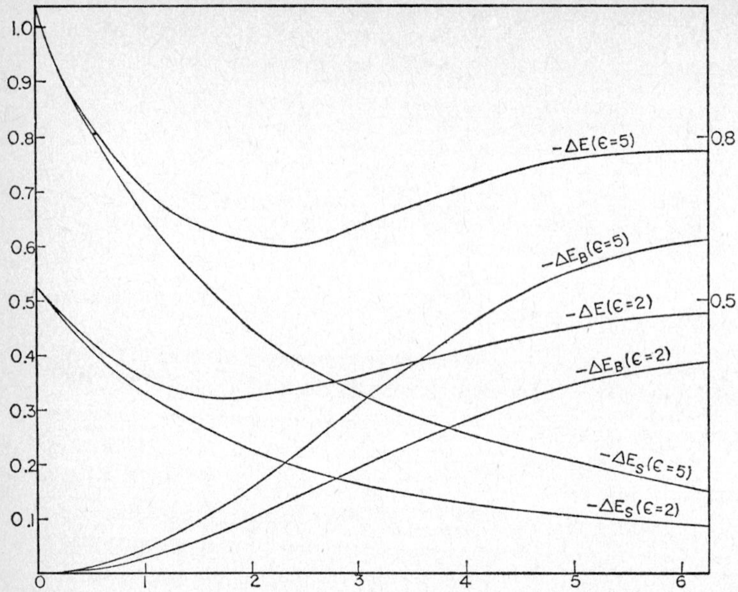

Fig. 7. *Abscissa:* $|x|$ (in unit of $1/2\ q_c$); *ordinate:* energy shifts (in unit of $e^2 q_c/\pi$).

Plots of the energy shift of the main line of the deep hole spectral density vs distance from the surface. Here ΔE_S is the contribution of the surface plasmons, ΔE_B is the contribution of the bulk plasmons and ΔE is the total. The sign of the shift is such that deep in the sample it would require less incident photon energy to produce a photoelectron of the same kinetic energy, than would be obtained if correlation effects were neglected. Note that this shift can be as much as several eV different from the bulk value if the core state is near the surface.

$$\Delta E_S(z) \to \frac{e^2}{4|z|}\left(\frac{\varepsilon-1}{\varepsilon+1}\right), \tag{29}$$

$$\Delta E_{\text{total}}(z) \to \beta\omega_B - \frac{e^2}{4|z|}\frac{1}{\varepsilon}\left(\frac{\varepsilon-1}{\varepsilon+1}\right). \tag{30}$$

This says simply that the energy shift is reduced from its bulk value $\beta\omega_B$ (denoted by $\Delta_r - \Delta$ in section II) by a z dependent potential which at large distances goes over into the classical "image potential" form, and that this image potential arises from the excitation of virtual bulk and surface plasmons, which are in the "cloud" around the deep hole. The quantities ΔE_B and ΔE_S and ΔE are plotted in fig. 7 for general values of z. Note that for sufficiently large ε, the shift is even larger near the surface than in the bulk and this effect can be as large as several eV. The quantities $\beta_S(z)$ and $\beta_B(z)$ can also be obtained from this plot, because in analogy with the simple bulk case discussed in section I, they satisfy

$$\hbar\omega_S\,\beta(z) = \Delta E_S(z) \tag{31}$$

$$\hbar\omega_B\,\beta_B(z) = \Delta E_B(z) \tag{32}$$

Although the above effects are interesting in their own right, it turns out that they are not too important in X-ray photoemission at the energies like 1 keV. This is because the contribution (27) to the photocurrent must be averaged over an escape function which decays exponentially in a distance in from the surface roughly equal to the inelastic mean free path λ. Therefore the relevant dimensionless parameter measuring the strength changes and shifts of the satellites due to the surface [see (27), (28), (29), (31), (32)]

$$\frac{e^2}{4\lambda\,\hbar\omega_p} \sim \frac{1}{4}\left(\frac{e^2}{\hbar v}\right)^2 \ln\left(\frac{m\cdot v^2}{\hbar\omega_p}\right) \ll 1 \tag{33}$$

At lower energies this effect of course gets more important, but then one will have to worry about similar effects on the escaping photoelectron as well as vertex corrections which as we know tend to produce cancellation.

To investigate the effect of the surface on the extrinsic satellites we made use of the approximate Green's function [23] for the photoelectron

$$G(z, z'; E)$$

$$\simeq (2ik_1)^{-1} \exp(-i(k_1 - i/2\bar{\lambda})(z - z'));$$

$$\text{for } z < 0, z' > 0$$

$$\simeq (-2ik_1^{\frac{1}{2}} k_2^{\frac{1}{2}})^{-1} \exp(i(k_1 - i/2\bar{\lambda}) z)\, e^{-ik_2 z'};$$

$$\text{for } z < 0, z' > 0. \quad (34)$$

where

$$\hbar k_1 = (2mE_1 - \hbar^2 k_{\|}^2)^2 \quad (35)$$

and k_1 is the component of the electron's wave vector parallel to the surface and E_1 is the electrons kinetic energy inside the material, $z < 0$; k_2 is the analogous quantity outside ($z > 0$). The quantity $\bar{\lambda} = \lambda[\hbar k_1/(2mE_1)^{\frac{1}{2}}]$ where λ is the inelastic mean free path. At the energies under consideration one can neglect the reflection by the change in potential at the surface. The other approximation made in writing (34) is to assume the self-energy is simply a local imaginary constant $i\hbar v/2\lambda$. This is not a valid approximation in the presence of a surface and the error introduced must be remedied in perturbation theory later.

Using (34) plus our model Hamiltonian (24), it is a straight-forward but tedious matter to calculate the escape function. I will not give the details here, but simply say that all the various surface effects (including the effect of the non-locality of the self-energy) combine to give a reduction in the first plasmon satellite, whose strength relative to the main line instead of α, is

$$\alpha - [(1 + \alpha)/2](e^2/\hbar v) F \quad (36)$$

where F is a slowly varying model dependent function of order unity, and where I have neglected to write other very small terms. Similarly a surface plasmon satellite of strength $\sim e^2/\hbar v$ appears in the escape function. The above formulae have been written for normal incidence; the effect of non-normal incidence is given roughly by replacing v by the normal component of v in the above, so that at a given energy the corrections become somewhat larger. Our treatment is of course not valid at grazing angle incidence. Finally it was shown that neither the second or subsequent bulk plasmon satellites nor the "fast–slow" interference term of fig. 6 is substantially affected by the surface at energies appropriate to our approximation ($e^2/\hbar v < 1$).

To summarize, then, the effect of the surface is of about the same magnitude as the fast–slow interference term which also tends to decrease the strength of the bulk satellites. Note also, because the surface weakens the first bulk satellite more than the second, it tends to counteract the intrinsic effect which strengthens the first satellite more than the second. This gives even more credence to the use of the numbers in table 1 to support the existence of an intrinsic effect.

VI. Thin Films

It has been suggested [5] that the use of thin films be used to separate the intrinsic from the extrinsic effects. Here we present the outlines for a theory.

Having established that surface effects are of limited importance at the energies under consideration, we are in a position to develop a simplified method for evaluating the diagrams for the escape function in photoemission in real space, which can easily be applied to thin films. The method can also be simply applied to the calculation of the extrinsic effects in other types of experiments.

Essentially what is involved is an extension of the methods I previously used [23] to derive probabalistic rate equations [30] for isotropic quasi-elastic scattering in low energy photoemission. There we took advantage of the fact that in a diagram like fig. 5*a* for the escape function, the equivalent of the wiggly line was local and therefore proportional to

$$\lambda_1^{-1} \delta(z_1 - z_1') \quad (37)$$

so that the oscillating parts of the product $G(z_0, z_1) G^*(z_0, z_1')$ cancelled out, neglecting any surface reflection. This implied in that case that the only effective z dependence of the Green's function inside (see eq. (34)) was proportional to

$$\exp(-|z_0 - z_1|/2\lambda) \quad (38)$$

The present case is slightly different: the scattering is still quasi-elastic as far as the motion of the electron goes, but instead of being isotropic is highly peaked in the forward direction. Said

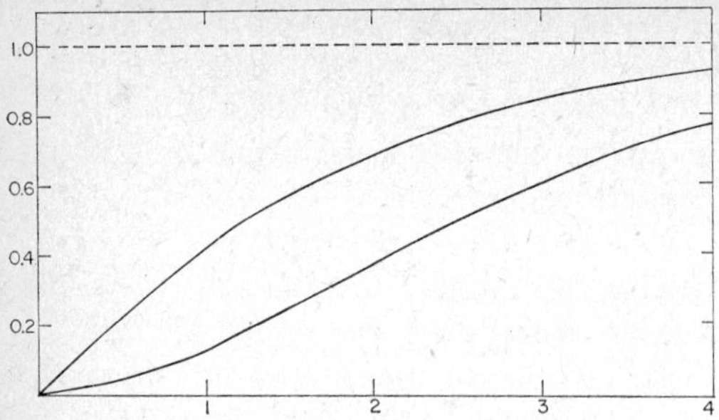

Fig. 8. *Abscissa:* $\xi = d/\lambda$; *ordinate:* units of α (*top curve*); units of α^2 (*bottom curve*).

Prediction of eqs (41) and (42) for the reduction of the strength of the first two extrinsic satellites as a function of film thickness d over the inelastic mean free path λ. These curves can be applied to non-normal escape provided that d is interpreted as the actual maximum path length of an electron through the film at the angle of the collimator.

another way, now we have $k \gg q_c \gg 1/\lambda$. This means that the interaction is still local on the scale of λ, but that the z dependence of the oscillating terms must still be integrated out to give approximate momentum conservation. What one must do is intuitively obvious, and is best verified by working out several examples, an exercise which is left to the reader. The result is that so far as the calculation of the weights of plasmon satellites due to extrinsic effects as exemplified for example by the diagram of fig. 5*a*, one may replace the z dependence of the fast electron Green's function by

$$\theta(z_1 - z_0)e^{-|z_0 - z_1|/2\lambda} \tag{39}$$

instead of (38); the z dependence of the wiggly line is still given by (37) because now the integrals over the oscillating terms have been performed to give the step function $\theta(z_1 - z_0)$, which says physically that a particle not originally headed for the surface will not get out, and that subsequent collisions have only a small effect on the electron's motion. Note, however that λ_1^{-1} is now the plasmon contribution to λ^{-1}. The above was written down for photoelectrons escaping normally to the surface; the simplest way to handle non-normal escape is to define the z direction to be parallel to the collimator (i.e. the measured electron's path) regardless of its angle to the surface; this is possible because this approximation does not include surface effects.

Using this prescription it is straightforward to evaluate the escape function for a film of thickness d, or electron path length d through the film for non-normal escape. The result is that the strength of the nth plasmon satellite[1] is (relative to the main line)

$$\alpha^n \int_{-d/\lambda}^{0} (e^{\xi}\xi^n/n!)\, d\xi \Big/ \int_{-d/\lambda}^{0} e^{\xi} d\xi, \tag{40}$$

where as before $\alpha = \Gamma_1/\Gamma = \lambda/\lambda_1$. Note that for $d \to \infty$ this reduces to what we had before—eq. (21). This may be straightforwardly worked out, yielding for the strength of the first satellite

$$\alpha\left[1 - \frac{\xi}{e^{\xi} - 1}\right], \tag{41}$$

and for the second satellite

$$\alpha^2\left[1 - \frac{\xi(\xi/2 + 1)}{e^{\xi} - 1},\right] \tag{42}$$

where $\xi = d/\lambda$. The expressions in (41) and (42) are plotted in fig. 8. Notice that the reduction in intensity of the extrinsic satellites is marked, even at film thickness substantially greater than the inelastic mean free path. Physically this occurs because the mean escape depth of an electron which contributes to the nth extrinsic plasmon satellite is equal to $(n+1)\lambda$, a fact which we pointed out [7] earlier, and which may be most simply proved by including an extra factor of ξ in the integrand in the numerator of eq. (40).

As the thickness becomes less and less the approximation (40) begins to break down even at very high electronic velocities, because the plasmon wave vectors become discreet. For escape normal to the surface, this effect becomes important when the film becomes so thin that the perpendicular wave vector of the lowest bulk mode becomes considerably less than the change in perpendicular wave vector of the photoelectron

[1] Of course what we have done really just constitutes a first principles derivation of what could have been written down intuitively from probabalistic considerations. Eq. (40) could also have been derived by the method of ref. [31] if an electron-hole production rate is introduced ad hoc. A similar comment applies to ref. [33] which appeared since this manuscript was prepared.

(this latter quantity is about ω_p/v). When this happens, the perpendicular component of momentum can no longer be conserved and the coupling of this electron to the bulk plasmon is reduced. Since the lost spectral weight is transferred to the excitation of surface modes, we expect this effect to reduce Γ_1 more than it changes Γ so that very crudely speaking this is an effect which reduces α. We may estimate its magnitude by using the calculations of Šunjić & Lucas [31] for transmission through thin films. Crudely we estimate that the reduction would be

$$\alpha \to \alpha[1 - D/B] \tag{43}$$

where D and B are given by eqs (43) and (45) respectively of ref. [31]. If we evaluate the integrals asymptotically for large d ($d = 2a$), we find

$$D/B \sim \frac{\pi}{2} \frac{v}{\omega_p d} \ln^{-1}\left(\frac{q_c v}{\omega_p}\right) \tag{44}$$

If we set $d = \lambda$, (43) becomes,

$$\alpha \to \alpha - \frac{\pi e^2}{2\hbar v} \tag{45}$$

Thus at 1.5 keV ($e^2/\hbar v < 0.1$) it is fairly safe to use (40) down to a thickness equal to the mean free path or perhaps slightly less; the quantization effects are, as might have been expected, of the same order of magnitude as other surface effects which in this section we are neglecting in assuming that $e^2/\hbar v \ll 1$.

This quantization also modifies the intrinsic satellites. In a bulk sample the intrinsic effect is isotropic; making the sample thinner simply means leaving a smaller number of modes with wave vectors less than q_c. Thus the coupling parameter β (see eq. (5)) decreases with decreasing film thickness. We can estimate this decrease in the model (24) in which case

$$\beta(z) = \frac{2e^2}{\hbar \omega_p d} \int q_\parallel \, dq_\parallel \sum \frac{\sin^2 (n\pi z/d)}{q_\parallel^2 + (n\pi/d)^2} \tag{46}$$

where the upper limit of the sum and integral is subject to the restriction $q_\parallel^2 + (n\pi/d)^2 < q_c^2$. If we average this over the sample, the $\sin^2$ term becomes 1/2 and the sum and integral are trivially performed, yielding

$$\langle \beta(z) \rangle = \frac{e^2}{\hbar \omega_p d} [n_c \ln (q_c d/\pi) - \ln n_c!] \tag{47}$$

where n_c is the largest integer in $q_c d/\pi$. For large d, the use of Stirling's approximation for the factorial function allows one to recover the normal bulk result $\beta = e^2 q_c/\pi\hbar\omega_p$. In fig. 9 we show a plot of $\langle \beta(z) \rangle/\beta$ as a function of d. The abscissa is $d/(\pi/q_c)$; for Al π/q_c is about 2.5 Å if we take $q_c \sim \omega_p/v_f$. Therefore a distance of about 6 or 8 units on the abscissa corresponds to a mean free path at 1.5 keV.

For Al films, for example, there remains a 10–20% reduction of β for thickness between 20 and 60 Å. Below 20 Å the intrinsic effect rapidly falls to zero, much faster than the extrinsic satellites. Therefore one will not be able to identify the existence of the intrinsic effect by looking for a quasi-intercept in a plot of satellite strength vs thickness. Possibly an identification could be made by fitting the extrinsic theory, eq. (40) and fig. 8 in the range $d = 2\lambda$ to $d = 4\lambda$ say, where this theory is valid, and then looking for the residual which would be considered the intrinsic contribution. It is clear that quite accurate data will be necessary, and that it will help to have an independent measure of the inelastic mean free path as may be obtained by measuring a spectral line from the substrate on which the sample film is evaporated as a function of thickness, as well as a line from the sample. I have fit preliminary data of Tracy [32] on Al films to

Fig. 9. *Abscissa:* $q_c d/\pi$; *ordinate:* $\langle\beta\rangle/\beta$.
Mean reduction in the value of the intrinsic coupling parameter $\langle\beta\rangle$ in units of the bulk value β as a function of film thickness d; the quantity q_c is the plasmon coupling parameter.

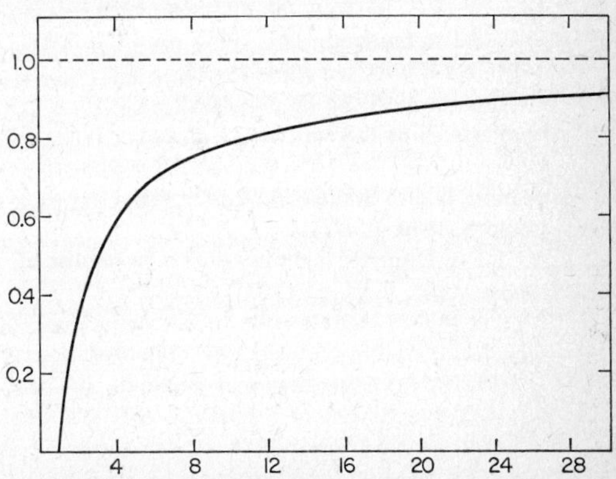

eq. (37) with no major anomalies; it is too early to say, however, whether the intrinsic effect can be isolated.

Note added in proof: Since this manuscript was submitted the following works [34, 35, 36] have come to my attention. These relate in varying degree to the topics discussed here.

I am indebted to the Nobel Foundation, to the Symposium Committee, and to all involved in the organization of this Symposium for making possible the presentation of this paper, and for making my stay in Sweden a scientifically profitable one. Much of the research reviewed or presented in this paper was partially supported by the National Science Foundation to whom I am grateful.

I would like to acknowledge helpful conversations and correspondence with S. Doniach concerning the sum rule, and with J. J. Chang concerning all aspects of this work. I thank Y. Baer, R. Pollak, J. C. Tracy and C. J. Vesely for sending me their still unpublished and in some cases preliminary data.

Finally, I am grateful to Mrs Mary Ann Niolu, not only for the typing of the manuscript, but for her conscientious and diligent work over the last three years in the handling of many of the administrative burdens which otherwise might have taken more of my time away from the research upon which a part of this paper is based.

References

1. Hedin, L, Lundqvist, B I & Lundqvist, S, Solid state comm 1967, 5, 237.
2a. Lundqvist, B I, Phys kondens Materie 1967, 6, 193.
2b. — Ibid 1967, 6, 206.
2c. — Ibid 1968, 7, 117.
3. — Ibid 1969, 9, 236.
4. Hedin, L & Johansson, A, J phys C, ser. 2, 1969, 2, 1336.
5. Hedin, L, Lundqvist, B I & Lundqvist, S, in Electron density of states (ed L H Bennett) p. 233. NBS Spec Publ 323, Washington, 1971.
6. Langreth, D C, Phys rev lett 1971, 26, 1229.
7. Chang, J J & Langreth, D C, Phys rev 1972, 5, 3512.
8a. — Bull Am phys soc 1973, 18, 55.
8b. — Phys rev 1973. To be published.
9. Nozières, P & de Dominicis, C J, Phys rev 1969, 178, 1097.
10. Langreth, D C, Phys rev 1969, 182, 973.
11. — Ibid B, 1970, 1, 471.
12. Yue, J T & Doniach, S, Phys rev. To be published.
13a. Mahan, G D, Phys rev 1967, 153, 882.
13b. — Ibid 1967, 163, 612.
14. Anderson, P W, Phys rev lett 1967, 18, 1049.
15. Doniach, S, in Computational methods in band theory (ed P M Marcus, J F Janak & A R Williams) Plenum Press, New York, 1971.
16. Doniach, S. Private communication.
17. Browers, F, Phys lett 1964, 11, 297.
18. Glick, A J & Longe, P, Phys rev lett 1965, 15, 589.
19. Shirley, D A (ed), Electron spectroscopy. North Holland, Amsterdam, 1972.
20. Du Bois, D F, Ann phys (New York) 1959, 8, 24.
21. Mahan, G D, Phys rev B, 1970, 2, 4434.
22. Ashcroft, N W & Schaich, W L, in Electronic density of states (ed L H Bennett) p. 129. National Bureau of Standards Spec. Publ. 323, Washington, 1971.
23. Langreth, D C, Phys rev B, 1971, 3, 3120.
24. Kadanoff, L P & Baym, G, Quantum statistical mechanics. Benjamin, New York, 1962.
25. Keldysh, V, Zh eksp i teor fiz 1964, 47, 1515. [Sov Phys JETP, 1965, 20, 1018).
26. Pollak, R, Ph.D. thesis, University of California, Berkeley, 1973. Unpublished.
27. Bear, Y. Private communication.
28. Bear, Y & Busch, G, Phys rev lett 1973, 30, 280.
29. Vesely, C J, Kingston, D L & Langer, D W, Phys rev. To be published.
30. Kane, E O, Phys rev 1966, 147, 335.
31. Šunjić, M & Lucas, A A, Phys rev 1971, 3, 719.
32. Tracy, J C. Private communication.
33. Feibelman, P I, Phys rev B 1973, 2305.
34. Mahan, G D, Phys stat sol (b), 1973, 55, 703.
35. Caroli, C, Lederer-Rozenblatt, D, Roulet, B & Saint-James, D, Phys rev. To be published.
36a. Gersten, J I & Tzoar, N, Phys rev. To be published.
36b. Tzoar, N & Gersten, J I, Phys rev. To be published.

Theory of High Energy Excitations in Solids

S. Doniach

Department of Applied Physics and Stanford Synchrotron Radiation Project,* W. W. Hansen Laboratories of Physics, Stanford University, Stanford, Calif. 94305, USA

The subject of high energy excitations in solids is of increasing practical interest, partly as a result of the pioneering work of Siegbahn [1] and collaborators in the development of ESCA (electron spectroscopy for chemical analysis).

For the most part it is convenient to classify such excitations as involving the core state of an atom in a solid, although valence band excitations induced by high energy photons are also of interest.

The main, rather obvious, point to be emphasized about the practical implications of core state excitations is the chemical specificity of the core state energy levels. It is for this reason that such excitations are important for the study of chemically complex systems, surface properties, inhomogeneous systems etc.

Theoretically, interest in core state excitations developed as a result of the work of Mahan [2], Nozières & de Dominicis [3] and others [4] on the singular response of the conduction electrons (in the metallic case) to the sudden potential created by the removal of a core election.

The resulting electron-hole pair excitations lead to an asymmetry [5] and relaxation shift of the resulting ESCA peak [6] whose nature is of importance in understanding the effect of chemical bonding on the ESCA spectrum. In this connection it is worth noting the recent work of Shirley and his collaborators [7] who have made considerable progress in the calculation of relaxation shifts following the theoretical method proposed by Hedin & Johansson [8]. However, the detailed effects of conduction band structure on such shifts still remains to be worked out, both theoretically and experimentally.

* Supported by the National Science Foundation through grant No. GH 39525.

In this brief summary we would like to turn attention to another aspect of high energy excitations which is of growing practical interest. This is the effect of local atomic structure on the shape of photoelectric X-ray absorption edges in solids and molecules. The so called "*E*xtended *X*-ray *A*bsorption *F*ine *S*tructure", or EXAFS, is a weak modulation of the ordinary X-ray absorption cross section for a few hundred electron volts above a core level photoelectric edge.

Recent work by Sayers et al. [9] has shown that a simple modified Born approximation in which an outgoing photoelectron wave is diffracted from neighboring atoms back to the source atom, where it interferes with the birth process can account very convincingly for the qualitative features of EXAFS spectra in simple solids.

From a simple one-electron viewpoint one can look at the process as very closely analagous to LEED. (In reality, the inelastic excitations associated with the core state formation should be folded into the one-electron matrix elements. This will probably not affect the basic periodicities in the EXAFS spectrum.) The photoelectrons are in the same energy range as LEED electrons, but in the EXAFS case consist of outgoing spherical waves from a central atom rather than plane waves as in the LEED case. Stated in those terms it is clear that multiple scattering effects, at least from near-neighbor shells, cannot be ignored. Recent work by Ashley & Doniach [10] and by Schaich [11], has shown how the EXAFS process can be reformulated in terms of electron multiple scattering theory:

The general X-ray absorption cross section

from a core state φ_{core} may be written in the form

$$\frac{d\sigma}{d\varepsilon} \propto \sum_{k,n} |\langle\varphi_{\text{core}}|x|\varphi_{k,n}\rangle|^2 \delta(\varepsilon - \varepsilon_{k,n}) \tag{1}$$

where $\varphi_{k,n}$ are the Bloch states for the photoelectrons in band n. In terms of the one electron Hamiltonian H in which the atomic potentials of the solid are treated in the non overlapping muffin tin approximation (1) may be re-written as

$$\frac{d\sigma}{d\varepsilon} \propto \frac{1}{\pi} \operatorname{Im} \left\langle \varphi_{\text{core}} \middle| x_{\text{op}} \frac{1}{\varepsilon - H - i\eta} \cdot x_{\text{op}} \middle| \varphi_{\text{core}} \right\rangle \tag{2}$$

We now follow Beeby [12], and expand (2) in terms of a complete set of one center basis functions $\varphi_{L,\varepsilon}(R_\alpha)$ at energy ε, angular momentum $L \equiv (l, m)$ to give

$$\frac{d\sigma}{d\varepsilon} \propto \sum_{LL'} M_{\text{core},LL'} \frac{1}{\pi} \operatorname{Im} G_{LL',\varepsilon}(R_\alpha, R_\alpha) \tag{3}$$

where

$$M_{\text{core},LL'} = \langle\varphi_{\text{core}}|x|\varphi_{L,\varepsilon}(R_\alpha)\rangle\langle\varphi_{L',\varepsilon}(R_\alpha)|x|\varphi_{\text{core}}\rangle$$

and G satisfies

$$G_{LL',\varepsilon}(R_\alpha, R_\beta) = G^0_{LL',\varepsilon}(R_\alpha, R_\beta)\,\delta_{LL'} + \sum_{L'', R_\gamma \neq R_\alpha} G^0_{LL',\varepsilon}(R_\alpha, R_\gamma)\,\tau_{L''}(\varepsilon)\,G_{L''L',\varepsilon}(R_\gamma, R_\beta) \tag{4}$$

Here $G^0_{LL',\varepsilon}(R_\alpha, R_\beta)$ are matrix elements of the free electron Green's function and $\tau_L(\varepsilon)$ is Beeby's t-matrix like function, related to the electron-atom scattering phase shift $\delta_L(\varepsilon)$ at angular momentum L by

$$\tau_L^{-1}(\varepsilon) = k(i - 2/\tan \delta_L(\varepsilon)) \tag{5}$$

On substituting (4) in (3), we see that $d\sigma/d\varepsilon$ may be written in the general form

$$\frac{d\sigma}{d\varepsilon} = \left.\frac{d\sigma}{d\varepsilon}\right]_{\text{atom}} + \frac{1}{\pi} \operatorname{Im} \sum_{\substack{LL'L'' \\ R_\gamma \neq R_\alpha}} M_{\text{core},LL'}\, G^0_{LL'',\varepsilon}(R_\alpha, R_\gamma) \times \tau_{L''}(\varepsilon)\, G_{LL',\varepsilon}(R_\gamma, R_\alpha) \tag{6}$$

This general formula has the form of the SLS Born approximation (since $G^0_{LL}(R) = Y_L(R)\,h_L(kR) \sim i^L Y_L(R) e^{ikR}/R$ at large R), but takes account, in principle, of all multiple scattering effects through eq. (4) for the full Green's function. The multiple scattering equations are essentially those of LEED theory where they have been successfully handled by numerical techniques [13, 14].

By different approximations to the above general formulation we show that for elements in the atomic number region of Ge:

(1) in a lower energy range (0–300 eV), multiple scattering effects are so large that they change the relative phase of waves scattered from first neighbor shells to those scattered from second neighbor shells by an amount which appears likely to be able to explain the anomalies observed by Stern & Sayers [15] in the close packed metal Cu, without the need to invoke the additional polarization of the local charge density due to the photoionized core hole which they bring in to account for the observations;

(2) in a higher energy region (300–800 eV) we show the importance of taking into account the energy dependence of the mean-free path of the photoelectrons and suggest a way in which this can be simply estimated in terms of electron-atom scattering cross sections. We believe the latter observation makes clear the importance of using this higher energy region, where the electron mean free path is increasing, for structural-diagnostic purposes. Our analysis leads us to suggest that the exploitation of EXAFS with high resolution (now possible using a synchrotron source) in this higher energy region will be capable of yielding detailed information on structural changes which, particularly for bio-molecules, are unobtainable by other means.

One particularly exciting possibility will be an attempt to observe subtle changes in bond distances in haemoglobin on oxygenation. Such changes have been discussed by Hopfield on pp 238–250. Here the chemical specificity of high energy excitations allows one to focus in on the iron atom in a complex molecule. Distances of neighboring C, N, O atoms in the molecule should be observable in the periodicity of the associated EXAFS spectrum.

In the case of C, N, O, estimates of the energy dependence of the photoelectron mean free path lead us to suggest [10] that the start of the high energy region will occur around 100 eV above the absorption edge. Thus we propose that above 100 eV there will be a great deal of fine structure, representing coordination lengths out to distant coordination shells, to be observed in EXAFS data, for organic systems containing heavy atoms such as Fe, of sufficiently high energy resolution. (A resolution of 5 eV would probe 20–50 shells.) The possibility of observing subtle changes in this fine structure as a result of biochemical changes is a particularly appealing one which could have far-reaching consequences.

Research supported by Army Research Office, Durham, N.C. and National Science Foundation through Center for Materials Research at Stanford University, Stanford, Calif. 94305, USA.

References

1. Siegbahn, K et al., Nova acta reg soc sci Upsaliensis ser IV, 1967, 20.
2. Mahan, G, Phys rev 1967, 163, 612.
3. Nozières, P & de Dominicis, D, Phys rev 1969, 178, 1097.
4. Hedin, L & Lundqvist, S Solid state physics (ed H Ehrenreich, F Seitz & D Turnbull) vol. 23, p. 2. Academic Press, New York, 1971.
5. Doniach, S & Sunjic, M, J phys C, 1970, 3, 172.
6. Doniach, S, Computational methods in band theory (ed P M Marcus, J Janak & A Williams) p. 500. Plenum Press, New York, 1971.
7. Shirley, D A, Chem phys lett 1972, 16, 220.
8. Hedin, L & Johansson, A, J phys B 1969, 2, 1336.
9. Sayers, D, Lytle, F & Stern, E A, Adv in X-ray analysis 1970, 13, 248.
10. Ashley, C & Doniach, S. In preparation.
11. Schaich, W. Preprint, University of California at San Diego, 1973.
12. Beeby, J L, Proc roy soc A 1964, 279, 82.
13. Duke, C B, Smith, D L & Holland, B W, Phys rev 1972, B5, 3358 (and references given there).
14. Jepsen, D W, Marcus, P M & Jona, E, Phys rev B6 1972, 3684.
15. Stern, E A & Sayers, D E, Phys rev lett 1973, 30, 174.

Liquid Crystals and Liquids

Recent Results in the Physics of Liquid Crystals

P. G. de Gennes

Collège de France, 11 Place Marcelin-Berthelot, 75231 Paris Cedex 05, France

Summary

Long organic molecules do not always show a direct transition between a crystal phase and an isotropic liquid. The intermediate phases are called "mesomorphic"; The main classes are: (*a*) nematics (anisotropic liquids with uniaxial optical properties); (*b*) smectics A and C: layered systems where each layer is a liquid; (*c*) smectics B: layered systems where, inside each layer, there appears to exist also some amount of two-dimensional crystalline order. A useful introduction to these classes can be found in [1, 6].

The main recent results inside the nematic phase concern

(1) convective instabilities under suitable external perturbations (electric fields or temperature gradients) [2];

(2) hydrodynamic instabilities inside the laminar flow regime, due to the coupling between orientation and flow [3].

The smectic A phase has a striking similarity with a superconductor. The order parameter includes both the amplitude and the phase of the density modulation describing the layers. Applications of this idea to the smectic-nematic transition are presented and compared with recent measurements on pretransitional anomalies.

The theory of elasticity in a layered phase such as a smectic A is unusual [4]: for certain deformations, the smectic behaves as a solid. For other deformations it behaves like a nematic liquid. This leads to a number of remarkable effects:

—a smectic single domain under mechanical tension shows instability towards a certain periodic distortion [5];

—under conflicting boundary conditions, the smectic deforms with a pattern of "focal conics" [6]. In the simplest case, we argue that the available space should be filled by tangent cylinders, the interstices being filled by smaller cylinders, etc. This iterative process should probably extend from the sample size down to something like 10 layer thicknesses (at least for cases of thermal equilibrium) [7]. The resulting arrangement is very different from familiar domain structure in magnets, superconductors, etc. and might possibly be probed by light scattering.

References

1. Brown, G, Doane, W & Neff, V, CRC critical reviews in solid state science 1972, 1, 303.
2. Pieranski, P, Dubois-Violette, E & Guyon, E, Phys rev lett 1973, 30, 736.
3. Guyon, E & Pieransky, P, Phys lett 1973.
4. De Gennes, P G, J phys 1969, 30, suppl. C4, 65.
5. Delaye, M, Durand, G & Ribotta, R. To be published.
6. Friedel, G, Ann phys 1922, 22, 273.
7. Bidaux, R, De Seze, L, Sarma, G, Boccara, N, Parodi O & de Gennes, P G, J phys 1973, 34, 661.

Discussion

Suhl: The time constant characterizing the instability of an *E* field due to anisotropic conductivity could be determined by a.c. modulation superposed on the *E* field. Has this possibility been considered?

de Gennes: This has been studied in particular by R. Kashnow at General Electric, and also to some extent in Orsay. It is possible to restabilise the system by a high frequency field.

Dzyaloshinsky: (1) In your discussion of smectic phases you used a long-range order parameter $|\psi|^2$, which possesses one-dimensional periodicity. But as Landau has shown such structures are unstable just because of the unusual elastic properties you have just described. What is your comment?

(2) Beside smectic structures which consist of liquid planes, distributed in space periodically, and which are, strictly speaking, thermodynamically unstable, in principle structures can exist which are liquid lines, periodically distributed. As Landau has shown these structures of threads are thermodynamically stable and it is a challenge to find such bodies in nature. Probably this tendency of thread-like structures to "melt" in longitudinal direction could explain a part of observed irregularities and peculiar behavior of conductivity.

de Gennes: (1) Yes, the phase fluctuations in the ordered phase of smectic A are divergent and smectic A represents a case of "quasi order", like the two-dimensional *xy* ferromagnet. The only case where I know the behavior is a two-dimensional model, where instead of layers we have non-intersecting lines. This can be mapped into a Fermion problem in one discussion, and the X-ray scattering intensity near the nominal Bragg position shows a logarithmic singularity.

(2) In usual nematics this type of order is probably ruled out by the X-ray data. But the "ordered rod" phases are indeed observed with lipids (hexagonal phases of the soaps).

Fisher: Could you say a little more about the nature of the B and C smectic phases and the evidence for it?

de Gennes: A very recent (unpublished) EPR experiment by C. Taupin, M. Dvolaitsky and F. Poldy shows that (at least in the case they studied) the solid$\leftrightarrow$B transition is a "melting" of the terminal aliphatic chains at both ends of the molecule. This picture also fits the transition enthalpy rather well. In the "molten chain" state the rigid parts in one layer are strongly coupled to the rigid parts of their neighbors in the same layer, but only weakly coupled to the rigid part in other layers. This leaves two possibilities for the B phase:

(1) If the interlayer coupling is still strong enough, the successive layers are locked and we have a solid with finite shear modulus, but with very low yield stress. The X-ray intensity near a Bragg position $\boldsymbol{\tau}$ then goes like $|\boldsymbol{q}-\boldsymbol{\tau}|^{-2}$.

(2) If the interlayer interactions are weak, the layers are uncoupled (the couplings can be treated by perturbation theory) and each layer essentially shows the anomalies of a two-dimensional crystal. The X-ray intensity near the (nominal) Bragg position would probably go like $|\boldsymbol{\tau}-\boldsymbol{q}|^{-X(T)}$, where $X(T)$ is a temperature dependent exponent.

Present data from Brillouin scattering (Harvard) and from X-ray diffuse scattering (Orsay) slightly favor explanation (1).

Fisher: You might then say that the B phases are partially melted solids?

de Gennes: Yes.

Electronic Correlation Functions in Liquid Metals and Alloys

N. H. March

Department of Physics, Imperial College of Science and Technology, London SW7 2BZ, UK

Summary

Evidence is presented in favour of relatively well defined collective modes of (*a*) phonon; (*b*) plasmon character in the simpler liquid metals. On the basis of a simple model which puts all the emphasis on these collective modes, the qualitative behaviour of the electronic correlation functions in a pure liquid metal is discussed. It is pointed out that the characteristic energy appearing in the asymptotic form of the partial direct correlation functions of a pure liquid metal is the zero-point energy of the plasmons.

For weak electron–ion interaction, a theory relating the partial dynamical structure factors in a binary metallic alloy is summarized, and shown to include the correct conditions of perfect screening of ions and electrons in the extreme long wavelength limit. The same screening conditions have consequences for triplet correlations, as emphasised by Parinello & Tosi.

Small angle neutron scattering experiments for binary metallic alloys, as a function of concentration, would now seem to be of considerable interest.

A summary is first given of experimental evidence on the simpler liquid metals which points to the existence of well defined collective modes, referred to below as 'phonons'.

The way in which such 'phonon' modes can be incorporated into a description of a pure liquid metal as a two-component system, electrons and positive ions, is considered. In the course of the discussion, it becomes clear that (*a*) the plasmon behaviour of the electron gas enters the long-wavelength (small q) form of the electron correlation functions; (*b*) the $q=0$ correlation functions must satisfy the condition of perfect screening.

Following the approach of Tosi and the writer [9], these collective motions, phonons and plasmons, are incorporated into a simple model which should be useful for describing the partial correlation functions.

A treatment of liquid binary alloys is then considered, which is valid for weak electron–ion interaction and which incorporates (*b*) above in a form appropriate to an alloy. The interest in the small-angle neutron scattering from binary alloys is stressed.

Finally, some remarks are made about electronic triplet correlations in pure liquid metals, following the work of Parrinello & Tosi [12], from which some numerical estimates for liquid Rb result in the extreme long wavelength limit.

There is, by now, a variety of evidence that in the simpler liquid metals such as Al, Pb and Ga, collective motions are much more sharply defined than in, for example, liquid argon. We can cite here the facts that for quite a few liquid metals, the ratio of the specific heats γ at the melting point is in the range 1.2–1.3, in contrast to liquid argon where $\gamma \sim 2.2$. Similarly, the specific heats at constant volume C_V are near to 3 R, and somewhat in excess of that value.[1] Again, for argon, C_V at the triple point is 2.6 R.

These results can be understood on the basis of a treatment starting from well defined collective modes. Thus, if we define a density fluctuation $\varrho_{\boldsymbol{q}}$ by

$$\varrho_{\boldsymbol{q}} = \sum_{\boldsymbol{R}_i} \exp(i\boldsymbol{q}\cdot\boldsymbol{R}_i) \tag{1.1}$$

where $\boldsymbol{R}_i$ denotes the atomic positions, then the assumption is that, roughly speaking, $\varrho_{\boldsymbol{q}}$ varies in time as

$$\varrho_{\boldsymbol{q}}(t) = \varrho_{\boldsymbol{q}}(0)\exp(i\omega_{\boldsymbol{q}} t). \tag{1.2}$$

Obviously there will be damping on the above expression for $\varrho_{\boldsymbol{q}}(t)$ but the understanding is that the oscillatory behaviour is sufficiently well defined to define $\omega_{\boldsymbol{q}}$. We shall loosely speak of such collective modes as 'phonons'.

[1] For transition and noble metals with d shells, the interionic potentials have evidently harder repulsive cores and, it turns out, also shorter range than for the simpler liquid metals. This leads to conditions rather more akin to liquid argon.

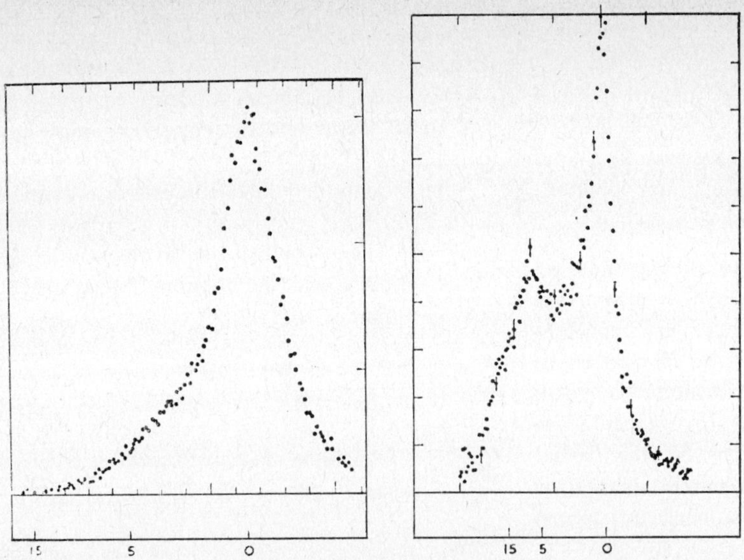

Fig. 1. *Abscissa:* energy transfer (meV); *ordinate:* intensity in arbitrary units.

Spectra of neutrons inelastically scattered by (*a*) liquid argon through 60° [4]; (*b*) liquid gallium through 58°. Temperatures are (*a*) 94 K; (*b*) 343 K.

If we have independent phonons as a zeroth order approximation, then we find immediately $\gamma = 1$ and $C_V = 3R$. Using a theory of such collective motions developed by Eisenschitz & Wilford [1], Bratby, Gaskell & March [2] have expressed the corrections (positive definite) to $C_V = 3R$ in terms of the structure factor $S(q)$. Reasonable numerical estimates can be made for potassium at the melting temperature and there is satisfactory agreement between theory and experiment.

Clearly, a zeroth order approximation $\gamma = 1$ and $C_V = 3R$ is a very bad starting point for liquid argon. This should be no surprise, for the above 'harmonic' model is completely inappropriate for hard cores and short range forces, such as we have in argon.

So far, this discussion has focussed on the thermodynamic quantities C_V and γ. But there is other evidence from neutron scattering experiments (see, for example, Barker et al. [3]) that collective motions are much more important in liquid metal Ga than in liquid argon. Thus, we show in fig. 1, taken from the above paper, the spectra of neutrons of incident wavelength 4 Å inelastically scattered by liquid argon through an angle of 60° [4] and by liquid gallium through 58°. The form for Ar is featureless, whereas for Ga an extra peak appears at a well-defined energy transfer. If ω at the peak is plotted against q, a $\omega(q)$ relation quite similar to the phonon dispersion curves of solid Ga is found.

Thus, in the simpler liquid metals, with relatively soft cores and long range forces, there is now substantial evidence for well defined collective modes, which, as explained above, we shall refer to, somewhat loosely, as phonon modes. Of course, low frequency collective modes are well defined in all liquids.

Two-component Theory of Liquid Metals

So far, of course, we have essentially been dealing with liquid structure through the static structure factor $S(q)$, which was involved in estimating deviations of C_V from $3R$, and the dynamic structure $S(q, \omega)$, which is essentially what is determined in neutron inelastic scattering. Needless to say, $S(q, \omega)$ as measured by neutrons refers to the dynamic structure factor of the ions.

Developments in liquid metal theory have been taking place in which, instead of dealing with ions coupled by effective interactions mediated by the conduction electrons, the liquid metal is treated as a two-component system, conduction electrons and positive ions [5–8]. We have already given evidence for well defined collective modes of the positive ions in the simpler liquid metals. There is, of course, no doubt that, in any dense assembly of conduction electrons, there are well-defined plasma modes. We shall therefore turn immediately to summarize a model for such a two-component system which starts out from the assumption that we have two types of well defined collective excitations: phonons and plasmons, following Tosi & March [9].

Phonon–plasmon model. On the assumption that, in the long wavelength limit, single-particle and multi-particle excitations are less important than collective modes, which we take to be: (i) Plasmon-like, with dispersion relation $\omega = \omega_p(q)$; (ii) phonon-like, with $\omega = v_s(q)q$, we can write down approximate expressions for the partial dissipation functions of the liquid metal (for ion–ion, ion–electron and electron–electron terms) as

$$\chi''_{ij}(q,\omega) = \frac{-\pi}{2}\{n_i n_j\}^{\frac{1}{2}}\{\alpha_{ij}(q)[\delta(\omega-\omega_p(q)) - \delta(\omega+\omega_p(q))] + \beta_{ij}(q)[\delta(\omega - qv_S(q)) - \delta(\omega + qv_S(q))]\} \tag{2.1}$$

We note at this point that the partial structure factors $S_{ij}(q)$, related to the usual pair functions $g_{ij}(\boldsymbol{r}_1-\boldsymbol{r}_2)$ by

$$S_{ij}(q) = \delta_{ij} + (n_i n_j)^{\frac{1}{2}}\int d(\boldsymbol{r}_1-\boldsymbol{r}_2)[g_{ij}(\boldsymbol{r}_1-\boldsymbol{r}_2)-1] \times \exp(i\boldsymbol{q}\cdot(\boldsymbol{r}_1-\boldsymbol{r}_2)) \tag{2.2}$$

are obtained from χ'' using the relation

$$S_{ij}(q) = -\int\frac{d\omega}{2\pi}\chi''_{ij}(q,\omega)\coth\left(\frac{\hbar\omega}{2kT}\right)(n_i n_j)^{-\frac{1}{2}}. \tag{2.3}$$

To satisfy the first moment theorem on χ''_{ij} in the long wavelength limit, we must choose $\alpha_{ij}(q) \propto q^2$ and $\beta_{ij}(q) \propto q$. It is then easy to see that the static structure factor $S_{ii}(q)$ at $q=0$ is entirely dominated by the phonon mode, and the compressibility sum rule gives the coefficients β_{ij} in $\beta_{ij}(q) \to \beta_{ij}q$ for small q as

$$\beta_{ii} = Z^{-1}\beta_{ee} = Z^{-\frac{1}{2}}\beta_{ie} = \hbar n_i K_T v_s \tag{2.4}$$

v_s being the velocity of sound.

Since we saw above that in many liquid metals at their melting points, the ratio of the specific heats $\gamma \sim 1{\cdot}2-1{\cdot}3$ the fact that the sound velocity entering above is identified with the isothermal sound velocity is not very important in practice.

Next we consider the coefficients s, s_i, and s_e defined by $\alpha_{ii}(q) \to sq^2$, $\alpha_{ie}(q) \to s_i q^2$, $\alpha_{ee}(q) \to s_e q^2$ as $q \to 0$. These coefficients involve both the plasmon mode and the phonons. More specifically, we have

$$Zs_e^{\text{plasmon}} = \frac{\hbar}{2m\omega_p}\frac{n_i M}{\mathcal{M}}$$

$$s_i^{\text{plasmon}} = \frac{\hbar}{2M\omega_p}\frac{n_e m}{\mathcal{M}} \tag{2.5}$$

and

$$s^{\text{plasmon}} = \frac{\hbar}{2m\omega_p}\frac{n_i m}{\mathcal{M}}$$

Here m and M are electron and ion masses, while $\mathcal{M}$ is defined by

$$\mathcal{M} = n_i M + n_e m \tag{2.6}$$

and ω_p is the reduced plasma frequency, given by

$$\omega_p^2 = 4\pi n_i Z^2 e^2\left(\frac{1}{M}+\frac{1}{Zm}\right). \tag{2.7}$$

In practice, because of the large difference between electron and ion masses, ω_p can be equated to the electronic plasma frequency in the homogeneous electron gas and $Zs_e^{\text{plasmon}} \sim \hbar/2m\omega_p$ whereas s_i^{plasmon} and s^{plasmon} are extremely small.

On the other hand, the phonon contribution to the same coefficients depends on differences between coefficients of q^3 terms in $\beta_{ij}(q)$ and those in $qv_s(q)$. The suggestion from this calculation is that these terms are usually small in magnitude and therefore negligible compared with the plasmon contribution in s_e. While the phonon terms are likely to be important relative to the plasmon terms in s_i and s, the conclusion is that these latter coefficients are very small compared with s_e. These conclusions can be shown, using hydrodynamic forms, to be relatively insensitive to phonon and to plasmon damping.

It is true, of course, that, at present, the only quantities directly accessible to experiment are the ionic structure factors.[1] Nevertheless, there is little doubt that a definitive basic theory of liquid metals must deal with both ionic and electronic excitations. The above theory represents a start on this problem and we can summarize the findings to date as follows:

(i) The static structure factor $S_{ee}(q)$ is expected to rise sharply from its value at $q=0$. The curvature is directly dependent on the zero-point energy of the plasmons.

(ii) The structure factor $S_{ii}(q) \equiv S(q)$ has little curvature at small q.

(iii) The cross-term $S_{ie}(q)$ is expected to be very small over the whole range of q.

(iv) The partial direct correlation functions, defined following Pearson & Rushbrooke [10], can be shown to behave asymptotically as the bare interaction energy, divided by a characteristic energy. This characteristic energy in argon is the thermal energy kT, but in the two-component theory of liquid metals is $\frac{1}{2}\hbar\omega_p$; the zero-point energy of the plasmons [8].

(v) The Bohm-Staver formula for the velocity of sound v_s comes back from the two-component theory, as does the customary expression for neutron scattering in the case of weak electron–ion interactions [8].

Having discussed the main results to date of such a two-component approach to liquid metals, it is of interest to consider binary alloys, where, as we shall see, there is a strong concentration dependence of the partial structure factors in the extreme long wavelength limit $q=0$. The somewhat more formal theory discussed below obviously can be applied immediately to the pure liquid metal: we have not given the formalism above to avoid repetition.

Generalization to binary alloys

In confronting theories of dynamical structure factors with experiment, recent work of March et al. [11] indicates that it is likely to be fruitful to study liquid metal alloys. The generalization of the above results to deal with partial dynamical structure factors in binary A–B alloys will now be sketched. The presentation follows that of Tosi & March [9] and is a generalization of the two-component theory for a pure metal.

We first apply an external potental V to the electrons. The corresponding density changes for the three components can then be written in the form

$$\delta\varrho_e(q,\omega)=\chi_{ee}(q,\omega)\,V_e(q,\omega) \tag{3.1}$$

$$\delta\varrho_A(q,\omega)=\chi_{Ae}(q,\omega)\,V_e(q,\omega) \tag{3.2}$$

$$\delta\varrho_B(q,\omega)=\chi_{Be}(q,\omega)\,V_e(q,\omega). \tag{3.3}$$

The density changes of the ionic components are caused by the potentials they feel due to the electronic polarization. These are:

$$\text{Potential seen by ion A} = V_{Ae}(q)\,\delta\varrho_e(q,\omega) \tag{3.4}$$

while

$$\text{Potential seen by ion B} = V_{Be}(q)\,\delta\varrho_e(q,\omega) \tag{3.5}$$

It follows by linear response theory that, for weak electron-ion interactions V_{Ae} and V_{Be}, the density changes of ions and electrons are

$$\delta\varrho_A(q,\omega) = \chi_{AA}(q,\omega)\,V_{Ae}(q)\,\delta\varrho_e(q,\omega) + \chi_{AB}(q,\omega)\,V_{Be}(q)\,\delta\varrho_e(q,\omega) \tag{3.6}$$

and

$$\delta\varrho_B(q,\omega) = \chi_{BA}(q,\omega)\,V_{Ae}(q)\,\delta\varrho_e(q,\omega) + \chi_{BB}(q,\omega)\,V_{Be}(q)\,\delta\varrho_e(q,\omega). \tag{3.7}$$

These equations then yield the response function

$$\chi_{Ae}(q,\omega) = V_{Ae}(q)\chi_{ee}(q,\omega)\chi_{AA}(q,\omega) + V_{Be}(q)\chi_{ee}(q,\omega)\chi_{AB}(q,\omega) \tag{3.8}$$

$\chi_{Be}(q,\omega)$ being obtained from this result by interchanging A and B.

If, for electrons, we neglect the frequency dependence of χ_{ee} i.e. we put

$$\chi_{ee}(q,\omega) \approx \chi_{ee}(q,0) \tag{3.9}$$

then we find, with n_A, n_B and n_e the number densities of ions A, ions B and electrons respectively,

[1] Very recently, Egelstaff & March (1973, to be published) have shown that the electronic correlation functions can be got from experiment by combining X-ray, electron and neutron scattering data.

$$S_{Ae}(q) = \left(\frac{n_A}{n_e}\right)^{\frac{1}{2}} \left\{ V_{Ae}(q)\,\chi_{ee}(q,0)\,S_{AA}(q) + V_{Be}(q)\,\chi_{ee}(q,0)\,S_{AB}(q)\left(\frac{n_B}{n_A}\right)^{\frac{1}{2}} \right\}. \quad (3.10)$$

S_{Be} follows by interchanging A and B.

This formula already contains the exact long wave limit results [8]

$$S_{Ae}(0) = \left(\frac{n_A}{n_e}\right)^{\frac{1}{2}} \left\{ zS_{AA} + ZS_{AB}\left(\frac{n_B}{n_A}\right)^{\frac{1}{2}} \right\} \quad (3.11)$$

and

$$S_{Be}(0) = \left(\frac{n_B}{n_e}\right)^{\frac{1}{2}} \left\{ ZS_{BB} + z\left(\frac{n_A}{n_B}\right)^{\frac{1}{2}} S_{AB} \right\}, \quad (3.12)$$

z and Z being the valencies of ions A and B respectively. These results are directly the consequence of perfect screening of an ion by electrons and the other ions, in a liquid metallic alloy: they generalize earlier results [6 and 7] for pure metals, given in equation (4.6) below for monovalent metals.

These relations (3.11) and (3.12) can be used in conjunction with thermodynamics to plot the electronic correlation functions at $q=0$ as a function of concentration. Results obtained for a Na–K alloy are shown in fig. 2 [11]. The striking feature is the strong concentration dependence of these electronic correlation functions. This is *not* described by the uniform electron gas theory, of course. However, the considerations of this paper strongly suggest that this model, and in particular the plasma mode, will correctly describe the small q behaviour, and in particular the term of $O(q^2)$, in $S_{ee}(q)$ in the alloy.

It would be of considerable interest now, in the light of these conclusions, to have small angle scattering data from a binary alloy such as Na–K or K–Hg.

Finally, we return to the case of the pure liquid metal, but to deal with triplet correlations, rather than the pair correlations which we have been discussing exclusively up to this point.

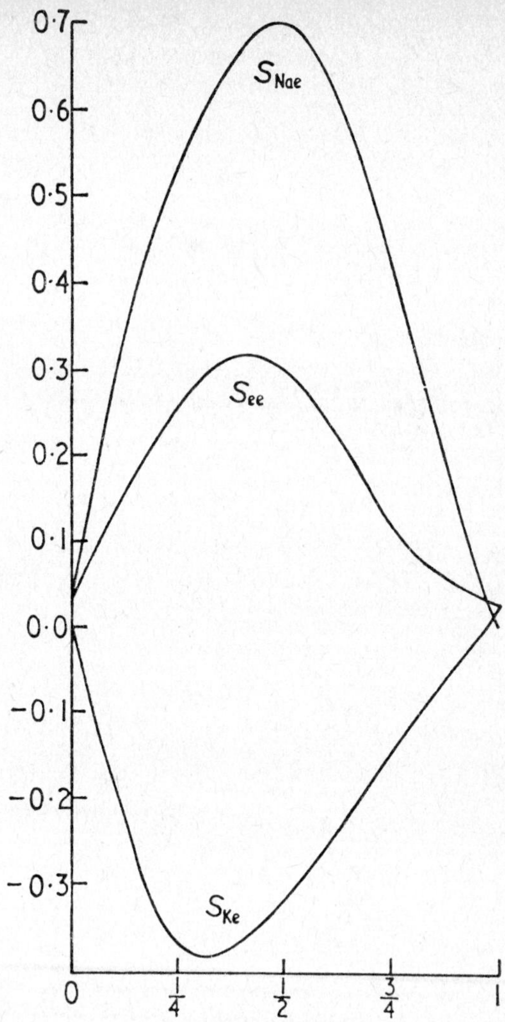

Fig. 2. *Abscissa:* conc.; *ordinate:* electronic correlation functions.

Electronic correlation functions S_{ee} (q), S_{Nae} (q) and S_{Ke} (q) at $q=0$ in a Na–K liquid alloy, as a function of concentration.

Triplet correlation functions

We wish to note that the same conditions of charge neutrality that lead to eqs (3.11) and (3.12) have been used by Parrinello & Tosi [12] to relate the ion–ion–electron correlation function to diffraction experiments under pressure. Furthermore, all the electronic–triplet correlations in the thermodynamic limit are related to the equation of state.

If we denote the n-particle correlation functions by $g_{(n)}$, with space coordinates referring in order to the particles indicated by the indices, then Parrinello & Tosi show that we can write for the relation between the triplet–ion correlation function g_{iii} and the ion–ion–electron correlation function g_{iie} for monovalent metals,

$$n\int \mathrm{d}\boldsymbol{r}_3[g_{\mathrm{iii}}(\boldsymbol{r}_1,\boldsymbol{r}_2,\boldsymbol{r}_3)-g_{\mathrm{iie}}(\boldsymbol{r}_1,\boldsymbol{r}_2,\boldsymbol{r}_3)]=-2g_{\mathrm{ii}}(\boldsymbol{r}_1,\boldsymbol{r}_2) \tag{4.1}$$

g_{ii} being the ion–ion pair function. This relation implies that the total charge density surrounding the two ions must integrate to $-2e$. Entirely similar relations hold by considering a pair of electrons, or an electron and an ion. By means of a relation due to Schofield [13] we can write

$$n^2\int \mathrm{d}\boldsymbol{r}_3[g_{\mathrm{iie}}(\boldsymbol{r}_1,\boldsymbol{r}_2,\boldsymbol{r}_3)-g_{\mathrm{ii}}(\boldsymbol{r}_1,\boldsymbol{r}_2)] = kT\left(\frac{\partial}{\partial p}[n^2 g_{\mathrm{ii}}(\boldsymbol{r}_1,\boldsymbol{r}_2)]\right)_T \tag{4.2}$$

Following Parrinello & Tosi (PT) we introduce the Fourier transforms of the correlation functions by

$$S_{\mathrm{ik}}(q)=\delta_{\mathrm{ik}}+n\int \mathrm{d}(\boldsymbol{r}_1-\boldsymbol{r}_2)\,[g_{\mathrm{ik}}(\boldsymbol{r}_1,\boldsymbol{r}_2)-1] \times\exp[\mathrm{i}\boldsymbol{q}(\boldsymbol{r}_1-\boldsymbol{r}_2)] \tag{4.3}$$

and

$$T_{\mathrm{jkl}}(\boldsymbol{q},\boldsymbol{q}')=n^2\iint \mathrm{d}(\boldsymbol{r}_1-\boldsymbol{r}_2)\,\mathrm{d}(\boldsymbol{r}_3-\boldsymbol{r}_1)\,t_{\mathrm{jkl}}(\boldsymbol{r},\boldsymbol{r}_2,\boldsymbol{r}_3) \times\exp[\mathrm{i}\boldsymbol{q}(\boldsymbol{r}_1-\boldsymbol{r}_2)+\mathrm{i}\boldsymbol{q}'(\boldsymbol{r}_3-\boldsymbol{r}_1)]. \tag{4.4}$$

Here the indices i, k and l take the value i for ions and e for electrons. $S_{\mathrm{ij}}(q)$ are the usual partial structure factors and $t_{\mathrm{jkl}}(\boldsymbol{r}_1,\boldsymbol{r}_2,\boldsymbol{r}_3)$ are the cluster-type triplet correlation functions

$$t_{\mathrm{jkl}}(\boldsymbol{r}_1,\boldsymbol{r}_2,\boldsymbol{r}_3)=g_{\mathrm{jkl}}(\boldsymbol{r}_1,\boldsymbol{r}_2,\boldsymbol{r}_3)-g_{\mathrm{jk}}(\boldsymbol{r}_1,\boldsymbol{r}_2)-g_{\mathrm{kl}}(\boldsymbol{r}_2,\boldsymbol{r}_3)-g_{\mathrm{lj}}(\boldsymbol{r}_3,\boldsymbol{r}_1)+2. \tag{4.5}$$

By using the relations [6] and [7] for monovalent liquid metals

$$\lim_{q\to 0}[S_{\mathrm{ii}}(q)=S_{\mathrm{ie}}(q)=S_{\mathrm{ee}}(q)]=nkTK_T, \tag{4.6}$$

K_T being the isothermal compressibility, it is readily shown from the perfect screening conditions that

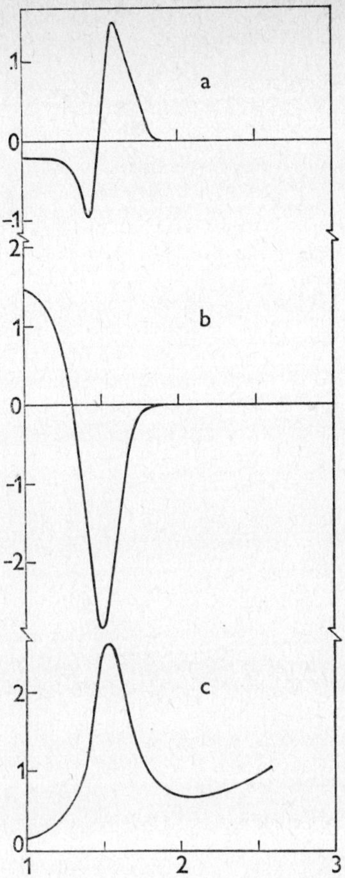

Fig. 3. *Abscissa:* q(Å^{-1}); *ordinate:* (*a*) $T_{\mathrm{iie}}(q,0)$; (*b*) $T_{\mathrm{iii}}(q,0)$; (*c*) $S_{\mathrm{ii}}(q)$.

(*a*) Ion-ion-electron triplet correlation function in liquid Rb at 330 K; (*b*) ion triplet correlation function; (*c*) ionic structure factor [14]; (after [12]).

$$T_{\mathrm{iie}}(q,0)=T_{\mathrm{iii}}(q,0)+2[S_{\mathrm{ii}}(q)-1] \tag{4.7}$$

$$T_{\mathrm{eei}}(\boldsymbol{q},\boldsymbol{q})=T_{\mathrm{iie}}(-\boldsymbol{q},-\boldsymbol{q}) \tag{4.8}$$

and

$$T_{\mathrm{eee}}(q,0)=T_{\mathrm{eei}}(q,0)-2[S_{\mathrm{ee}}(q)-1] \tag{4.9}$$

Thus, as first shown by PT, in the thermodynamic limit the electronic triplet functions are all related to the ionic triplet function, which is, in turn, determined by $(\partial K_T/\partial p)_T$.

For liquid Rb, PT show from data reported by Egelstaff et al. [14] that $T_{\mathrm{eee}}(0)=T^{(0)}_{\mathrm{iii}}\simeq 1.9$ and $T_{\mathrm{eei}}=T_{\mathrm{iie}}\simeq -0.03$. For non-zero momentum transfer q, we can write

$$T_{\text{iie}}(q, 0) = nkT\left\{K_T(S_{\text{ii}}(q) - 1) + \left(\frac{\partial S_{\text{ii}}(q)}{\partial p}\right)_T\right\}. \tag{4.10}$$

PT's results for $T_{\text{iie}}(q, 0)$ are shown in fig. 3 compared with the ionic correlation functions.

It is a great pleasure to acknowledge the fact that the work reported here was stimulated by interaction with Professors Mario Tosi and Avadh Bhatia and, of course, draws heavily on work that has resulted from our mutual collaboration, as cited in the references. Special thanks are due to Dr Parrinello and Professor Tosi for making their work on triplet correlations available to me, before publication.

References

1. Eisenschitz, R & Wilford, M J, Proc phys soc 1962, 80, 1078.
2. Bratby, P, Gaskell, T & March, N H, Phys chem liquids 1970, 2, 53.
3. Barker, M I, Johnson, M W, March, N H & Page, D I, The properties of liquid metals (ed S Takeuchi). Proc Tokyo conf, Sept. 1972, p. 99. Taylor & Francis, London, 1973.
4. Sköld, K & Larsson, K E, Phys rev 1967, 161, 102.
5. Cowan, R D & Kirkwood, J G, J chem phys 1958, 29, 264.
6. Watabe, M & Hasegawa, M (cf [3], p. 133).
7. Chihara, J, (cf [3], p. 137).
8. March, N H & Tosi, M P, Ann phys. In press 1973.
9. Tosi, M P & March, N H, Nuovo cimento 1973, 15B, 308 .
10. Pearson, F J & Rushbrooke, G S, Proc roy soc Edinburgh 1957, A64, 305.
11. March, N H, Tosi, M P & Bhatia, A B, J phys C 1973, 6, L59.
12. Parrinello, M & Tosi, M P, J phys C 1973, 6, L254.
13. Schofield, P, Proc phys soc 1966, 88, 149.
14. Egelstaff, P A, Page, D I & Heard, C R T, J phys C 1971, 4, 1453.

Biology

The Physical Basis of Cooperativity in the Binding of Oxygen by Mammalian Hemoglobin[1]

J. J. Hopfield

Department of Physics, Princeton University, Princeton, N.J. 08540
and Bell Laboratories, Murray Hill, N.J. 07974, USA

For organisms having dimensions of 10^{-4} cm diffusion is an adequate means of the transport of oxygen, metabolites, and products of metabolism. Mammals have linear dimensions 10^5 times larger than this, and diffusion times will be 10^{10} longer. What was an adequately fast process for transport in elementary organisms is clearly ineffective for the transport of materials necessary to sustain large animals. To cope with this problem a circulatory system has developed, and hemoglobin plays an essential role in the transport of oxygen in that system. Water is not a good circulatory fluid for oxygen transport because the solubility of oxygen in water is too low. The function of hemoglobin is to serve as a molecule, circulating with the blood, to which oxygen can weakly bind. Such a molecule enhances the apparent solubility of oxygen in the circulatory fluid, thus making water containing hemoglobin an effective fluid for the transport of oxygen. This paper is chiefly concerned with the physical mechanism by which the unusual physical chemistry of hemoglobin is obtained. The architect Mies von der Rohe remarked, "form follows function," and this should be equally true for the architecture of biological molecules.

The fundamental structure of proteins is a short linear polymer of amino acids, with a chain length of typically 150 amino acid units. Each amino acid unit has a nitrogen-carbon-carbon backbone, and the main chain of the polymer is a ...NCCNCCNCC... sequence. There are twenty different amino acids which are characterized by twenty different side chains from the central carbon atom as is shown in fig. 1. A particular protein consists of a precisely defined sequence having a particular amino acid in the first position, another particular amino acid in the second position, etc., through all ~150 positions, and is a well defined chemical compound. The so-called primary structure of a protein is the specification of the sequence of the amino acids along the polymer chain.

Under physiological conditions proteins are not just randomly coiled in solution. Particular hydrogen bonds tend to form, cross-linking one part of the molecule with another, and regular superstructures develop within the polymer chain. One of the commonest superstructures is the α-helical structure first described by Pauling & Corey [1]. Such local superstructures of parts of the protein chain are called secondary structure. There is still further folding of the molecules into a complicated spatial pattern which allows the interaction of local regions of secondary structure with other regions. Such folding is stabilized by additional hydrogen bonds, by disulfide bridges, by salt bridges, and by Van der Waals forces. This final three-dimensional form of the secondary structure is called the tertiary structure of the protein.

The full structure of a globular protein looks rather like a knotted helical spring. Fig. 2 shows a schematic drawing of the molecule myoglobin. Only the nitrogen-carbon-carbon backbone is drawn, all the side groups being omitted for clarity. Myoglobin is a protein used for temporary oxygen storage in muscle and is in structure and in function related to hemoglobin. Located in the middle of the folded helical structure of fig. 2 is one heme group shown in detail in fig. 3. The

[1] This research was supported in part by US Air Force Office of Scientific Research under contract F46620-71-0-0108.

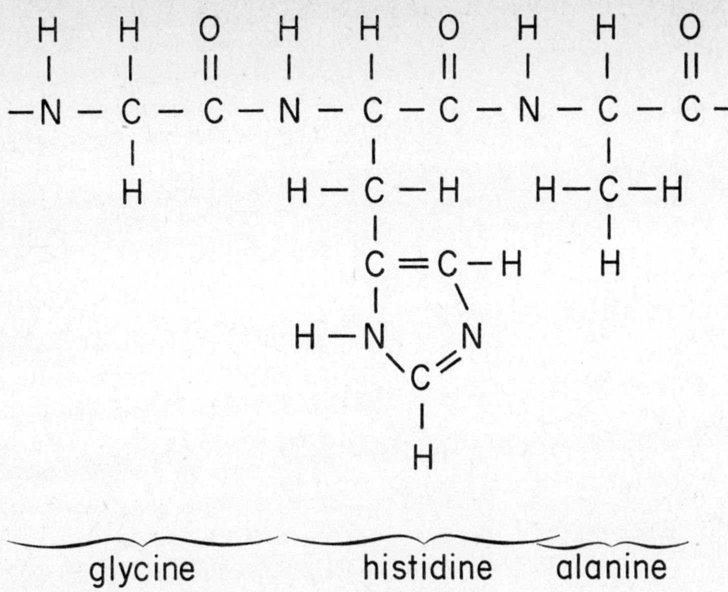

Fig. 1. A short section of an amino acid polymer (primary sequence) showing the linking between the amino acids glycine, histidine, and alanine. The polymer of a typical protein would be of the order of 150 amino acids long.

heme is covalently bonded to the protein (globin) by a bond from the central iron atom (directed perpendicular to the plane of the paper) to a nitrogen of an imidazol ring of a side-chain of a histidine group in the protein. The Fe atom is then surrounded at 5 of 6 octahedral positions by nitrogen atoms. The oxygen binding-site is the 6th octahedral position, directly above the iron atom out of the plane of the paper in fig. 3. There is only one oxygen binding-site per myoglobin molecule.

The structure of hemoglobin, as determined by Perutz and coworkers [3, 4] from X-ray diffraction studies, is similar to four myoglobin molecules put together. Myoglobin has a molecular weight of about 17 000, and hemoglobin has a molecular weight of about 67 000. Each of the four subunits of hemoglobin contains one heme group with one iron atom, in an environment very similar to that of the heme group and iron atom in myoglobin. A picture of the electron density model of hemoglobin is shown in fig. 4. One can readily identify the subunits, some columns of high electron density from α-helices, and discs of high electron density representing the heme groups. The iron atoms are approx. 35 Å apart. The way in which the subunits are put together in a multiple subunit protein is called quaternary structure. One important detail emerging from the X-ray studies is the fact that the quaternary structure of hemoglobin without oxygen (deoxyhemoglobin) and hemoglobin with oxygen bound

Fig. 2. The tertiary structure of myoglobin. The side-chains of the amino acids are not shown. The planar group in the upper center of the molecule is the heme group (after Dickerson [2]).

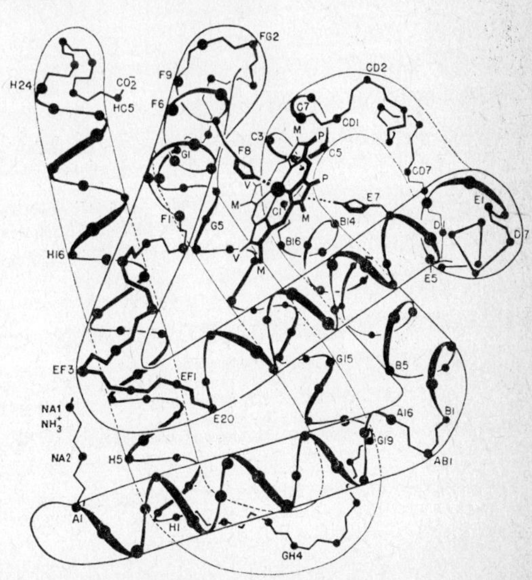

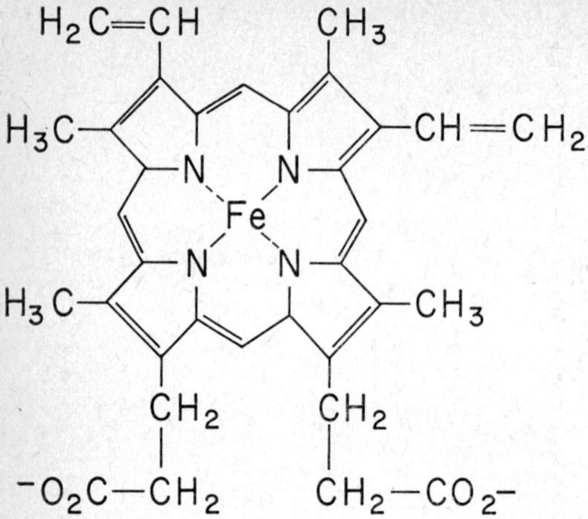

Fig. 3. The heme group. A carbon atom lies at each unlabelled vertex. The group is covalently bonded to a nitrogen atom directly below the iron, and the oxygen molecule binds directly above the iron.

on each of the four iron atoms (oxyhemoglobin) are different [4, 5].

The physical chemistry [6] of hemoglobin relevant to oxygen transport is its oxygen binding as a function of the partial pressure of oxygen. Let Y be the fraction of oxygen binding sites having oxygen bound to them. For myoglobin, with one oxygen binding site per molecule, the reversible reaction $Mb + O_2 \rightleftharpoons MbO_2$ yields from elementary physical chemistry a hyperbolic binding curve,

$$Y = \frac{Kp}{1 + Kp} \tag{1}$$

where K is a constant and p is the partial pressure of oxygen in equilibrium with the solution. If myoglobin were to be used as a transport molecule one would want its binding constant K to be fixed in an optimum fashion. Transport molecules will pick up oxygen in the lungs at one partial pressure and deliver it somewhere else in the body where the partial pressure of oxygen is lower. The molecule operates around some ambient partial pressure and transport is accomplished if, when it moves from a place of one pressure to a place with a partial pressure difference of Δp, there is an amount of oxygen ΔY given up. Clearly the greater the slope $\Delta Y/\Delta p$ at the ambient pressure, the greater the effectiveness of oxygen transport. For a one-site transport mole-

Fig. 4. A picture of the hemoglobin electron density model of Perutz and coworkers. The two α subunits are modeled in white, and the two β subunits in black (from [3]).

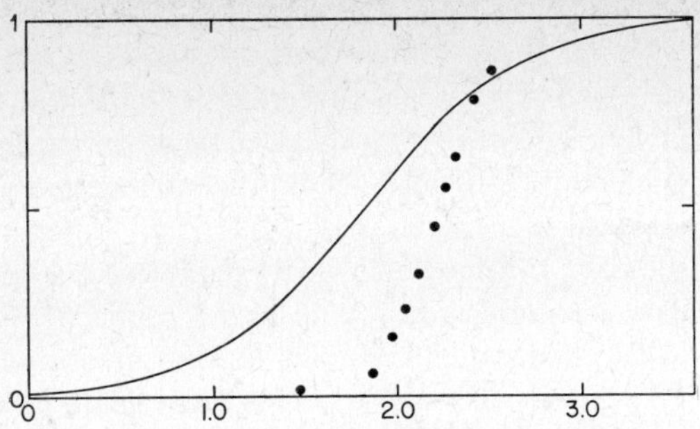

Fig. 5. *Abscissa:* log oxygen pressure; *ordinate: Y.* ——, myoglobin (expt & theory); ●, hemoglobin (expt).

The oxygen coverage of binding sites of myoglobin and hemoglobin as a function of oxygen pressure. Different chains of K would slide the theoretical curve horizontally with invariant shape.

cule the optimum is achieved by fixing $K^{-1} = p_{\text{ambient}}$. A similar conclusion is reached even if Δp is a finite fraction of p_{ambient}.

Fig. 5 shows a comparison of the shape of the binding curve of eq. (1) with experimental data on hemoglobin. The plot is of Y as function of $\log p$ so that the value of the constant K simply shifts the theoretical curve with no change of shape. Experimental points for myoglobin would lie on a curve identical in shape to that of the theory. The experimental points for hemoglobin actually plotted in this figure clearly do not follow the predictions of eq. (1). The experimental curve has a slope which is about three times as steep as the theoretical curve. Since slope is of the essence, the hemoglobin molecule is in fact about three times as effective a transporter of oxygen (per oxygen binding-site) as a simple myoglobin-like molecule would be.

It is known from the physical chemistry of hemoglobin that the hemoglobin molecules act independently in the transport of oxygen under physiological conditions in mammals. If the binding sites within a single molecule acted independently the discrepancy between the experimental points and eq. (1) shown in fig. 5 could not exist. There must be interactions between the oxygen binding sites within a hemoglobin molecule, as is easily exhibited with a so-called Hill plot [7]. This plot shown in fig. 6 of $\log(Y/(1-Y))$ vs $\log p$ would give a straight line if a single site were reacting according to eq. (1). The intercept of the line would be related to the free energy of binding of oxygen to that site. The experimental data for hemoglobin show a sigmoid curve whose asymptotes are straight lines. The four binding reactions are

$$
\begin{aligned}
&\mathrm{Hb} + \mathrm{O_2} \rightleftharpoons \mathrm{HbO_2} \quad 4K_1 \\
&\mathrm{HbO_2} + \mathrm{O_2} \rightleftharpoons \mathrm{Hb(O_2)_2} \quad 3/2\ K_2 \\
&\mathrm{Hb(O_2)_2} + \mathrm{O_2} \rightleftharpoons \mathrm{Hb(O_2)_3} \quad 2/3\ K_3 \\
&\mathrm{Hb(O_2)_3} + \mathrm{O_2} \rightleftharpoons \mathrm{Hb(O_2)_4} \quad 1/4\ K_4
\end{aligned}
\qquad (2)
$$

The low-pressure asymptote (which corresponds to a reaction in which the first oxygen molecule is bound) has a small intercept and low binding energy describing K_1. The high oxygen pressure asymptote corresponds to a process in which all but one site already has an oxygen molecule bound and examines the binding energy of a fourth oxygen molecule to the last vacant site in hemoglobin with binding constant K_4. The statistical factors in eq. (2) are chosen so that $K_1 = K_2 = K_3 = K_4$ in the case of no interaction

Fig. 6. *Abscissa:* $\log p$; *ordinate:* $\log(Y/(1-Y))$.

The Hill plot of the binding of oxygen to human hemoglobin. The difference between the asymptotic straight lines is 0.15 eV per (oxygen) molecule. The dashed line is experimental data from Tyuma, Shimizu & Imai [8].

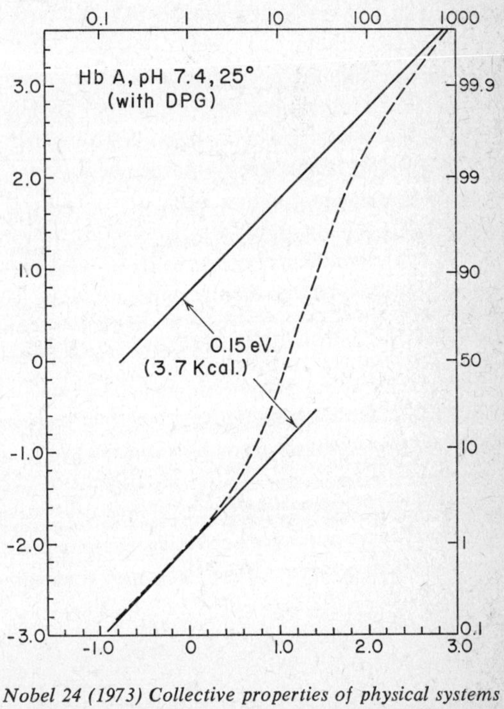

between binding sites. The difference between the binding energy of the first and of the last molecule (the two asymptotic lines) is the free energy of cooperation and is about 3.6 Kcal or about 0.15 eV for O_2 binding to Hb. This number does not seem very large, but in biological systems all energies are measured on a scale of kT, and generally occur at $\exp(E/kT)$. Population ratios (the relevant parameter) on the order of 200 are produced by such energies. This free energy of cooperation is responsible for the increased slope in the oxygen binding-curve. As one can see in the Hill plot, the slope of Y is a function of pressure, becoming steeper in the central region when the molecule switches from a low energy binding (for earlier oxygen molecules) to a high energy binding (for late oxygen molecules). Notice that this is exactly the opposite of what would happen if there were simply different non-interacting sites having different binding energies.

The physics problem posed by the hemoglobin physical chemistry, physiology, and structure can be summarized as follows. The free energy of cooperation is the source of effective oxygen transport by the hemoglobin molecule. The design of hemoglobin must have been tailored to result in this free energy of interaction between the binding sites. What is the mechanism of interaction which generates this free energy of cooperation? Considering that the binding sites are 35 Å apart, and the molecule operates in a 0.1 molar salt solution, direct "action at a distance" from one oxygen binding site to another seems ruled out as an explanation. We must look to the structure of hemoglobin as providing a mechanism of interaction between binding sites.

It is convenient for describing the physics of the action of hemoglobin to divide the problem into two separable steps. We will first discuss a phenomenological model of hemoglobin which explains and classifies the many physical properties of hemoglobin in terms of a few phenomenological parameters. Second, having examined the model and shown that it is verifiable in reasonable detail, we will try to understand the physical origin and magnitude of the parameters of the phenomenological model.

The model which most nearly explains all known hemoglobin physical chemistry and connects and interprets most physical experiments on hemoglobin is an allosteric model first discussed by Monod, Wyman & Changeux [9]. They began with the X-ray structure observation that there were two quaternary structures of the hemoglobin molecule and assumed that in either quaternary structure there was no energy of interaction between oxygenation of different subunits. If hemoglobin were fixed in one quaternary structure there would be no cooperativity in the oxygenation according to this supposition. They presumed that the two different quaternary structures had different oxygen binding energies. This description can be summarized in the energy level scheme shown in fig. 7. Each quaternary structure has five states, namely having 0, 1, 2, 3 or 4 oxygen molecules bound to the hemoglobin. The separation between these energy levels is the binding free energy of oxygen to the hemoglobin molecule. The different separation between the R-structure and the T-structure energy levels is due to the different binding energies attributed to these two structures. Within this energy level diagram the molecule is in equilibrium both with the oxygen partial pressure and between the two quaternary structures.

The cooperative oxygenation can be understood as follows. When no oxygen is bound, the hemoglobin molecules are primarily in the T structure, because T_0 is a lower energy state than R_0. The first oxygen binds to hemoglobin in the T structure and thus has a low binding energy. By the time three oxygen molecules are bound the molecule has switched chiefly to the R quaternary structure because R_3 has lower energy than T_3. The fourth oxygen goes on with a large binding energy characteristic of the R structure. Within this description the free energy difference ΔG obtained from the Hill plot is closely related to the difference in the binding free energy of oxygen between the R and T quaternary structures. It is clear that this model builds in the change of quaternary structure between deoxy hemoglobin and oxy hemoglobin.

Many experiments bear on this model both quantitatively and qualitatively. This model gives an adequate account of the binding curve of oxygen to hemoglobin in terms of three parameters [10] rather than the four parameters which would be possible for the most general models. While this is not a very sensitive test of the model, it is certainly a *sine qua non* for any model to be considered further.

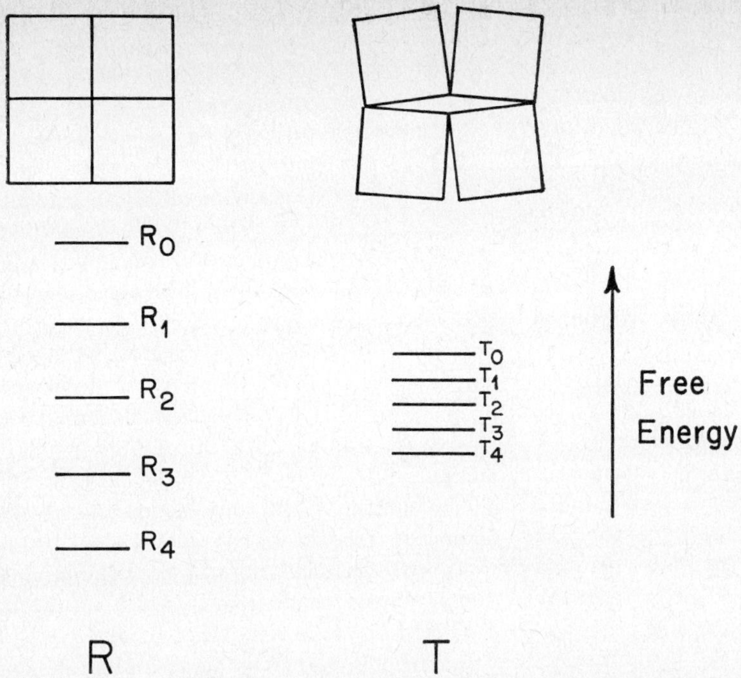

Fig. 7. The energy levels for the two different quaternary structures R and T within the MWC model. State T_n has oxygen molecules bound to n subunits. The subunit arrangement is denoted schematically at the top.

A great variety of kinetic experiments have also been shown to agree with the predictions of this model [11]. Kinetic experiments can be a particularly useful bridge between physical chemistry and models. On the one hand, ratios of kinetic constants are equilibrium constants. On the other hand, "on rates" can be properties of particular states, not simply thermodynamic averages over all states. A typical example of such agreement is the kinetics of the binding of carbon monoxide to hemoglobin (carbon monoxide being convenient to use for kinetic studies). Equilibrium studies show that the difference between the cooperative bindings of oxygen and of carbon monoxide is simply the difference in the scale of the absolute binding constants, the binding constants for carbon monoxide being approx. 200 times larger than those for oxygen and the cooperativity of CO binding being essentially identical with that of oxygen.

Carbon monoxide can be removed from hemoglobin or myoglobin by flash photolysis with a quantum efficiency near unity. A small flash of light shown on $Hb(CO)_4$ hemoglobin will remove at random a few molecules of carbon monoxide yielding $Hb(CO)_3$. The rate of recombination (i.e. the second-order rate constant) of the fourth CO molecule after the flash can be directly measured. The rate of binding of this molecule is about 60 times the rate of binding of the first carbon monoxide molecule to be bound to deoxy hemoglobin. This latter can be determined in an experiment in which deoxy hemoglobin is rapidly mixed with a solution of carbon monoxide in buffer. Within a kinetic description of the MWC model, the first of these rate-constants applies to the binding $R_3 + CO \rightarrow R_4$, while the second (in its early stages) is a measure of the process $T_0 + CO \rightarrow T_1$.

What should be expected after an intense flash of light when all the hemoglobin molecules have all four CO molecules removed? Deoxy hemoglobin results, but a prediction of the allosteric kinetic model is that immediately after flash photolysis from R_4 the molecule will be left in R_0, still in the R quaternary structure. Two decay routes toward $Hb(CO)_4$ are now possible,

$$R_0 \rightarrow R_1 \rightarrow R_2 \ldots$$

$$R_0 \rightarrow T_0 \rightarrow T_1 \ldots$$

Immediately after the flash one therefore might

expect the carbon monoxide rate of binding to be characteristic of the R structure (as long as $R_0 \to T_0$ is not overwhelmingly fast) rather than the slow rate first carbon monoxide bound to normal deoxy hemoglobin. Exactly this result was found by Gibson [12] who discovered the so-called quickly reacting form of hemoglobin in flash photolysis. When all the carbon monoxide is removed by flash some of the hemoglobin recombines rapidly with a rate which is characteristic to the R structure, and the rest of the hemoglobin recombines slowly with a rate characteristic of the T structure. The branching ratio is determined by the relative rates for the first steps of the two paths listed above.

There are many kinds of kinetic experiments which have been done and are in accord with the predictions of a kinetic allosteric model. The flash photolysis experiment was chosen here as a simple example which very directly seems to verify the existence of a state with the properties to be expected for the state R_0. In normal hemoglobin this state is negligibly populated in equilibrium studies of oxygen binding, while in a transient experiment almost 100% of the molecules can be put in state R_0.

Nuclear magnetic resonance is a physical tool which has proved extraordinarily useful in investigating the properties of particular states of hemoglobin. One application of such experiment by Schulman & Ogawa [13] concerns the verification of the existence of two quaternary structures for hemoglobin having two of its four binding sites occupied. In these experiments use is made of the fact that the four subunits of hemoglobin are not identical. Two of the subunits are of one kind and the other two subunits of a slightly different kind, the so-called α and β subunits. The hemoglobin molecule can be decomposed into monomers and the subunits separated. Cyanide (in place of oxygen) is then bound (essentially irreversibly) to one of the kinds of subunits and not to the other kind. The molecule can then be reassembled, and has cyanide on two binding sites and nothing on the other two binding sites. Ogawa and Shulman demonstrated the existence of some particular proton resonances in high resolution NMR which were sensitive to quaternary structural change, the positions of which can be used to determine the quaternary structure of the molecule in solution. The experiments on the "mixed state" (hemoglobin with two sites occupied) demonstrated the simultaneous existence of both quaternary structures for such molecules. A glance at the energy leve diagram for the MWC model in fig. 7 shows why this happens.

Mutant and chemically modified hemoglobins provide another useful test of the allosteric description of hemoglobin. For example, those mutants (typically having one amino acid changed from normal hemoglobin) which show the same quaternary structure for oxy and deoxy hemoglobin do not exhibit cooperativity in oxygen binding, an obvious prediction of the MWC model. Those mutants which are fixed in the R quaternary structure have oxygen affinities for all the subunits which are characteristic of the binding of oxygen to the last binding site ($Hb(O_2)_3 + (O_2) \rightleftharpoons Hb(O_2)_4$) in normal hemoglobin and exhibit kinetic characteristics which are again characteristic of the last oxygen binding.

Hemoglobin Kansas is an interesting mutant whose kinetics and nuclear magnetic resonance were recently studied [14, 15]. This mutant has one amino acid different from normal hemoglobin (HbA), at the interface between subunits, a position crucial to the equilibrium between the two quaternary structures. In only one of the two quaternary structures this amino acid makes a hydrogen bond across to another subunit. In Hb Kansas the amino acid substitution does not allow this hydrogen bond to form.

In the MWC model the most obvious effect of this mutant, which occurs at a location well removed from the heme group, should be to alter the equilibrium between the two quaternary structures [16], but to leave the physical properties of the two quaternary structures (as far as oxygen binding is concerned) very similar to what they were in HbA. A comparison of the energy level diagram for Hb Kansas and Hb A based on this supposition is shown in fig. 8.

Let us consider the effects of this change of the energy levels on the binding of carbon monoxide by deoxy hemoglobin. For Hb A this can be described by the first carbon monoxide binding at a rather slow rate (the first arrow shown in the path of fig. 8), the second also binding slowly (the second arrow), the third carbon monoxide also binding slowly followed by a rapid quaternary structure change (because R_3 is lower than T_3)

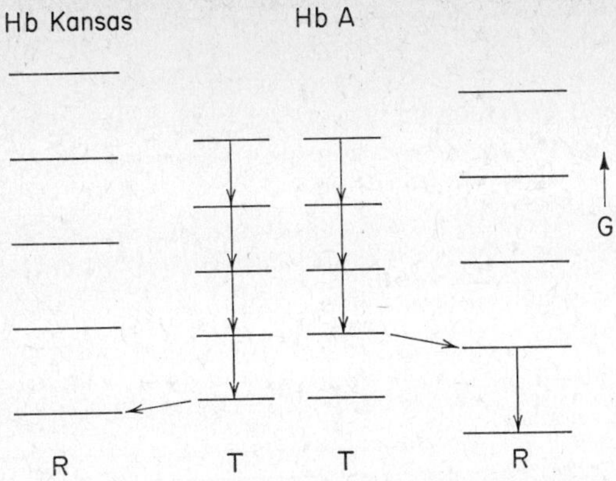

Fig. 8. A comparison of the energy levels of Hb A with those of Hb Kansas. The energies of the R states of Hb Kansas have been raised by an additive constant with respect to Hb A. The dominant pathway for the combination of CO with deoxy hemoglobin in a kinetic experiment is shown by the arrows.

and the fourth CO binding rapidly. This path results in an acceleration of the apparent bimolecular binding rate of carbon monoxide to hemoglobin because effectively two CO molecules are bound in the third slow step. This qualitative description can easily be given quantitative meaning [11]. Such acceleration is well known in hemoglobin kinetic studies. For Hb Kansas, on the other hand, exactly the same description will be true for the first two CO molecules bound, but the R state is not at a comparable energy to the T state until all the binding sites are occupied with carbon monoxide. All four CO molecules should bind equally slowly when the energy levels are as shown. Thus CO binding to Hb Kansas should show no increase in the apparent bimolecular rate. Its initial bimolecular rate should be the same as that of normal hemoglobin. Fig. 9 shows that the initial rates of Hb Kansas and Hb A are the same within experimental error, but Hb A exhibits an accelerating rate of binding while Hb Kansas shows an apparent rate-constant independent of time (or equivalently, of Y).

Nuclear magnetic resonance studies of Hb Kansas [15] showed that when CO was bound to all sites on Hb Kansas both quaternary structures could be simultaneously observed. Clearly such a result is allowed for Hb Kansas from the near equality of the energy levels T_4 and R_4 in fig. 8. Only the level R_4 is observed in equilibrium in Hb A because the R_4 state lies so much lower in energy than T_4.

There is agreement between the predictions of the phenomenological allosteric model of Hb and a wide variety of physical experiments which characterize properties of particular states. Solution conditions—pH, temperature, the presence of organic phosphates, the ionic strength—are important, but seem chiefly to renormalize the MWC parameters while preserving the phenomenology. Some degree of subunit inequivalence exists, but it does not seem to be large enough to be essential in a model which seeks to describe only the dominant phenomena. The agreement between experiments and the predictions of the allosteric model is sufficiently good that it makes sense to go one step further and inquire about the physical origin and meaning of the parameters of the MWC model.

There are three equilibrium parameters of the allosteric model, K_R (the binding constant of the R quaternary structure), K_T (the binding constant of the T quaternary structure) and L (the population ratio between the two quaternary structures $[T_0]/[R_0]$ when no oxygen is bound. It is useful to think of the ratio $K_T/K_R \equiv c$ as one of the three parameters, eliminating K_T as an independent parameter, because the value of $-kT \ln c$ is essentially the free energy of cooperation for Hb A.

K_R can be thought of as describing a property similar to the small molecule reaction of oxygen binding to iron in a heme group. This binding-constant is not itself intrinsically interesting from the point of view of biophysics, nor can I possibly calculate this number. L determines the equilibrium ratio between the two quaternary structures. It is clear what controls L, namely, all the bonds between the subunit interfaces which are somewhat changed when the quaternary structure changes. This interfacial energy, though conceptually simple, is contributed to by many Van der Waals contacts, salt bridges and some hydrogen bonds, and cannot be evaluated with precision although its origins are conceptually obvious. The essence of cooperativity lies in the constant c, which is closely related to the free energy of cooperation and which involes what are effectively interactions between different oxygen binding sites. The central physical problem is therefore to obtain an understanding of the origin of the parameter c.

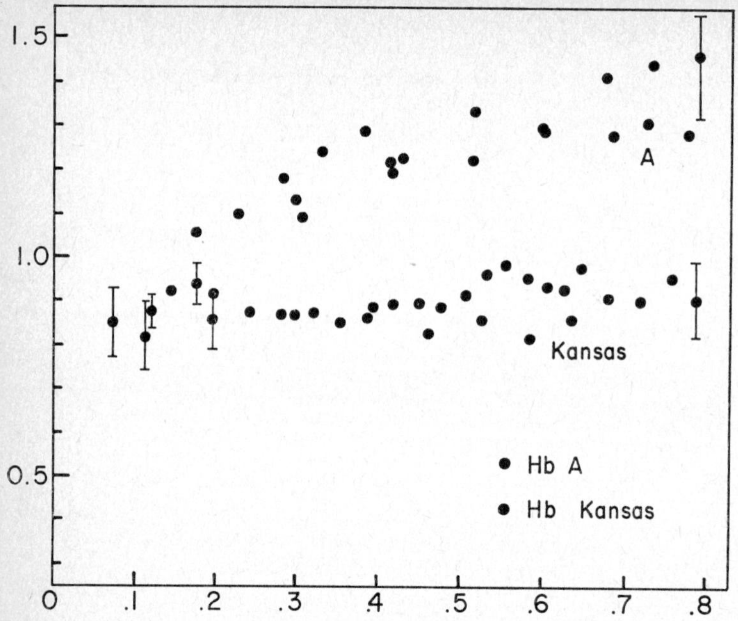

Fig. 9. *Abscissa: Y; ordinate:* rate (10^5 M^{-1} sec^{-1}).
The apparent second order rate constant as a function of the degree of ligation Y for the binding of CO by Hb A and Hb Kansas. Solutions at pH 7.0, 0.1 M phosphate, 0.01 M dithionite, and at a temperature of 20°C [14].

We will focus our attention on c by asking where the free energy $\Delta G = -kT \ln c$ is stored in the molecule [17]. There are three extreme descriptions of this location. First, this energy could be localized in some particular bond. Such bonds might for example be present in the T quaternary structure and not in the R, and present in a subunit of a T quaternary structure molecule only when oxygen is not bound to that subunit and not present when oxygen is bound. A bond of such systematics would be capable of explaining the free energy of cooperation, for the bond would break when oxygen was bound to the T structure but not to the R. The free energy ΔG could be localized in such a bond. (Of course, a bond need not be literally made or broken but need only be distorted.) Are there particular bonds in the molecule whose systematics really seem to demonstrate that those bonds store a significant amount of the free energy of cooperation? Obvious candidates include the direct bond between the oxygen and the iron (this bond might be different in the two quaternary structures), and particular interface bonds.

It is not possible to investigate each bond to see whether it stores the free energy ΔG, but some have been investigated in detail. Ogawa & Shulman [18] investigated the bonds between iron and the molecule bound to iron (typically cyanide) by NMR studies of protons on the heme to try to see the nature of the change in that bond when the quaternary structure changes. These studies demonstrate that no appreciable part of the free energy of cooperation is directly stored in the bond between the oxygen (or CN) and the iron atom. The wide-spread small spectral changes seen in the NMR studies suggested instead that the energy might be more widely distributed than in a particular bond [19].

The X-ray structural studies analyzed by Perutz in 1970 [20] suggested the localization of energies of cooperation might lie in particular bonds in the interface. It has not been possible to directly test this hypothesis. At present, the sytematics of X-ray structures suggest that the bonds Perutz considered may be important in the Bohr effect but are probably not as important in the free energy of cooperation. X-ray structural studies are again indicative of wide-spread small structural changes in the subunits due to quaternary structure change. As a result of such considerations, I have recently been investigating the consequences of presuming ΔG is widely distributed—perhaps among 10 or 20 bonds

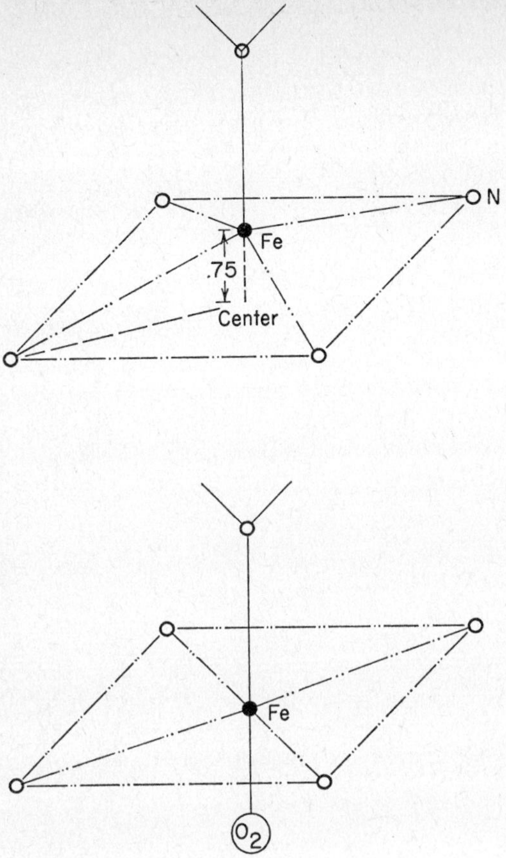

Fig. 10. The geometry of the iron atom with respect to its four neighboring heme nitrogens in deoxy hemoglobin and in oxy hemoglobin.

rather than chiefly concentrated in 1 or 2. Fortunately the physics of this situation is simpler than if the free energy of cooperation were concentrated in one or two bonds. I will briefly describe how the energy of cooperation could be widely spread, and show some new connections which should be present between different kinds of experiments if the ΔG is not localized.

Let us start, with Hoard [21] and Perutz [20], with the idea of a trigger, asking how does the protein part of the molecule know when oxygen is bound to the iron? The X-ray structural studies show that in deoxy Hb the iron atom lies 0.75 Å above the mean heme plane [20] as is shown in fig. 10. When oxygen binds to hemoglobin the iron atom moves down into the heme plane. The oxygen is not known, in the case of iron hemoglobin, to form bonds to other parts of the protein. As a result the motion of the iron with respect to the heme plane seems the chief means by which the protein knows of the oxygenation of the subunit.

If not large part of the free energy of cooperation is stored in any particular bond, one can think of the protein as responding in a linear fashion to the effect of the motion of the iron atom with respect to the heme plane. This gives rise to a description of the free energy of cooperation in terms of stored elastic energy. The position of the iron atom with respect to the heme plane is controlled by two competing forces, the covalent bond force between the iron atom and the heme, and the force via an indirect path from the iron atom to the nitrogen on the proximal histidine, through the protein, and back to the heme through Van der Waals bonds between the protein and outer parts of the heme ring. There are thus two "springs" connecting the iron atom to the heme plane, one from the direct iron-nitrogen bond, and one which represents the path through the globin (protein). These two springs do not have the same equilibrium length, i.e., they do not individually produce zero force at the same iron atom position. The spring system therefore stores strain energy in equilibrium, the amount of strain free energy stored depending upon the equilibrium length difference of the two springs. Most of the energy is stored in the protein spring, since this is the weaker spring, while the direct-bond spring chiefly determines the position of the iron atom.

After making approximations which are probably appropriate to the parameters of the hemoglobin molecule one obtains the results

$$G = G_0 + \delta_L F_p. \tag{3}$$

The free energy G of oxygen binding to hemoglobin in a fixed quaternary structure is equal to an ideal number G_0 plus a correction term which comes from the influence of the protein. This correction free energy $\delta_L F$ is the product of the distance δ_L which the iron moves with respect to the heme plane when the oxygen binds to the iron atom and the force F with which the protein pulls on the iron atom. This force will be dependent upon the quaternary structure. In this description the free energy of cooperation ΔG is given by

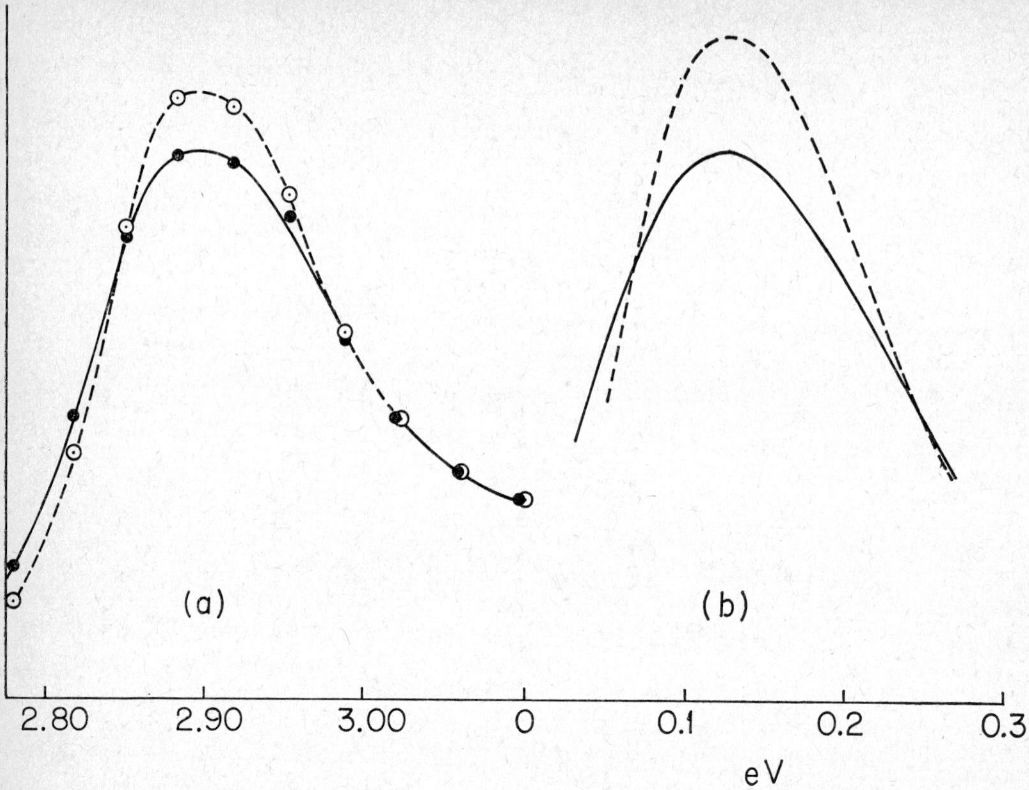

Fig. 11. *Abscissa:* photon energy (eV); *ordinate:* molar extinction coefficient (arbitrary units).
(*a*) The absorption spectrum of ○, deoxy hemoglobin and ●, the quickly reacting form of Hb [12]; (*b*) calculation from a configuration-coordinate model. Only the scale of the changes can be legitimately compared [17].

$$\Delta G = \delta_L(F_R - F_T). \qquad (4)$$

The force difference $F_R - F_T$ can be obtained from the known value of ΔG and the known displacement of the iron atom on binding, and is about 0.2 eV/Å.

An estimate of the restoring force-constant for iron motion with respect to the heme plane and the magnitude of $F_R - F_T$ combine to indicate that less than 10% of the free energy of cooperation is stored in the heme, the rest of it being stored in the protein. One should remember that in the case of a protein the "spring-constant" is determined by complicated factors, the energy involved being not internal energy but free energy including all the entropic effects of protein configuration.

The importance of this simple eq. (4) is that it can be checked. The energy of cooperation is a product of two terms, one of which depends on the iron motion and on the heme (but is essentially independent of the protein), and the other which depends on the protein but is independent of what is being bound to the iron atom. Eq. (4) relates an energy determined from physical chemistry to forces which can perhaps be measured in optics or NMR studies, and to a distance scale determinable from structural studies. A certain scale of forces is predicted in the equilibrium state, and we next examine some evidence for such a force.

There are small spectroscopic differences [12] between the absorption spectrum of the heme when deoxy hemoglobin is in the high affinity form (the R structure) or when it is a low affinity form (T structure) as is shown in fig. 11. We have made an unsophisticated configuration-coordinate analysis of the absorption spectrum to determine the effect of a particular size of force on the iron atom. Fig. 11 shows that the configuration-coordinate model predicts the observed general scale of the absorption changes due to a force whose magnitude is supplied by eq. (4)

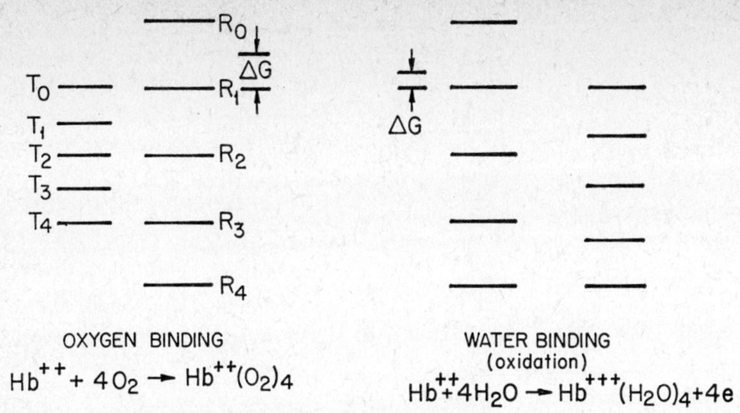

Fig. 12. The energy levels of hemoglobin for binding oxygen and for binding water, illustrating the effect of halving ΔG in the latter case.

and data from physical chemistry and structure studies.

While this result cannot be made quantitative certain symmetry considerations work out well. Deoxy hemoglobin shows spectral changes between the two quaternary structures, but oxy hemoglobin does not. The reason for this can be seen from a symmetry argument. In the case of deoxy hemoglobin the iron atom lies 0.75 Å above the heme plane. A force on the iron atom directed normal to the heme plane will produce a change in the spectrum linear in that force. In the case of oxy hemoglobin the iron atom lies in the heme plane. There should be no spectral changes linear in the force because the heme plane is essentially a plane of symmetry. Thus the force change due to changing quaternary structures should theoretically produce much larger spectral changes for deoxy hemoglobin than for oxy hemoglobin.

An alternative route to the checking of eqs (3) or (4) is to try to vary the parameter δ_L. When water binds in the same position to which oxygen binds at pH 6.5 the iron atom does not move all the way into the heme plane, but remains 0.3 Å above it. Thus in the case of the binding of water δ_L is only about half as large as it is in the binding of oxygen. The free energy of cooperation ΔG should therefore be half as great, changing the MWC parameter *c*. *L* is the same as for oxygen binding, since it is a property of deoxy hemoglobin. The energy levels for water binding and for oxygen binding are compared in fig. 12. For the case of water binding, the levels T_4 and R_4 are nearly the same in energy (just as they were for oxygen binding to Hb Kansas). The failure of water to bind cooperatively at pH ~6.5 is thus understood in terms of the MWC model and the observed differences between the iron motion for water and for oxygen binding. Recently Perutz [22] and coworkers have shown that in the case of the binding of water (so-called acid methemoglobin) both optical studies and SH-reactivity studies indicate that the energy levels T_4 and R_4 are so close together that the molecule can be switched between them by changing solution conditions.

The parameters of the MWC model are altered by solution conditions, particularly important parameters being the pH of the solution and the concentration of organic phosphates. We surmized in eq. (3) that *G* is equal to *G* ideal plus a correction term. δ_L should be chiefly a property of the electronic binding of oxygen to hemoglobin to the heme (or of any other particular binding), whereas the force of the protein will change with solution conditions, for protein structure is sensitive to solution conditions. The optical changes of the deoxy heme are hypothesized to be linear in *F* so this provides a possible experimental means of trying to check the basic equation of the distributed-energy model. When changing the affinity of a particular subunit in a fixed quaternary structure by changing solution conditions the spectral changes for deoxy hemoglobin should be linear in the change in the free energy of oxygen binding. The constant of proportionality is already known from the spectral change between the two quaternary structures.

Experiments to check this point are underway. Unfortunately the prediction needs the three suppositions of the basic model, the linearity of optical changes, and the supposition that the Bohr effect changes subunit affinity by the same pathway as does the quaternary structure change.

The pH dependence of the oxygen binding curves is of particular interest to physicists for two reasons. First, it is used by the body to increase the transport effectiveness of hemoglobin. At a place where oxygen is being used carbon dioxide is being generated which, when it enters in the bloodstream, changes the pH in a direction such that it reduces the oxygen affinity of the hemoglobin. This feedback dumps more oxygen in the place where carbon dioxide is being generated, thus helping the transport of oxygen. That such devious schemes should be resorted to suggests how important effective oxygen transport is in physiology. The second interesting aspect of the pH dependence of the oxygenation curve is historical. This bit of physical chemistry is called the Bohr effect, named for its discoverer, Christian Bohr, the father of Niels.

The MWC allosteric model seems to give a good phenomenological description of hemoglobin action. I have described a possible simple view of how the most important parameter comes about. The free energy of cooperation in this description is stored in strain energy, distributed widely enough that the system can be viewed as responding linearly. There are enough detailed predictions of relations between structural studies, NMR, optics, and physical chemistry that the description can be checked in detail. It is doubtless an oversimplification, but may be useful for organizing purposes even if it is only in part valid. If it seems to be verified, one can go one layer deeper and look in a microscopic structural sense for how the strain propagates from the proximal histidine to the interface between subunits and see what parts of the interface are important in communicating the strain information.

At a meeting chiefly devoted to collective phenomena, the question of why hemoglobin has only four subunits will logically arise. The larger the number of cooperating subunits, the steeper the central section of the Hill plot, and the more effective the transport per binding site. In the limit of large numbers, the MWC model becomes a first-order phase transition. The penalty for better transport, however, is a well-defined transition pressure, and if the ambient pressure in the lungs were to drop below this value, all transport would cease. The number four seems a compromise between efficient transport and catastrophe prevention.

If hemoglobin were the only biological molecule having an allosteric transition it would be an isolated and uninteresting case. There is however a class of allosteric enzymes which are often controlled by a quaternary structure change. Some of these enzymes have separate control subunits and catalytic subunits with the activity of the catalytic subunit controlled by a binding at the control sites. Aspartate transcarbamylase is an example of such an enzyme for which the gross structure has been recently determined by X-ray diffraction [23]. Hemoglobin may also serve as a test case for understanding how a protein affects the rate of chemical reaction. Hemoglobin is unusual in the sense that its equilibrium chemistry is chiefly important to its biological function. It differs from enzymes for which the rate of reaction is biologically important. In hemoglobin one is trying to understand how the protein affects the equilibrium of a particular "small molecule" reaction. If this equilibrium alteration can be understood in hemoglobin, a next step to study is the effect of the protein on the hemoglobin kinetics, certainly simpler than most of the problems in enzyme kinetics.

The study of silicon in detail generated the understanding of phenomena in a wide class of semiconductors and greatly enhanced the understanding of metals and insulators as well. The detailed study of hemoglobin. which might be regarded as the physicists biological molecule, should teach us a great deal about the global control of the chemical properties of "local" parts of enzymes and other biological molecules.

References

1. Pauling, L & Corey, R B, Proc natl acad sci US 1951, 37, 235.
2. Dickerson, R E, in The proteins, vol. 2, p. 603. Academic Press, New York, 1964.
3. Perutz, M F, J mol biol 1965, 13, 646.

4a. Muirhead, H, Cox, J M, Mazzarella, L & Perutz, M F, J mol biol 1967, 28, 117.

4b. Perutz, M F & Lehmann, H, Nature 1968, 219, 902.

5. Muirhead, H & Perutz, M F, Nature 1963, 199, 633.
6. An excellent review of the chemistry of myoglobin and hemoglobin is available in Antonini, E & Brunori, M, Hemoglobin and myoglobin in their reactions with ligands. North Holland/American Elsevier, Amsterdam/New York, 1971.
7. See Wyman, J, Adv protein chem 1965, 19, 233.
8. Tyuma, I, Shimiza, K & Imai, K, Biochem biophys res commun 1971, 43, 423.
9. Monod, J, Wyman, J & Changeux, J P, J mol biol 1965, 12, 88.
10. Imai, K, Biochemistry 1973, 12, 798.
11a. Hopfield, J J, Shulman, R G & Ogawa, S, J mol biol 1971, 61, 425.
11b. Shulman, R G, Ogawa, S & Hopfield, J J, Arch biochem biophys 1972, 151, 68.
12. Gibson, Q H, Biochem j 1959, 71, 293.
13a. Shulman, R G, Ogawa, S, Mayer, A & Castillo, C L, Proc NY acad sci 1973. In press.
13b. Ogawa, S & Shulman, R G, Biochem biophys res commun 1971, 42, 9.
13c. Cassoly, R, Gibson, Q H, Ogawa, S & Shulman, R G, Biochem biophys res commun 1971, 44, 1015.
13d. Shulman, R G, Ogawa, S & Hopfield, J J, Cold Spring Harbor symp quant biol 1971, XXXVI, 337.
14. Hopfield, J J, Ogawa, S & Shulman, R G, Biochem biophys res commun 1972, 49, 1480.
15. Ogawa, S & Shulman, R G, Biochem biophys res commun 1972, 49, 1485.
16. Edelstein, S, Nature 1971, 230, 224.
17. The following viewpoint is described in detail in Hopfield, J J, J mol biol 1973, 77. In press.
18. Ogawa, S & Shulman, R G, J mol biol 1972, 70, 315.
19. Shulman, R G, Ogawa, S, Wuthrich, K, Yamane, T, Peisach, J & Blumberg, W E, Science 1969, 165, 251.
20. Perutz, M F, Nature 1970, 228, 726.
21. Hoard, J L, in Structural chemistry and molecular biology (ed A Rich & N Davidson). Freeman, San Francisco, 1968.
22. Perutz, M F, Biochem soc trans 1973, 1, 42.
23. Evans, D R, Warren, S G, Edwards, B F P, McMurray, C H, Bethge, P H, Wiley, D C & Lipscomb, W N, Science 1973, 179, 683.

Discussion

Wilson: How do you know that the distortion δ_L due to binding H_2O is half the distortion when binding O_2?

Hopfield: There are direct X-ray structure determinations of the position of the iron atom when water is bound, and it lies 0.3 Å above the heme plane. The case of O_2 bound to hemoglobin has not in fact been measured, but indications from CO as a liquid and from model system studies both show that a low spin ferrous iron atom (such as when O_2 is bound) will be in the plane.

A Possible Organization of Animal Memory and Learning[1]

L. N. Cooper

Brown University, Providence, R.I. 02912, USA

Summary

A brief account is given of some of the properties of a neural model which displays on a primitive level features which suggest some of the mental behavior associated with animal memory and learning. The model as well as the basic passive procedure by which the neural network modifies itself with experience is consistent with known neurophysiology as well as with what information might be available in the neuron network. One must, however, assume that there exists a means of communication (electrical or chemical) between the cell body and dendrite ends—communication in a direction opposite to the flow of electrical signals. This same modification procedure could also lead to the formation of cells of the type observed by Hubel & Wiesel in the cat's visual cortex. It is suggestive that a network modification procedure that could produce such early processing cells might also be responsible, in cortex, for 'higher' mental processes. The explicit mathematics employed here is that of linear transformations on a vector space. However, as only certain topological properties are used, it is possible to construct a more general non-linear theory.

We have been analyzing a class of neural models that display, on a primitive level, features such as recognition, association and generalization, which suggest some of the mental behavior associated with animal memory and learning [1]. The mechanisms employed seem to be biologically plausible and are not inconsistent with known neurophysiology. In addition the networks that result seem to be a reasonable outcome of evolutionary development under the pressure of survival. Some of the ideas discussed are related to or are generalizations of earlier concepts such as perceptrons or similar models [2–4]. In addition non-local memories have been explored previously among others by Longuet-Higgins [5, 6].

[1] This work was supported in part by the US National Science Foundation.

Quasi-Random Network

Because of the enormous complexity of the neural network, it seems reasonable to assume that an animal's central nervous system is not completely predetermined genetically. This view is reinforced by the observed fact that in mammals a functioning system is relatively resistant to the death or malfunction of individual units. A system is required, therefore, which is either highly redundant, in which the exact placement of individual units is not critical, or in some other way can continue functioning in spite of the failure of individual units.

A quasi-random network is in some ways analogous to a many-body system as treated in statistical mechanics. There the number of degrees of freedom is large compared to the number of parameters such as temperature or pressure, usually specified to characterize any given ensemble. The specification of the few parameters ensures that any actual configuration is a member of a particular ensemble.

We consider the possibility that genetic information determines a small number of overall parameters governing the growth of neurons. These might be, for example, density or types of neurons in various regions, approximate regions of projection, general direction of growth, extent of dendritic arborization, number and type of synaptic connections, etc. There is some evidence that this may be the case, particularly in cerebral cortex. In visual cortex there has long been known to be strong topographic organization, but there also appears to be local randomness of connection [7]. Auditory cortex seems to show even less obvious organization [8], and the olfactory and taste systems show no easily understood organization at all [9]. The model we consider accepts such local randomness as a fundamental principle of organization. Such a system could arise naturally in evolution by allowing progressive modification of previously completely specified neural systems.

Linear Mappings

Neuron behavior is complex and highly nonlinear. However, in spite of the short time nonlinear behavior of neurons, there is presently justification for considering linear or quasi-linear models in which the neuron potentials are averaged over short periods of time as surprisingly good approximations of some aspects of neural response [10]. Anderson explored the properties of a simple neural model based on a linear mapping between one set of neurons and another [11, 12]. This mapping contains a non-local memory with some of the properties of the holographic-type memory discussed by Longuet-Higgins [5, 6], but in contrast to the model of Longuet-Higgins, this mapping seems easily realizable with known physiology. Anderson showed that his model would act as a matched filter and be capable of recognition of events to which it is previously exposed. We have developed further the consequences of such systems and have shown that they are capable of recognition, association and a form of generalization which seems possible to relate in a primitive way to some animal mental behavior.

Network Modification

In addition we have been able to introduce a method of network modification in which learning takes place in an effortless or what we call a passive manner. The system is placed in an environment and, without any search procedure, forms an internal representation of its external world—an internal representation which enables it to recognize and associate.

The means of modifying the neural network employed here has a long history. It involves changing the strength of the synaptic junction (the ratio of output to input spiking frequencies) according to products of pre- and postsynaptic activity. This is physiologically possible (as will be discussed in more detail) if there is a means of communicating information between the cell bodies and dendrite ends—communication in a direction opposite to the flow of electrical signals. If such communication exists, alterations in the synaptic junctions could be made according to what information is locally available. (One can easily add to this the possibility of overall (global) controls affecting all of the network.)

Though there is no difficulty imagining a variety of possible mechanisms by which such changes might take place, experimental evidence bearing on which of these might be operative (or whether in fact synaptic modifications occur) is very sparse.

The modification procedure could also account for the formation of cells which respond to particular patterns as observed, for example, by Hubel & Wiesel in visual cortex of the cat [13]. The suggestive possibility emerges that the action of a single adaptive mechanism designed originally, perhaps, for the formation of cells to facilitate processing in the outer portions of the brain, can account also for 'higher' mental processes such as memory.

Space of Events and Representations

The duration and extent of an 'event' should be defined self-consistently by the interaction between the environment and the system itself. We proceed initially, however, as though an event is a well-defined objective happening and envision a space of events, E, labelled e^1, e^2, e^3 ... Imagine that these are mapped by the sensory and early processing devices of the system through the mapping P (processing) into signal distributions in the neuron space f^1, f^2, f^3 ... The mapping, P, is denoted by the double arrow, fig. 1. For the moment we maintain the fiction that this mapping is not modified by experience. (What seems actually to be the case is that such early processing systems are at least partially constructed in the youth of the animal and become 'hardened' at some relatively early stage in its development [14]).

Although we do not discuss the mapping P in any detail, it can be very complex; it must be rich and detailed enough so that a sufficient amount of information is preserved to be of interest. *We assume that the mapping P from E to F has the fundamental property of preserving in a sense (not yet completely defined) the closeness or separateness of events.*

Two events e^ν and e^μ map into f^ν and f^μ whose separation is related to the separation of the original events. In a vector representation we imagine that two events as similar as a white cat and a grey cat map into vectors which are close to parallel while two events as different as the sound of a bell and the sight of food map into vectors which are close to orthogonal to each other.

Given the signal distribution in F which is the result of an event in E, we imagine that the signal distribution f is mapped onto another set of neurons (or onto the same set) by a mapping, A, denoted by the single arrow, fig. 1. This latter type of mapping is modifiable, and we propose that it is in such mappings that animal memory is contained.

The cortex of higher animals is of course very complex, and if any such systems of neurons exist they would be expected not only to be complicated but to be interspersed with neuron assemblies which do specific processing. Further one would imagine that large numbers of such systems arranged serially or in parallel would be required to reproduce even a portion of the complexity and variety of mental processes.

As will be seen later, the behavior of such neuron sub-networks is dependent on parameters which could easily be imagined to vary from sub-network to sub-network resulting in quite different characteristics. It is possible that sequences of such sub-networks with different values of the biological parameters could be involved in actual mental behavior.

In what follows we construct an idealized model of a network which incorporates a modifiable mapping and explore some of its properties.

Fig. 1. The N neurons in the F bank are connected via synaptic junctions to the N neurons of the G bank.

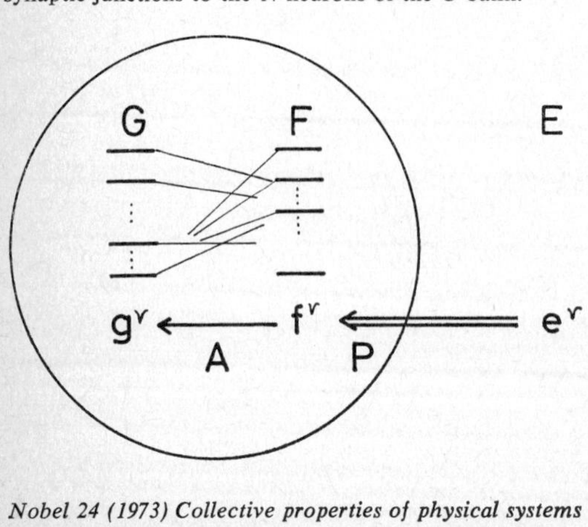

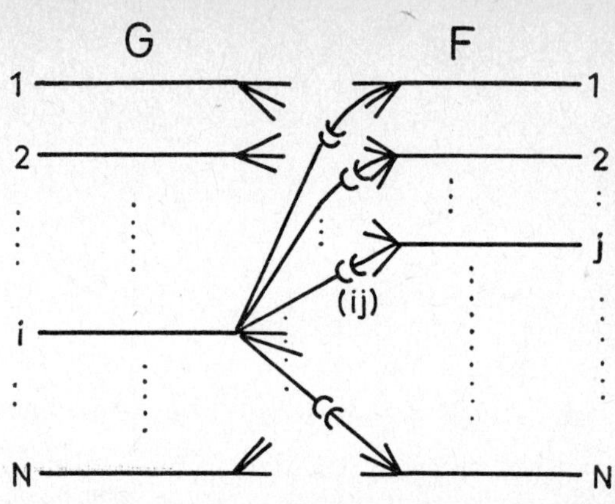

Fig. 2. The ideal associator unit. Each of the N incoming neurons in F is connected to each of the N outgoing neurons in G by a single ideal junction. (Only the connections to i are drawn.) We assume that the firing rate of neuron i in G, g_i, is mapped from the firing rates of all of the neurons in F by: $g_i = \Sigma_j A_{ij} f_j$.

Consider N neurons 1, 2 ... N, each of which has some spontaneous firing rate r_{j0}. (This need not be the same for all of the neurons nor need it be constant in time.) We can then define an N-tuple whose components are the difference between the actual firing rate r_j of the jth neuron and the spontaneous firing rate r_{j0}.

$$f_j \equiv r_j - r_{j0}$$

By constructing two such banks of neurons connected to one another (or even by the use of a single bank which feeds signals back to itself), we arrive at a simplified model as illustrated in fig. 1.

The actual connections between one neuron and another are generally complex and redundant; we idealize the network by replacing this multiplicity of connections between axon and dendrites by a single ideal junction which summarizes logically the effect of all of the synaptic contacts between the incoming axon branch from neuron j in the F bank and the dendrite tree of the outgoing neuron i in the G bank (fig. 2). Each of the N incoming neurons, in F, is connected to each of the N outgoing neurons, in G, by a single ideal junction. We then assume that: *the firing rate of neuron i in G, g_i, is mapped* *from the firing rates of all of the neurons, f_j, in F by:*

$$g_i = \sum_{j=1}^{N} A_{ij} f_j$$

This is the fundamental linear assumption which gives the influence of firing rates in F on those in G. Although most of the results we obtain below do not require so strong an assumption, the simplicity of this hypothesis makes worthwhile an exploration of its consequences. In making this hypothesis, we are focusing our attention on firing rates, on time averages of the instantaneous signals in a neuron (or perhaps a small population of neurons). We are further using the known integrative properties of dendrite branches. Although we have available both excitatory or inhibitory synaptic junctions, such a module could be built if necessary of excitatory neurons alone since a decrease of incoming signal (if $r_j - r_{j0} < 0$) would decrease the output (less excitation).

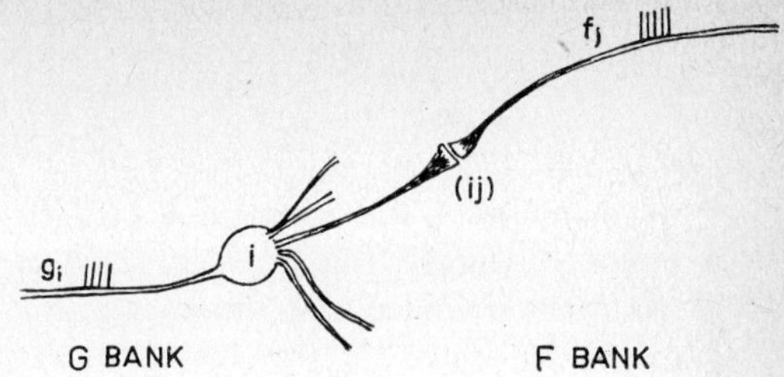

Fig. 3. The ideal junction.

The Associative Mapping, Memory and Mental Processes

It is in modifiable mappings of the type A that the experience and memory of the system are stored. In contrast with machine memory which is at present local (an event stored in a specific place) and addressable by locality (requiring some equivalent of indices and files) animal memory is likely to be distributed and addressable by content or by association. In addition for animals there need be no clear separation between memory and 'logic'. We show below that the mapping A can have the properties of a memory that is non-local, content addressable and in which 'logic' is a result of association and an outcome of the nature of the memory itself.

The suggestion that animal memory is non-local goes back at least to Lashley [15]; it is implied in Perceptron-like devices [2–4]. The holographic memory of Longuet-Higgins [5, 6] is non-local and content addressable but difficult to realize physiologically. The form we employ here was introduced by Anderson [11, 12].

Anderson's associative mapping is most easily written in the basis of the mapped vectors the system has experienced:

$$A = \sum_{\mu\nu} c_{\mu\nu} g^{\mu} \times f^{\nu}$$

Although this is a transparent mathematical form, its meaning as a mapping among neurons deserves some discussion. The ij'th element of A gives the strength of the ideal junction between the incoming neuron j in the F bank and the outgoing neuron i in the G bank (fig. 3).

Thus, if only f_j is non-zero

$$g_i = A_{ij} f_j.$$

Since

$$A_{ij} = \sum_{\mu\nu} c_{\mu\nu} g_i^{\mu} f_j^{\nu}$$

the ij'th junction strength is composed of a sum of the entire experience of the system as reflected in firing rates of the neurons connected to this junction. Each experience or association ($\mu\nu$), however, is stored over the entire array of $N \times N$ junctions. This is the essential meaning of a distributed memory: Each event is stored over a large portion of the system, while at any particular local point many events are superimposed.

Recognition and Recollection

The fundamental problem posed by a distributed memory is the address and accuracy of recall of the stored events. Consider first the 'diagonal' portion of A

$$(A)_{\text{diagonal}} \equiv \mathcal{R} \equiv \sum_{\nu} c_{\nu\nu} g^{\nu} \times f^{\nu}$$

An arbitrary event, e, mapped into the signal, f, will generate the response in G

$$g = Af$$

If we equate recognition with the strength of this response g, say the value of

$$(g, g),$$

then the mapping A will distinguish between those events it contains (the f^{ν}, $\nu = 1, 2, \ldots, \mathrm{k}$) and other events separated from these.

The word 'separated' in the above context requires definition. In a type of argument given previously [11, 12], the vectors f^{ν} are thought to be independent of one another and to satisfy the requirements that on the average

$$\sum_{i=1}^{N} f_i^{\nu} = 0$$

$$\sum_{i=1}^{N} (f_i^{\nu})^2 = 1.$$

Any two such vectors have components which are random with respect to one another so that a new vector, f, presented to $\mathcal{R}$ above gives a noise-like response since on the average (f^{ν}, f) is small. The presentation of a vector seen previously, f^{λ}, however, gives the response

$$\mathcal{R}f^{\lambda} = c_{\lambda\lambda} f^{\lambda} + \text{noise}$$

It is then shown that if the number of imprinted events, k, is small compared to N, the signal to noise ratios are reasonable.

If we define separated events as those which map into orthogonal vectors, then clearly a recognition matrix composed of k orthogonal vectors $f^1 f^2, \ldots, f^{\mathrm{k}}$

$$\mathcal{R} = \sum_{\nu=1}^{\mathrm{k}} c_{\nu\nu} g^{\nu} \times f^{\nu}$$

will distinguish between those vectors contained, $f^1 \ldots f^{\mathrm{k}}$, and all vectors separated from (perpendicular to) these. Further the response of $\mathcal{R}$ to a vector previously recorded is unique and completely accurate

$$\mathcal{R}f^{\lambda} = c_{\lambda\lambda} g^{\lambda}$$

In this special situation the distributed memory is as precise as a localized memory.

In addition this type of memory has, as has been pointed out before [5, 6], the interesting property of recalling an entire associated vector g^{λ} even if only part of f^{λ} is presented. Let

$$f^{\lambda} = f_1^{\lambda} + f_2^{\lambda}$$

If only part of f^{λ}, say f_1^{λ}, is presented, we obtain

$$\mathcal{R}f_1^{\lambda} = c_{\lambda\lambda}(f_1^{\lambda}, f_1^{\lambda}) g^{\lambda} + \text{noise}$$

The result is thus the entire response to the full f^{λ} with a reduced coefficient plus noise.

Association

If we now take the point of view that presentation of the event e^{ν} which generates the vector f^{ν} is recognized if

$$\mathcal{R}f^{\nu} = c g^{\nu} + \text{noise}$$

Then the off-diagonal terms

$$\mathcal{A} \equiv \sum_{\mu \neq \nu} c_{\mu\nu} g^{\mu} \times f^{\nu}$$

may be interpreted as leading to association of events initially separated from one another

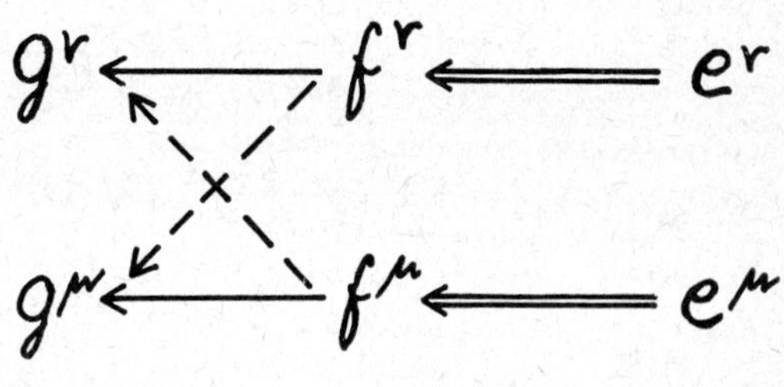

where $(f^{\nu}, f^{\mu}) = 0$.

For with such terms the presentation of the event e^{ν} will generate not only g^{ν} (which is equivalent to recognition of e^{ν}) but also (perhaps more weakly) g^{μ} which should result with the presentation of e^{μ}. Thus, for example, if g^{μ} will initiate some response (originally a response to e^{μ}) the presentation of e^{ν} when $c_{\mu\nu} \neq 0$ will also initiate this response.

We, therefore, can write the association matrix

$$A = \sum_{\mu\nu} c'_{\mu\nu} g^{\mu} \times f^{\nu} = \mathcal{R} + \mathcal{A}$$

where

$$\mathcal{R} = (A)_{\text{diagonal}} \equiv \sum_{\nu} c_{\nu\nu} g^{\nu} \times f^{\nu} \quad \text{[recognition]}$$

and

$$\mathcal{A} = (A)_{\text{off-diagonal}} \equiv \sum_{\mu \neq \nu} c_{\mu\nu} g^{\mu} \times f^{\nu} \quad \text{[association]}$$

The $c_{\mu\nu}$ are then the 'direct' recognition and association coefficients.

"Logic"

In actual experience the events to which the system is exposed are not in general highly separated nor are they independent in a statistical sense. There is no reason, therefore, to expect that all vectors, f^ν, printed into A would be orthogonal or even very far from one another. Rather it seems likely that often large numbers of these vectors would lie close to one another. Under these circumstances a distributed memory of the type contained in A will become confused and make errors. It will 'recognize' and 'associate' events never in fact seen or associated before.

To illustrate, assume that the system has been exposed to a class of non-separated events $\{e^1 \dots e^k\}:\{e^\alpha\}$ which map into the k vectors $\{f^1 \dots f^k\}:\{f^\alpha\}$.

The closeness of the mapped events can be expressed in a linear space by the concept of community. We define the community of a set of vectors such as $\{f^\alpha\}$ above as the lower bound of the inner products (f^s, f^t) of any two vectors in this set. *The community of the set of vectors* $\{f^\alpha\}$ *is* Γ, $C(f^\alpha) = \Gamma$, *if* Γ *is the lower bound of* (f^s, f^t) *for all* f^s *and* f^t *in* $\{f^\alpha\}$.

If each exposure results in an addition to A (or to $\mathcal{R}$) of an element of the form

$$c_{\nu\nu} g^\nu \times f^\nu,$$

then the response to an event f^s from this class $f^s \in \{f^\alpha\}$ is

$$\mathcal{R} f^s = g = \sum_\nu c_{\nu\nu} g^\nu (f^\nu, f^s) = c_{ss} g^s + \sum_{\nu \neq s} (f^\nu, f^s) c_{\nu\nu} g^\nu$$

where $(f^\nu, f^s) \geq \Gamma$. If Γ is large enough the response to f^s is, therefore, not very clearly distinguished from that of any other f contained in $\{f^\alpha\}$. (In the next section we discuss how such an A might be constructed.)

If a new event, e^{k+1}, not seen before is presented to the system and this new event is close to the others in the class α (for example, suppose that e^{k+1} maps into f^{k+1} which is a member of the community $\{f^\alpha\}$) then $\mathcal{R} f^{k+1}$ will produce a response not too different from that produced for one of the vectors $f^s \in \{f^\alpha\}$. Therefore, the event e^{k+1} will be recognized though not seen before.

This, of course, is potentially a very valuable error. For the associative memory recognizes and then attributes properties to events which fall into the same class as events already recognized. If in fact the vectors in $\{f^\alpha\}$ have the form

$$f^\nu = f^0 + n^\nu$$

where n^ν varies randomly, f^0 will eventually be recognized more strongly than any of the particular f^ν actually presented.

We have here an explicit realization of what might loosely be called 'animal logic'—which, of course, is not logic at all. Rather what occurs might be described as the result of a built-in directive to 'jump to conclusions'. The associative memory by its nature takes the step

$$f^0 + n^1, f^0 + n^2 \dots f^0 + n^k \dots \to f^0$$

which one perhaps attempts to describe in language as passing from particulars: cat[1], cat[2], cat[3] ... to the general: cat.

How fast this step is taken depends (as we will see in the next section) on the parameters of the system. By altering these parameters, it is possible to construct mappings which vary from those which retain all particulars to which they are exposed, to those which lose the particulars and retain only common elements—the central vector of any class.

In addition to 'errors' of recognition, the associative memory also makes errors of association. If, for example, all (or many) of the vectors of the class $\{f^\alpha\}$ associate some particular g^β so that the mapping A contains terms of the form

$$\sum_{\nu=1}^{k} c_{\beta\nu} g^\beta \times f^\nu$$

with $c_{\beta\nu} \neq 0$ over much of $\nu = 1, 2, \dots, k$, then the new event e^{k+1} which maps into f^{k+1} as in the previous example will not only be recognized

$$\mathcal{R} f^{k+1}, \mathcal{R} f^{k+1}) \quad \text{large}$$

but will also associate g^β

$$A f^{k+1} = c g^\beta + \dots$$

as strongly as any of the vectors in $\{f^\alpha\}$.

If errors of recognition lead to the process described in language as going from particulars to the general, errors of association might be described as going from particulars to a universal: cat[1] meows, cat[2] meows ... → all cats meow.

There is, of course, no 'justification' for this animal process. It is performed as a consequence of the nature of the system. Whatever efficacy it has will depend on the order of the world in which the animal system finds itself. If the world is properly ordered, an animal system which 'jumps to conclusions' in the sense above may be better able to adapt and react to the hazards of its environment and thus survive. The animal philosopher sophisticated enough to argue 'the tiger ate my friend but that does not allow me to conclude that he might want to eat me' might then be a recent development whose survival depends on other less sophisticated animals who jump to conclusions.

By a sequence of mappings of the form above (or by feeding the output of A back to itself) one obtains a fabric of events and connections

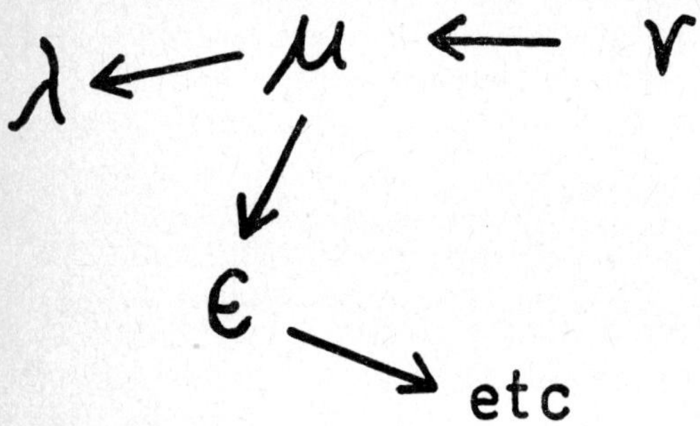

which is rich as well as suggestive. One easily sees the possibility of a flow of electrical activity influenced both by internal mappings of the form A and the external input. This flow is governed not only by direct association coefficients $c_{\mu\nu}$ (which can be explicitly learned as described next) but also by indirect associations due to the overlapping of the mapped events as indicated in fig. 4. In addition one can easily imagine situations arising in which direct access to an event, or a class of events, has been lost ($c_{\gamma\gamma} = 0$ in fig. 4) while the existence of this event or class of events in A influences the flow of electrical activity.

One serious problem in making the identifications suggested above is a direct consequence of the assumption of the linearity of the system. Any state is generally a superposition of various vectors. Thus one has to find a means by which events—or the entities into which they are mapped are distinguished from one another.

There are various possibilities; neurons are so non-linear that it is not at all difficult to imagine

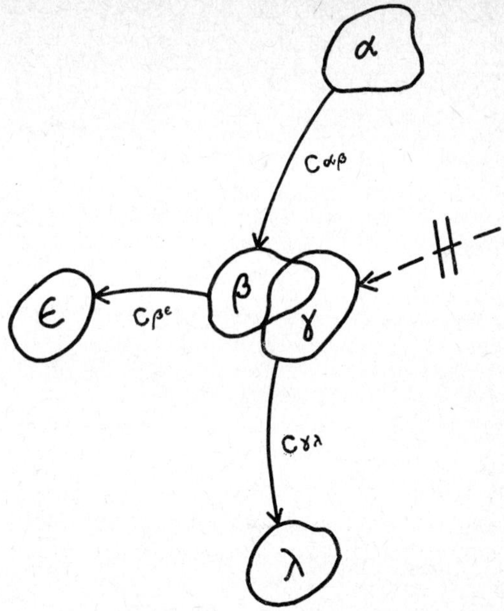

Fig. 4. There will be a flow from $\{\alpha\}$ to $\{\lambda\}$ even though no directly learned coefficient $c_{\alpha\lambda}$ exists, due to the overlap between the $\{\beta\}$ and $\{\gamma\}$ classes. In the particular situation above direct access to $\{\gamma\}$ has been lost.

non-linear or threshold devices that would separate one vector from another. But the occurrence of a vector in the class $\{f^{\alpha}\}$ in a distributed memory is a set of signals over a large number of neurons each of which is far from threshold. A basic problem, therefore, is how to associate the threshold of a single cell or a group of cells with such a distributed signal. How this might come about will be described in a later section.

In addition to the appearance of such 'pontifical' cells, there will be a certain separation of mapped signals due to actual localization of the areas in which these signals occur. For example, optical and auditory signals are subjected to much processing before they actually meet in cortex. It is possible to imagine that identification of optical or auditory signals (as optical or auditory) occurs first from where they appear and their immediate cluster of associations. Connections between an optical and an auditory event might occur as suggested in fig. 5.

I need hardly mention that even assuming that the physiological assumptions that underlie these constructions are possible (or even correct), there is a distance to be travelled before it is

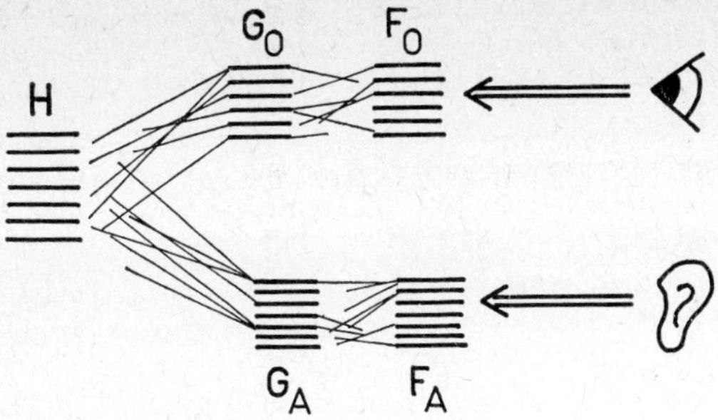

Fig. 5. A Model Optical-Auditory System.

shown that combinations of such elements as introduced above, working together could really reproduce animal mental behavior. The most important step is to make contact between theory and experiment. Some attempts in this direction will be discussed below.

Network Modification, Learning

We now ask how a mapping of the type A might be put into the network. The ij'th element of the associative mapping A

$$A_{ij} = \sum_{\mu\nu} c_{\mu\nu} g_i^\mu f_j^\nu$$

is a weighted sum over the j components of all mapped signals, f^ν, and the i components of the responses, g^μ, appropriate for recollection or association. Such a form could be obtained by additions to the element A_{ij} of the following type:

$$\delta A_{ij} \sim g_i f_j$$

This δA_{ij} is proportional to the product of the differences between the actual and the spontaneous firing rates in the pre and post synaptic neurons i and j[16]. The addition of such changes in A for all associations $g^\mu \times f^\nu$ results finally in a mapping with the properties discussed in the previous section.

For such modifications to occur, there must be a means of communication between the cell body and dendrite ends in order that the necessary information be available at the appropriate junctions; this information must move in a direction opposite to the flow of electrical signals [17]. The junction ij, for example, must have information of the firing rate f_j (which is locally available) as well as the firing rate g_i which is somewhat removed (fig. 6). One possibility would be that the integrated electrical signals from the dendrites produce a chemical or electrical response in the cell body which controls the spiking rate of the axon and at the same time communicates to the dendrite ends the information of the integrated slow potential.

There are a variety of means by which the coefficient A_{ij} might be modified, given that the necessary information is available at the ij'th junction. Among these might be growth of additional dendritic spines, adding new synaptic junctions, activation of synaptic junctions previously inactive, changes in membrane resistivity and/or changes in the amount of transmitter in a synapse. Although some structural changes have been observed, there is little evidence yet to choose among the possibilities mentioned above or in fact little evidence that such processes take place at all in the cortex of an adult animal.

To make the modifications

$$\delta A \sim g^\mu \times f^\nu$$

by any of the mechanisms suggested above, the system must have the signal distribution f^ν in its F bank and g^μ in its G bank. It is easy to obtain f^ν since this is mapped in from the event e^ν by P. But to get g^μ in the G bank is more difficult since this in effect is what the system is trying to learn.

In what we denote as active learning (which has been much explored in the past) the system is presented with some f^λ, searches for a response, and is given some indication of when it is coming

Fig. 6. In order that the junction (ij) be modified in proportion to $g_i f_j$, a means is needed for communicating the firing rate g_i which is the result of signals incoming from all the dendrites $g_i = \Sigma_j A_{ij} f_j$ back to the junction (ij).

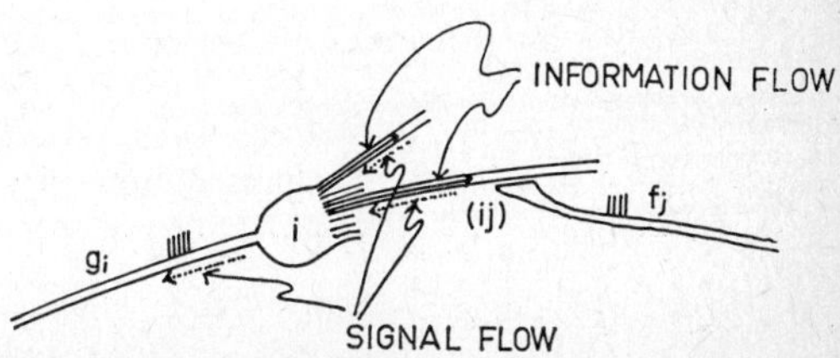

closer. When (after what could be a long time) by some procedure or another it finds the "right" response, say g^{ω}, it is "rewarded" and responds to the reward by printing into A the information:

$$\delta A_{ij} = \eta g_i^{\omega} f_j^{\lambda}$$

(The information is available at the time of the reward since at that time the system is mapping f^{λ}, responding g^{ω}, and thus has just the desired spiking frequencies in the F and G banks of neurons.) Active learning probably describes a type of learning in which a system response to an input is matched against an expected or desired response and judged correct or incorrect.

However, there is a type of animal learning which does not seem from visible external indications to require this type of a search procedure. It is the type of learning in which, as far as can be seen, an animal is placed in an environment and seems to learn to recognize and to recollect in a passive manner.

To arrive at an algorithm which produces passive learning, we utilize a distinction between forming an internal representation of events in the external world as opposed to producing a response to these events which is matched against what is expected or desired in the external world.

The simple but important idea is that *the internal electrical activity which in one mind signals the presence of an external event is not necessarily (or likely to be) the same electrical activity which signals the presence of that same event for another mind.* There is nothing that requires that the same external event be mapped into the same neural patterns by different animals. The event e^{ν} which for one animal is mapped into the signal distributions f^{ν} and g^{ν}, in another animal is mapped into f'^{ν} and g'^{ν}. What is required for eventual agreement between animals in their description of the external world is not that electrical signals mapped be identical but rather that the relation of the signals to each other and to events in the external world be the same (fig. 7).

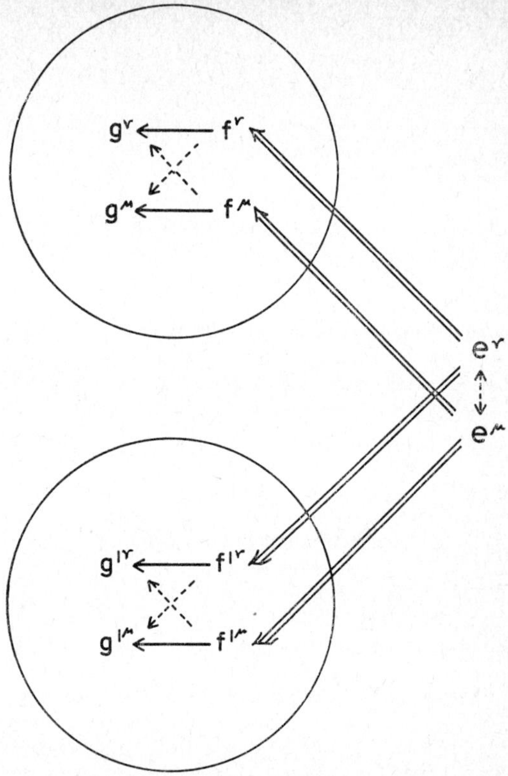

Fig. 7. Representations in two different systems of the same external fabric of events. The two representations are not identical but they each stand in a one-to-one relation to the external fabric and to each other.

Passive Learning

Call $A^{(t)}$ the A matrix after the presentation of t events. We write

$$A^{(t)} = \gamma A^{(t-1)} + \eta g^{t} \times f^{t}$$

In the equation γ is dimensionless and is a measure of the uniform decay of information at every site (a type of forgetting). One would expect that

$$0 \leqslant \gamma \leqslant 1$$

It turns out that values of γ close to one are of most interest. Since f or g are firing rates (spikes/second), η or $\varepsilon = \eta/\gamma$ have the dimensions of s^2. In what follows we normalize all vectors $(f,f) = (g,g) = 1$ so that η and ε become dimensionless.

If we now say that g^{t} is

$$g^{t} = A^{(t-1)} f^{t} + g_{R}^{t} + g_{A}^{t}$$

We see that the total post-synaptic potentials are composed of three terms: a passive response, $A^{(t-1)} f^{t}$, an active but random term, g_{R}^{t}, and an active response, g_{A}^{t}. For purely passive learning we consider only the first term so that

$$\delta A = \eta g^{t} \times f^{t} = \eta A^{(t-1)} f^{t} \times f^{t}.$$

Here the post synaptic potentials are just those produced by the existing mapping, $A^{(t-1)}$, when the vector f^{t} in F is mapped into G.

$$g^{t} = A^{(t-1)} f^{t}$$

The passive learning algorithm is then

$$A^{(t)} = A^{(t-1)}(\gamma + \eta f^{t} \times f^{t})$$

$$= \gamma A^{(t-1)}(1 + \varepsilon f^{t} \times f^{t})$$

where in general $\varepsilon = n/\gamma$ is presumably much smaller than one. Before any external events have been presented, A has the form $A^{(0)}$ which could be random. This will contain among other things the connectivity of the network.

With this algorithm, after k events A has the form

$$A^{(k)} = \gamma^{k} A^{(0)} \prod_{0\,\nu=1}^{k} (1 + \varepsilon f^{\nu} \times f^{\nu})$$

where $\prod_0$ is an ordered product in which the factors with lower indices stand to the left:

$$\prod_{0\,\nu=1}^{k} f(\nu) = f(1) f(2) \ldots f(k)$$

This can also be written:

$$A^{(k)} = \gamma^{k} A^{(0)} \Big[1 + \varepsilon \sum_{\nu=1}^{k} |f^{\nu}\rangle\langle f^{\nu}|$$

$$+ \varepsilon^{2} \sum_{\nu<\mu} |f^{\nu}\rangle\langle f^{\nu}|f^{\mu}\rangle\langle f^{\mu}| + \ldots$$

$$+ \varepsilon^{k} |1\rangle\langle 1|2\rangle\langle 2|3\rangle\langle 3|4\rangle \ldots \langle k-1|k\rangle\langle k| \Big]$$

It is striking that the passive learning algorithm generates its own response $A^{(0)} f^{\nu}$ to the incoming vector f^{ν}, a response that depends on the original configuration of the network through $A^{(0)}$ and on the vector f^{ν} mapped from the event e^{ν}. For example if f^{ν} is the only vector presented, A eventually takes the form

$$A \sim g^{\nu} \times f^{\nu}$$

where

$$g^{\nu} = A^{(0)} f^{\nu}.$$

Special cases of A

We now display the form of A in several special cases; in all of these ε is assumed to be constant and small.

(1) If the k vectors are orthogonal, A becomes

$$A^{(k)} = \gamma^{k} A^{(0)} \left(1 + \varepsilon A^{(0)} \sum_{\nu=1}^{k} f^{\nu} \times f^{\nu} \right).$$

Letting $A^{(0)} f^{\nu} = g^{\nu}$, the second term takes the form of the "diagonal" part of A

$$(A)_{\text{diagonal}} \equiv \mathcal{R} = \sum_{\nu=1}^{k} g^{\nu} \times f^{\nu}$$

and will serve for the recognition of the vectors $f^{1} \ldots, f^{k}$. (It should be observed that the associated vectors g^{ν} are not given in advance; they are generated by the network.) If ε is small however this seems inadequate for recognition since the recognition term would be weak. Further one would expect recognition to build up only after repeated exposure to the same event.

(2) The following example demonstrates that the passive learning algorithm does build up recognition coefficients for repeated inputs of the same event. If the same vector f^{0} is presented l times, A becomes eventually

$$A^{(l)} \simeq \gamma^{l}_{A^{(0)}} (1 + e^{l\varepsilon} f^{0} \times f^{0}).$$

If l is large enough so that $e^{l\varepsilon} \gg 1$, the recognition term will eventually dominate. Presumably when $e^{l\varepsilon}$ becomes large enough there should be no further increase in the coefficient. This can be accomplished by making ε a function of the response to the incoming vector in some way so that beyond some maximum value there is no further increase of the coefficient.

(3) The presentation of m orthogonal vectors $l_1, l_2, \ldots l_m$ times results in a simple generalization of the second result. When $\gamma = 1$ (for simplicity)

$$A^{(l_1 + l_2 + \ldots + l_m)} = A^{(0)} \left(1 + \sum_{\nu=1}^{m} e^{l_\nu \varepsilon} f^{\nu} \times f^{\nu} \right)$$

which is just a separated associative recognition and recall matrix

$$A \simeq \sum_{\nu=1}^{m} c_{\nu\nu} g^{\nu} \times f^{\nu}$$

if

$$e^{l_\nu \varepsilon} \equiv c_{\nu\nu} \gg 1.$$

(4) Some of the effect of non-orthogonality can be displayed by calculating the result of an input consisting of l noisy vectors distributed randomly around a central f^0

$$f^\nu = f^0 + n^\nu$$

Here n^ν is a stochastic vector whose magnitude is small compared to that of f^0. We obtain

$$A^{(l)} \simeq \gamma^l A^{(0)} \exp\left(l\varepsilon \frac{n^2}{N}\right) (1 + e^{l\varepsilon} f^0 \times f^0)$$

where n is the average magnitude of n^ν. We see that the generated $A^{(l)}$, with the additional factor due to the noise, is just of the form for recognition of f^0. Thus the repeated application of a noisy vector of the form above results in an A which recognizes the central vector f^0.

Association Terms

Off-diagonal or associative terms can be generated as follows. Assume that A has attained the form

$$A = \sum_{\nu=1}^{k} A^{(0)} f^\nu \times f^\nu = \sum_{\nu=1}^{k} g^\nu \times f^\nu.$$

Now present the events e^α and e^β so that they are associated so that the vectors f^α and f^β map together. (The precise conditions which result in such a simultaneous mapping of f^α and f^β will depend on the construction of the system. The simplest situation to imagine is that in which $(f^\alpha + f^\beta)/\sqrt{2}$ is mapped if e^α and e^β are presented to the system close enough to each other in time.) We may assume that e^α and e^β are separated so that $(f^\alpha, f^\beta) = 0$. In the F bank of neurons we then have $(f^\alpha + f^\beta)/\sqrt{2}$, where the vector is normalized for convenience.

After one such presentation of e^α and e^β, A becomes:

$$A^{(1)} \simeq \sum_{\nu=1}^{k} g^\nu \times f^\nu + \frac{\eta}{2}(g^\beta \times f^\alpha + g^\alpha \times f^\beta).$$

The second term gives the association between α and β with the coefficient

$$c_{\alpha\beta} = c_{\beta\alpha} = \eta/2$$

which presumably (except in special circumstances) would be small. If f^α and f^β do not occur again in association, $c_{\alpha\beta}$ or $c_{\beta\alpha}$, although they do grow upon the presentation of f^α or f^β separately, always remain small compared to the respective recognition coefficients $c_{\beta\beta}$ or $c_{\alpha\alpha}$. However if $(f^\alpha + f^\beta)/\sqrt{2}$ is a frequent occurrence (appearing for example l times) the coefficient of the cross term becomes

$$c_{\alpha\beta} \simeq \frac{\gamma^l}{2} e^{l\varepsilon}$$

as large as the recognition coefficient.

We are at present persuing analytical calculations and machine simulations to explore further the development of A in a variety of situations.

Hubel-Wiesel Type Cells

It is most important to make some contact between theory and experiment. We have been working in several directions in such attempts; I will describe one below.

In a classic series of experiments Hubel & Wiesel [13] have shown that cells in the cat's visual cortex respond in a very specific way to external visual patterns to which the cat is exposed. Such patterns can for example be vertical or horizontal lines of specific sizes, in specific positions or with definite motions. Recently Hirsch & Spinelli [18] have shown that depriving the cat of visual experience (such as the experience of horizontal lines) results in the absence of Hubel-Wiesel cells which respond to such lines in the adult. We outline below one means by which such cells might develop, assuming that network modification can occur in a similar manner to that employed previously for the construction of the mapping A.

Since these cells occur in the early processing regions we divide the mapping P so that

$$P = M_2 \times M_1$$

$$c^\alpha \underset{M_2}{\leftsquigarrow} d^\alpha \underset{M_1}{\Leftarrow} e^\alpha$$

where now M_1 is taken to be prewired but M_2 is modifiable (such modification being possible presumably only in the early development of the animal). As M_2 is to be modified by the same mechanism used previously we write:

$$\delta M_2 \sim g^{\mu} \times f^{\nu}.$$

We now assume that the spontaneous firing rate of the cells in the left hand bank of M_2 (other than the random spikes we admit below) is very low or zero and that $M_2^{(0)}$ (the initial value of the mapping M_2) is small, so that the cells in the left hand bank, at least in the development stages, are in general below threshold and do not respond strongly to external signals. These cells therefore are in a pre-threshold and highly non-linear region.

We suppose further (as is in fact the case) that the cells in the left hand bank fire occasional spikes spontaneously and at random. We can then write:

$$M_2^{(t)} = \gamma M_2^{(t-1)} + \eta P_n M_2^{t-1} d^t \times d^t + \eta c^n \times d^t$$

where $P_n M_2$ is the projection of M_2 onto the n'th outgoing cell

$$P_n M_2 = \sum_i \delta_{ni}(M_2)_{ij} \quad \text{for all j.}$$

As before the first term gives the uniform loss of memory, the second the passive response, while the third gives active additions to M_2 due to the simultaneous arrival of the mapped vector d^t and the spontaneous and random firing of the n'th cell in the left hand bank

$$c_i^n = 0 \quad i \neq n$$

$$c_i^n = 1 \quad i = n$$

Since M_2 is initially thought to be small, an incoming vector d^t produces no change in M_2 unless one of the cells in the left hand bank is firing simultaneously. If, however, the entry of the pattern d^α coincides with the firing of the n'th cell we add a term to M_2 of the form

$$\delta M_2 = \eta P_n M_2 d^\alpha \times d^\alpha + \eta c^n \times d^\alpha$$

It is reasonable to assume that a single such entry is not sufficient to fire the n'th cell upon a second presentation of d^α. However, with l coincidences of d^α and the firing of c^n we obtain

$$M_2^{(l)} \simeq \gamma^l e^{l\varepsilon} c^n \times d^\alpha$$

Therefore, there is a build-up of such recognition cells and, depending upon the parameters, they will eventually fire upon the presentation of the pattern with which they have been associated.

There are several problems. One would expect several cells to respond to the same pattern. This is very likely to be the case in fact. However, the addition of a mechanism, such as exists in the visual processing system of Limulus, by which a firing cell suppresses the activity of its neighbours would reduce the number of cells which would respond to the same pattern. In addition it might occur that the same cell would pick up several patterns. How likely this is is being tested by computer simulation with various values of the parameters. Whether this happens in fact is an open question which could be answered experimentally. Such multiple patterns could also eventually be diminished by the addition of a Limulus-like suppression.

All of the prior considerations of picking central vectors out of noisy entries would apply; the details would depend on the parameters chosen. Such cells, linked to F or G type neuron banks, might also be of use for the separation of one vector from another as suggested at the end of the second section.

If these considerations in some measure correspond to facts, it becomes intriguing to speculate that a form of network modification which arose originally in evolutionary development to assist in the formation of special cells for early processing—to increase the flexibility of systems so that they required less genetic pre-programming—might (functioning in a region of cortex which retains for much or all of a lifetime its ability for modification) be responsible also for what we like to call "higher" mental processes.

The author wishes to express his appreciation for the hospitality offered to him by the Laboratoires de Physique Théorique et Hautes Energies, Universités Paris VI and Paris XI (Orsay) where part of this article was written. He would also like to thank those colleagues, in particular Professors Jack Cowen, Charles Elbaum and Bruce Knight, who have offered helpful criticism.

References

1. Anderson, J, Cooper, L, Nass, M, Freiberger, W & Grenander, U, AAAS symposium, theoretical biology and bio-mathematics (1972).
2. Block, H D, Rev mod phys 1962, 34, 123.
3. Block, H D, Knight, B W, Jr & Rosenblatt, F, Rev med phys 1962, 34, 135.

4. Minsky, M & Papert, S, Perceptrons: An introduction to computational geometry. MIT Press (1969).
5. Longuet-Higgins, H C, Nature, London 1968, 217, 104.
6. — Proc roy roc London B 1968, 171, 327.
7. Creutzfeldt, O D & Ito, M, Exptl brain res 1968, 6, 324.
8. Goldstein, M, Jr, Hall, J L, II & Butterfield, B O, J acoustical society of America 1968, 42, 444.
9. Lettvin, J Y & Gesteland, R C, Cold Spring Harbor symp quant biol 1965, 30.
10. Mountcastle, V B, The neurosciences (ed G C Quarton, T Melnechuk & F O Schmitt) p. 393. Rockefeller University Press, New York, 1967.
11. Anderson, J A, Math biol-sci 1970, 8, 137.
12. — Ibid 1972, 14, 197.
13. Hubel, D H & Wiesel, T N, J physiol 1962, 160, 106.
14. Wiesel, T N & Hubel, D H, J neurophysiol 1965, 28, 1029.
15. Lashley, K S, Arch neurol psychiat, Chicago 1924, 12, 249.
16. Alterations in junction strengths proportional to f_j or the immediate dendrite response to f_j would also seem to be physiologically possible and in some situations might be useful. However, such modifications do not result in the various properties discussed here.
17. Such 'retrograde signalling' has also been postulated by J P Changeux & A Danchin. Private communication.
18. Hirsch, H V & Spinelli, D N, Exptl brain res 1971, 12, 509. See also Blakemore, C & Cooper, G F, Nature 1970, 228, 477.

Discussion

Anderson: Some theories of memory I have seen (not understood!) discuss the question of capacity. Is there corresponding discussion here?

Cooper: Yes, one can discuss the capacity. There seems to be a lot of it. Whether it is enough, I do not know.

Wilson: If you stick to your linear theory the capacity of the brain in your theory seems to be limited, namely there are at most 10^8 linearly independent f's if there is one component of f for each neuron. This would limit the brain to much less than 10^8 distinct remembered patterns.

Cooper: I am not sure whether or not 10^8 distinct patterns (if that is the number) would or would not be enough. One should also recall that each of the "distinct" remembered patterns can be very complex. Further one does not expect a complete absence of non-linearity.

Conference Summary

Conference Summary

If past experience at this conference is any guide at all, I will still be talking to you at least an hour from now. And what's more after I sit down somebody will ask a question and I will suggest that I have about three more preprints or reprints to describe; therefore, so that I will bore you on my own time and not anybody else's, I think I'd like to start by acting as a second representative of the visitors here to express our appreciation for what, for many of us I'm sure, is the most magnificent scientific meeting we've ever attended. And I think we all owe a very hearty vote of thanks to our great leaders Stig and Bengt (Lundqvist)[1]. Everything as far as I could see was done perfectly. I think I should mention particularly that I've never seen a projection setup that worked as consistently well as this one. Harry Suhl referred to many other things last night, but of course Harry could not comment on last night's banquet because he hadn't had it yet. I think I must say of the banquet that at just about the time when I said to Gerry Mahan "they've laid on just about everything but dancing girls," in came the dancing girls.

Of course we have also to give a hand to the other Swedish members of the committee as well, to Professors (Lamek) Hulthén and (Lars) Hedin and to all the young people whom we've come to rely on so heavily for practically everything.

And then I think we do have to say something about the power behind the throne, i.e. John Wilkins. Actually he was sometimes to be found well in front of it with his neck stuck out. But I think it's obvious to all of us how valuable John's help has been, not least in insisting in a break after every major talk. I don't know when I've met in more frequently changed air and that's been surprisingly helpful. I have to mention that John's effect (and that of several of the others I have mentioned), on the meeting has not been entirely positive, because while certain aspects of the late night hospitality have been very pleasant, at the moment I'm suffering from at least some of them and perhaps some of the rest of you are. I was still reading my notes at 2: 30 last night. So if this seems a bit incoherent that's the reason why.

I would also like to thank John for always being the man to ask the question I wish I'd thought of to make things clear in my mind, though I'm afraid there were times when things weren't clear nonetheless.

Now, at this time I think I'd like to get about the business of awarding a number of prizes. Considering the title of the Symposium one can hardly object if these become Nobel prizes in the minds of their holders in the future. I think the first prize is the prize for Top Banana that should have gone to John Wilkins, but he gave it away to Bill Brinkman so we'll have to make Bill Brinkman the Top Banana.[2] I was going to give a prize for the most clearly, concisely and beautifully presented talk to Seb Doniach but unfortunately he just disqualified himself.[3] It was a clearly and beautifully presented talk but so were most of the others. Therefore that prize has got to be shared between John Hubbard and I. Dzyaloshinsky. And I suspect there were various other people who copped out (or were pushed out) whose names never have formally appeared on the program and I express our appreciation for the forbearance of all of them. A memorable remark on this subject not on the list (which you will soon see) was by some chairman or organizer who said "Well, we've put that headache off for awhile". I won't remind you which paper that was except to say it was a good one.

The prize for the most artistic doodler unquestionably goes to L. P. Gorkov. I don't know if all of you have seen any of his productions. Last night Michael Fischer and I were scrambling about among the papers up here looking for an example to show you, but unfortunately either the students have stolen them or he's taken them home with him. Michael came in second.

Next I wanted to award prizes for a few of the

[1] Editors remarks.

[2] John Wilkins presented an object billed as a Triplet Banana to Bill before his repeatedly postponed talk.

[3] Doniach's talk had been cancelled but time was found for it in the last session.

memorable things people said during the meeting. I noted a few of them myself, and others I collected last night wandering around asking what had struck people one way or another, as amusing but also with some thematic application to what we have been hearing. I'll present the list first:

Quotes—sometimes out of context

"In this calculation I have everything going for me." (K.G.W.)

"RNG is as general and vigorous as the partition function approach to statistical mechanics was." (M.E.F.)

"Right for partly the wrong reason." (P.W.A.)

"When Einstein speaks, I listen." (J.R.S.)

"Like South America, the Kondo effect will always have a great future." (H.S.)

"What did Smetana do in Göteborg?" "He suffered." (a Lundqvist)[1]

"Five years in superconductivity and five years in liquid crystals—this field was *made* for me!" (P. G. de Gennes)

The first one struck a lot of people as funny; I suppose because of the overtones of God being on the side of the big battalions. I was sorry I didn't get more votes for Ken's (Wilson)[2] remark about his role being to replace a 30% by a 10% theory.

The second one was said with such conviction and caused such discussion that it just had to be put down here.

The third—I do remember in the early days of the BCS theory and a bit before there was a very eminent theorist, who will be nameless, who was invariably wrong for absolutely the right reason. I have been accused of preferring the other way around.

The next two are self-explanatory. Harry Suhl provided a problem of selection—we will mention at least one other of his.

The next one—I should hasten to explain my reference here.[1] I have a very poor attendance record at most meetings—I believe I average about 25% of the sessions—but here I've been at absolutely every session, suffering from the responsibility for this talk. But I didn't suffer actually all that much because there certainly was no single session that I didn't feel had been very much worth my going to.

Again, the last one is self-explanatory.

Now, finally, I'd like to run very informally through the program and say a little about what we actually did hear, try to put some flesh on the bones of Harry Suhl's remarks last night that amounted to essentially saying that Many-Body Theory is alive and well and living right here in Lerum.

It was an incredibly rich meeting from a scientific point of view. There were four developments of absolutely major significance. I hardly need to tell you what they were: the renormalization group on the first day, the ^{3}He experimental data and all its consequences on the second day, the solution of the Kondo problem on the third day, and the announcement of the results of Heeger on the third day as well. These were all so exciting that a lot of things may have slipped by that were really pretty major, without people noticing their importance and weight. Of course the monumental work that Walter Kohn described is an example, but there were other things that were just described in seconds.

Here are some examples picked at random: in most meetings the announcement of the first reasonably practical calculations of LEED intensities would have been of major importance. De Gennes' discussion remark on smectic B is a very important, exciting development. I thought Gorkov's very exciting new ideas on V_3Si and on one-dimensional systems are very much worth our serious consideration, though perhaps later I'll say things about what remains unsatisfying about them.

Now to take the program day by day, more or less systematically:

The first day, we were of course very grateful for the road show that blew in from Temple in the States, (a very similar sequence of talks had been given at a full three-day meeting in Philadelphia two weeks before) and we also realized it may have been a little dull for them because they were saying, one presumes, many of the same things they'd just been saying two weeks before. But it was not dull, by any means, for any of us who have not followed the field all that carefully.

[1] Those who attended the Symposium will remember that we spent quite a bit of time milling about the Smetana room in the Town Hall of Gothenburg on Wednesday night.

[2] Editors remarks.

What that day really was about was the application of the renormalization group techniques, the new renormalization group as opposed to the old renormalization group, of course, to the problem for which they were invented, and especially the epsilon expansion and the $1/n$ expansion.

I guess I forgot to put down a particularly memorable thing Harry Suhl said in his talk which was the remark that if a theory satisfies the fundamental principles of conservation of momentum and energy and is mathematically correct, it always seems to be worth doing. I would have, a few years ago, felt that the study of critical points was somewhere near that category. It should be done of course, but only because it satisfies momentum and energy and is correct. But now I think Harry's remark has come home to us very seriously in this case.

I think that we may or may not have something that is as rigorous as equlibrium statistical mechanics. I think we have to keep in mind that these fixed points are at least the limit points of lines of essential singularities, namely the droplet singularities; and God knows what other singularities of peculiar Thouless-Kosterlitz type or something like that might often be associated with them. But it certainly is a classification, a way of thinking, and a quantitative approach that can be thought of in terms of expansions of various kinds, i.e. it's got an enormous potential in its own field. But I think what one will see more and more of is an extension of these ideas into other fields, an expansion of them into doing different kinds of things of the general sort that we saw in the Kondo problem. There are very wide possibilities.

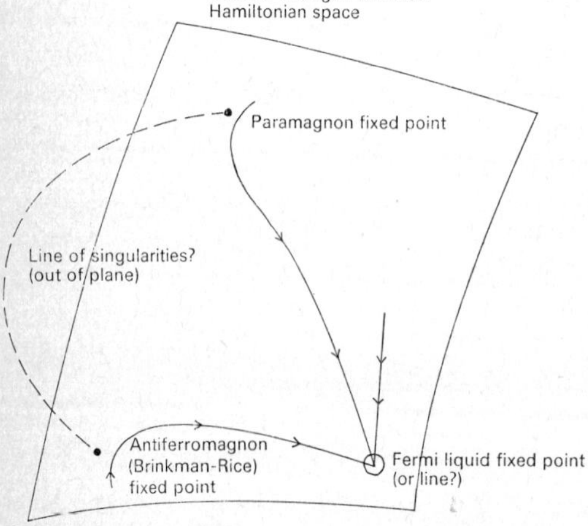

Bob Schrieffer said something privately to me (I don't know if he wants to have it quoted) which opened my eyes to some of these possibilities. He said that this is the kind of qualitative idea that may be of value in very distant fields—information handling, sociology, psychology. It's the kind of thing that can apply to a very much wider range of situations—the basic ideas of scale change, irrelevant variables, limit points, etc. are very general.

Now another application: renormalization group is one of these things like speaking prose we've been doing all our life. One special example, for instance, is the Landau Fermi liquid theory which is an effective Hamiltonian valid in a certain range. It's an effective Hamiltonian which is invariant to the form and behavior of the starting parameters and always gives you the same final Hamiltonian with perhaps different values of the coefficients. So one thinks that the Landau Fermi liquid theory must be the equivalent, in the electron gas, of the Gaussian fixed point, the ordinary fixed point that everything goes to in the end. Then there are obviously at least two other fixed points. I have made a little diagram of Fermi Hamiltonian space with its singular points. There's the Landau Fermi liquid theory fixed point, and as you see everything runs through Fermi sea Hamiltonian space and eventually arrives at that. But one can very well tell if in the course of its trajectory through Fermi sea space it has passed near what one might call the Doniach-Schrieffer or paramagnon fixed point. There's undoubtedly at least one other point, the Brinkman-Rice antiferromagnon point you might call it, and I wonder if perhaps there isn't a line of singularities connecting these two. But unquestionably that is something like the structure of the Hamiltonian space as you cut off renormalized Fermi sea theories. That's the full three-dimensional version of the same diagrams that Ken Wilson talked about on the third day, where, as someone was saying this morning at breakfast, the localized state Hamiltonian is the single atom version of the Hubbard Hamiltonian.

To see how things might go with the renormalization of the strong-coupling Fermi sea, let's rephrase the Kondo solution. The Kondo renormalization starts from a localized, strong coupled spin fluctuation, renormalizes it down, and eventually gets, at the low temperature end, to a weak coupling localized spin fluctuation which is nonmagnetic and apparently uninteresting, kind of the equivalent of the Landau Fermi liquid theory, but by no means in all cases has that Kondo center forgotten that it went very close to a singularity, to an unstable fixed point. I don't think I'll take time to speculate further except to remark that this program is in fact perhaps underway already in the case of the paramagnon, that what Hertz and Edwards have done looks very much like a beginning of the full renormalization group approach to the paramagnon problem.

This perhaps then suggests what I hoped would be another theme of my talk, which is universality and breadth of techniques. The fact is that the techniques which were developed for this at first apparently very specialized problem of a rather restricted class of special phase transitions and their behavior in a restricted region are turning out to be something which is likely to spread over not just the whole of physics but the whole of science. A second example can perhaps be introduced by talking about the ^{3}He work. I guess we all felt that was a very exciting day. I found the most exciting moment was when I finally began to realize that (a) I understood Tony (Leggett's) longitudinal resonance and (b) that he maybe understood what I was saying and then (c) that we weren't in any real disagreement. That's the kind of thing that meetings like this are for. There have been many other discussions but that perhaps was the most fruitful that actually took place on the meeting floor.

Well, the point about ^{3}He is the very exciting experimental results. I was asked by (Robert)[1] Richardson to announce one which just arrived in a telegram from Wheatley. The fact that the B phase, Wheatley's B phase, is the same as the Cornell group's B phase is now experimentally confirmed at least on the basis that what Wheatley has been calling the AB transition has a first order magnetization change. So, yet more experimental data are coming in.

It seems very likely that this incredibly unlikely phase that's been postulated is the correct description. There are quite a number of pretty solid experimental vs theoretical checks, and several further indications that one can see qualitatively are strongly in the right direction, and I know of nothing which is seriously contradictory at this point. So, I think, perhaps, this will have been the first conference at which such a wide spectrum of experiment and theory have seemed to come into agreement that one begins to be able to suspend disbelief in the superfluid phase.

Lars Hedin asked me, "Well, all right, so it's superfluid. What are you going to do with it?" And the answer is, "Absolutely nothing." I don't think that that's irrelevant nonetheless. As Harry says: "It satisfies energy and momentum, it must have some use." One remembers that back in 1957, after Bob (Schrieffer)[1], Leon (Cooper)[1] and John Bardeen made their advances, the theoretical developments in understanding superconductivity eventually leaked over into high energy physics; the most exciting modern developments in high energy physics are based on one or another of those concepts that we had then: The concept of broken symmetry, the concept of Goldstone bosons, the concept of what we now call Higgs bosons—the plasmons, in fact. ^{3}He is a nice example of a field with a broken internal symmetry, of non-Abelian type, which is just exactly what people are doing with the renormalization group in high energy physics, at this point. One must assume that having an actual physical model is going to serve again as a guide for that field of physics.

The Kondo effect: I think one must associate Ken Wilson's magnificent solution ... to me it's an incredible intellectual feat ... with the other uses of the renormalization group and again emphasize the point I think he's been making that while, yes, it is very important, in doing things with the renormalization group, that one understands the classification of singular points, in the end what one will want to be doing it for is actually getting from a bare Hamiltonian to an effective Hamiltonian, and knowing really how to get down from A to B and then calculate some numbers and do some things with them. We really basically understood the Kondo problem from a classification point of view. It hadn't been proved by any means, but more than speculations certainly existed, and yet this new solution is infinitely more valuable.

Then Heeger's new data and all the new questions about what kind of one-dimensional fluctua-

[1] Editors remarks.

ting states can have this conductivity which appears to be of a superconducting type. What was most exciting to me, because I guess I'd been a bit familiar with these data, was the new data on the thermoelectric power which are extremely striking, and I think Alan (Luther)[1] is quite right to claim those are something which one really has to contend with. They show ... this is another Harry Suhl quote (or is it C. Kittel?) they show that Forces are at Work.

I would like to run through the days of the meeting and remind you, aside from those four very major things, of some of the things that went on.

On the first day I think I have one more prize that I want to award and that is for maximum overload on my information handling capacity. I award that to Gerard Toulouse. The main problem of course was that he had an enormous amount to say, and I think that this anatomy of the n–d plane is a delightful and fascinating game, and more than a game. Again it satisfies energy conservation and momentum conservation (although I'm not sure that negative dimensionality satisfies anything). I do wish that one of the questions John Wilkins had asked was, what has zero dimensionality got to do with percolation theory and self-avoiding walks. I had it explained to me once but it's completely gone.

Nonetheless, it was a very good, very interesting day, not just the three major talks, really, but all the remarks that were made in addition. I was glad Dzyaloshinsky reminded some of these phase transition buffs that there are n's greater than 3, lots and lots of them.

^{3}He: I guess I really have covered that pretty much. Perhaps I should mention the feeling of intellectual excitement that Pierre de Gennes gave us all in his "anatomy of the liquid ^{3}He horse". That's perhaps why I have quoted that last sentence from his talk. Then the Kondo problem. We've talked about that, but can add that Fred Zawadowski, of course, has pointed out that that particularly simple logarithmic renormalization group is not confined to that one problem at all. There are two simpler problems which it does completely: the so-called $n=2$ one-dimensional Ising model and the ordinary X-ray problem. So there are quite a few problems to which this can be applied, even though it doesn't quite in its simplest form successfully do the Kondo problem.

There were the few-dimension sessions and various things ... another of those things that was whipped past very fast is this very strange puzzle of the $(H_c)_2$ of layered superconductors which Alan Luther mentioned, which suggests again that Forces are at Work.

I did mention the Gorkov talk which I thought was very important, and of course we all were very excited by the discussion of the Heeger session.

In surfaces there was the monumental paper and sequence of work that Walter Kohn has assembled on this subject. I personally had not realized how closely one could get agreement with experiment. It seems the basic equilibrium processes that happen on surfaces are beginning to be very straightforwardly understood. There do of course remain problems.

I think the discussion session will be valuable on the question of the plasmon contribution.

Yet another of these problems that passed over our heads rather rapidly, in an extremely exciting field of physics, is the question of superstructures that André Blandin talked about. I know of at least four groups that are making progress in the understanding of various kinds of surface superstructures. It's a very spectacular phenomenon from time to time. I have seen a diffraction pattern, and nothing seemed wrong about it to me, that claimed to be a 43 by 43 superstructure on a triangular layer. There are certainly 13 by 13 superstructures on TaS_2 and, of course, on germanium and silicon there are 7 by 7s and 8 by 8s that have been known for many years. These are very interesting phenomena. They may well end up being understood as excitonic transitions. One may hope so.

And then, finally, of course, there are the threshold effects. Perhaps that session seemed a little tame in comparison with the rest of the conference, not very tame I must say, because that was the problem we solved last year instead of this year, and all the rest of the conference has been about the problems we either solved this year or are only about to solve. And there people have done some pretty magnificent things, but they seem to have a lot of it very much under control. One mustn't forget, however, that they are important and exciting many body effects.

---

[1] Editors remarks.

Now I'd like to say some things in summary and conclusion about open questions and especially about the biology talks. We are fortunate in that the "cat" has by no means lapped up all the "cream" in our field, there is more and more coming along. One can think of problem after problem. In Harry Suhl's talk the peculiar problem of the compensation of chemical activations was mentioned. John Hopfield has also worked on that—an incredible puzzle. One may pick out many others. Certainly the one-dimensional structures are full of puzzles that one has no particular hope of understanding. For instance, I would certainly like to suspend my disbelief in Gorkov's theory of the Nb_3Sn, V_3Si transitions, but I find myself very unhappy with the fact that one uses a purely one-dimensional approach. I know that in the layer metals an actual band calculation gives one interlayer hopping integrals of the order of 1 eV which is two orders of magnitude higher than 50 K, which was being assumed I think in this calculation, and those layers are very far apart and are separated by van der Waals' bonds, whereas Nb_3Sn is a very, very rigid tough metal with obvious three-dimensional character. And what's more, one is going out of one dimension in order to get the right relevant states, which I am sure Gorkov has got. So, can one suggest that perhaps it renormalizes towards a one-dimensional fixed point of some sort? Perhaps if one starts with two pieces of Fermi surface which are parallel within 1 eV and also a strongly interacting system, when one renormalizes they become more parallel. That's a wild suggestion, but it's the only thing that I might understand. There is evidence after evidence that these crystals are one-dimensional. That one of them could be is perfectly reasonable. That all of them could be, I will believe when I see a band structure calculation that I absolutely believe. But they could be renormalized one-dimensional systems... The other thing about them of course is that the transitions do take place and real one-dimensional transitions don't take place. To what extent is one justified in eating one's cake, and having it too, using one-dimensionality at one point, and not using it at another? So there are some very interesting problems ... I mentioned that at such length because this is not by any means an isolated problem. The existence of anomalous quasi-one-dimensional and quasi-two-dimensional fluctuations is a very widespread phenomenon that occurs both in insulators and in metals, in ferroelectrics and in soft mode systems: and remains one of our major puzzles. So we have a great future ahead of us, like South America.

We also have a future ahead of us travelling outside our fields as Harry Suhl and Leon Cooper and John Hopfield showed us so ably. All of these talks I think illustrate why it is that many body theorists can and should invade wider fields of science, why they can do so well in it. I think we are uniquely those physicists who are completely eclectic in our methodology, and eclectic in our point of view. We are not unwilling to take methods and ideas from the most abstract mathematical level, as a couple of these papers make quite obvious. (I certainly was lost at some points.) We're also many of us not by any means unwilling to get our hands dirty and to get down and really look at the phenomena, really look not just at the phenomenology, but at the electrons and atoms as they move around, and with one technique or another, dive in and really understand it. And if it's a matter of just doing kinetics right as opposed to the incredible hash that the literature was in before, that's something we can at least do ... if we're as good as John Hopfield ... and something which was not done until many-body theorists moved into that particular field.

So, we have a good field to work in. We also have an excellent opportunity, particularly for those of us who are somewhat younger than I am, to move out of the field and look at a wider spectrum of things. I think we must consider ourselves very fortunate.

P. W. Anderson

Nobel Symposia 1–22

Publishers:

Almqvist & Wiksell, Stockholm, Sweden
John Wiley & Sons, New York, USA

Nobelsymposium 1, 1966
Muscular afferents and motor control
(ed Ragnar Granit)

Nobelsymposium 2, 1967
Prostaglandins
(ed Sune Bergström & Bengt Samuelsson)

Nobelsymposium 3, 1967
Gamma globulins
(ed Johan Killander)

Nobelsymposium 4, 1968
Current problems of lower vertebrate phylogeny
(ed Tor Ørvig)

Nobelsymposium 5, 1967
Fast reactions and primary processes in chemical kinetics
(ed Stig Claesson)

Nobelsymposium 6, 1968
Problems of international literary understanding (ed Karl Ragnar Gierow)

Nobelsymposium 7, 1968
International protection of human rights
(ed Asbjörn Eide & August Schou)

Nobelsymposium 8, 1968
Elementary particle theory
(ed Nils Svartholm)

Nobelsymposium 9, 1968
Mass motions in solar flares and related phenomena
(ed Yngve Öhman)

Nobelsymposium 10, 1969
Disorders of the skull base region
(ed Carl-Axel Hamberger & Jan Wersäll)

Nobelsymposium 11, 1969
Symmetry and function of biological systems at the macromolecular level
(ed Arne Engström & Bror Strandberg)

Nobelsymposium 12, 1970
Radiocarbon variations and absolute chronology
(ed Ingrid U Olsson)

Nobelsymposium 13, 1970
Pathogenesis of diabetes mellitus
(ed Erol Cerasi & Rolf Luft)

Nobelsymposium 14, 1970
The place of value in a world of facts
(ed Arne Tiselius & Sam Nilsson)

Nobelsymposium 15, 1971
Control of human fertility
(ed Egon Diczfalusy & Ulf Borell)

Nobelsymposium 16, 1973
Frontiers in gastrointestinal hormone research
(ed Sven Andersson)

Nobelsymposium 17, 1971
Small states in international relations
(ed August Schou & Arne Olav Brundtland)

Nobelsymposia 18 and 19 (Postponed)

Nobelsymposium 20, 1972
The changing chemistry of the oceans
(ed David Dyrssen & Daniel Jagner)

Nobelsymposium 21, 1972
From plasma to planet
(ed Aina Elvius)

Nobelsymposium 22, 1972

ESR applications to polymer research
(ed Per-Olof Kinell & Bengt Rånby)

Nobel Symposia 23–25

Publishers

Nobel Foundation, Stockholm, Sweden
Academic Press, New York, USA

Nobel Symposium 23, 1972
Chromosome identification—technique and applications in biology and medicine
(ed Torbjörn Caspersson & Lore Zech)

Nobel Symposium 24, 1973
Collective properties of physical systems
(ed Bengt Lundqvist & Stig Lundqvist)

Nobel Symposium 25, 1973
Chemistry in botanical classification
(ed Gerd Bendz & Johan Santesson)